ADVANCED TEXTS IN ECONOMETRICS

General Editors

C. W. J. GRANGER G. E. MIZON

Nonstationary Time Series Analysis and Cointegration

Edited by Colin P. Hargreaves

Oxford University Press
1994

Oxford University Press, Walton Street, Oxford OX2 6DP
Oxford New York
Athens Auckland Bangkok Bombay
Calcutta Cape Town Dar es Salaam Delhi
Florence Hong Kong Istanbul Karachi
Kuala Lumpur Madras Madrid Melbourne
Mexico City Nairobi Paris Singapore
Taipei Tokyo Toronto
and associated companies in
Berlin Ibadan

Oxford is a trade mark of Oxford University Press

Published in the United States
by Oxford University Press Inc., New York

British Library Cataloguing in Publication Data
Data available

Library of Congress Cataloging in Publication Data
Data available
ISBN 0-19-877391-9
ISBN 0-19-877392-7 (Pbk)

1 3 5 7 9 10 8 6 4 2

Printed in Great Britain
on acid-free paper by
Biddles Ltd.,
Guildford & King's Lynn

Contents

List of Figures vii
List of Tables xiii
List of Contributors xv
Foreword xvii

1. Introduction 1

2. Towards a Theory of Economic Forecasting 9
MICHAEL P. CLEMENTS and DAVID F. HENDRY

3. Bayes Models and Forecasts of Australian Macroeconomic Time Series 53
PETER C. B. PHILLIPS

4. A Review of Methods of Estimating Cointegrating Relationships 87
COLIN HARGREAVES

5. A Test of the Null Hypothesis of Cointegration 133
DAVID HARRIS and BRETT INDER

6. Modelling Seasonal Variation 153
SVEND HYLLEBERG

7. Cointegration, Seasonality, Encompassing, and the Demand for Money in the United Kingdom 179
NEIL R. ERICSSON, DAVID F. HENDRY and HONG-ANH TRAN

8. Evaluating a Real Business Cycle Model 225
FABIO CANOVA, MARY FINN and ADRIAN R. PAGAN

9. Misspecification versus bubbles in the Cagan hyperinflation model 257
STEVEN N. DURLAUF and MARK A. HOOKER

10. Regime Switching with Time-Varying Transition Probabilities 283
FRANCIS X. DIEBOLD, JOON-HAENG LEE and
GRETCHEN C. WEINBACH

Name Index 303

Subject Index 307

List of Figures

Chapter 2.

1. Ratio of trace MSFE of the DV model to correctly specified case 45

Chapter 3.

1. C: 1959(3) – 1991(4) 66
   (a) Log-levels, (b) Prediction errors, (c) Evolving Best Bayes Model,
   (i) AR(p) + Trend (r) parameters, (ii) Unit Root present or not,
   (d) Bayes Model Forecast Encompassing Test Statistic: dQ^B/dQ^F
2. RC: 1959(3) – 1991(4) 67
   (a) Log-levels, (b) Prediction errors, (c) Evolving Best Bayes Model,
   (i) AR(p) + Trend (r) parameters, (ii) Unit Root present or not,
   (d) Bayes Model Forecast Encompassing Test Statistic: dQ^B/dQ^F
3. GDP: 1959(3) – 1991(4) 68
   (a) Log-levels, (b) Prediction errors, (c) Evolving Best Bayes Model,
   (i) AR(p) + Trend (r) parameters, (ii) Unit Root present or not,
   (d) Bayes Model Forecast Encompassing Test Statistic: dQ^B/dQ^F
4. RGDP: 1959(3) – 1991(4) 69
   (a) Log-levels, (b) Prediction errors, (c) Evolving Best Bayes Model,
   (i) AR(p) + Trend (r) parameters, (ii) Unit Root present or not,
   (d) Bayes Model Forecast Encompassing Test Statistic: dQ^B/dQ^F
5. PGDP: 1959(3) – 1991(4) 70
   (a) Log-levels, (b) Prediction errors, (c) Evolving Best Bayes Model,
   (i) AR(p) + Trend (r) parameters, (ii) Unit Root present or not,
   (d) Bayes Model Forecast Encompassing Test Statistic: dQ^B/dQ^F
6. CPI: 1959(3) – 1991(4) 71
   (a) Log-levels, (b) Prediction errors, (c) Evolving Best Bayes Model,
   (i) AR(p) + Trend (r) parameters, (ii) Unit Root present or not,
   (d) Bayes Model Forecast Encompassing Test Statistic: dQ^B/dQ^F

Chapter 3 continued.

7. U: 1959(3) – 1991(4) 72
(a) Log-levels, (b) Prediction errors, (c) Evolving Best Bayes Model,
(i) AR(p) + Trend (r) parameters, (ii) Unit Root present or not,
(d) Bayes Model Forecast Encompassing Test Statistic: dQ^B/dQ^F
8. WR: 1959(3) – 1991(4) 73
(a) Log-levels, (b) Prediction errors, (c) Evolving Best Bayes Model,
(i) AR(p) + Trend (r) parameters, (ii) Unit Root present or not,
(d) Bayes Model Forecast Encompassing Test Statistic: dQ^B/dQ^F
9. RWR: 1959(3) – 1991(4) 74
(a) Log-levels, (b) Prediction errors, (c) Evolving Best Bayes Model,
(i) AR(p) + Trend (r) parameters, (ii) Unit Root present or not,
(d) Bayes Model Forecast Encompassing Test Statistic: dQ^B/dQ^F
10. S&P500: 1959(1) – 1991(12) 75
(a) Log-levels, (b) Prediction errors, (c) Evolving Best Bayes Model,
(i) AR(p) + Trend (r) parameters, (ii) Unit Root present or not,
(d) Bayes Model Forecast Encompassing Test Statistic: dQ^B/dQ^F
10'. S&P500: 1959(1) – 1991(12) 76
(a) Log-levels, (b) Prediction errors, (c) Evolving Best Bayes Model,
(i) AR(p) + Trend (r) parameters, (ii) Unit Root present or not,
(d) Bayes Model Forecast Encompassing Test Statistic: dQ^B/dQ^F
11. INT1: 1969(7) – 1991(12) 77
(a) $(\text{Levels})^{-1}$, (b) Prediction errors, (c) Evolving Best Bayes Model,
(i) AR(p) + Trend (r) parameters, (ii) Unit Root present or not,
(d) Bayes Model Forecast Encompassing Test Statistic: dQ^B/dQ^F
11'. INT1: 1969(7) – 1991(12) 78
(a) Levels, (b) Prediction errors
12. INT2: 1969(7) – 1991(12) 79
(a) $(\text{Levels})^{-1}$, (b) Prediction errors, (c) Evolving Best Bayes Model,
(i) AR(p) + Trend (r) parameters, (ii) Unit Root present or not,
(d) Bayes Model Forecast Encompassing Test Statistic: dQ^B/dQ^F
12'. INT2: 1969(7) – 1991(12) 80
(a) Levels, (b) Prediction errors
13. INT3: 1969(7) – 1991(12) 81
(a) $(\text{Levels})^{-1}$, (b) Prediction errors, (c) Evolving Best Bayes Model,
(i) AR(p) + Trend (r) parameters, (ii) Unit Root present or not,
(d) Bayes Model Forecast Encompassing Test Statistic: dQ^B/dQ^F
13'. INT3: 1969(7) – 1991(12) 82
(a) Levels, (b) Prediction errors

Chapter 6.

1. Logarithm of the Quarterly Industrial Production in the Netherlands, 1960.1 to 1986.4 155
2. Logarithm of the Quarterly Production in Australia, 1963(1) to 1986(4) 157
3. Logarithm of the Monthly Industrial Production in Canada, 1960(1) to 1986(12) 159
4. Logarithm of Quarterly Real Disposable Income in Japan, 1961(1) to 1987(4). 162
5. Logarithm of Real Total Consumption in Japan, 1961(1) to 1987(4) 167
6. The Residuals from the Cointegrating Regression $c_{3,t}$ on I, SD, $y_{3,t}$ and $y_{3,t-1}$ 173

Chapter 7.

1. Nominal Money Stock 195
   (a) Logs of NSA and SA data [m_t and m_t^a],
   (b) The seasonal component [$m_t - m_t^a$]
2. Prices 195
   (a) Logs of NSA and SA data [p_t and p_t^a],
   (b) The seasonal component [$p_t - p_t^a$]
3. Real Money Stock 196
   (a) Logs of NSA and SA data [$(m-p)_t$ and $(m^a - p^a)_t$],
   (b) The seasonal component [$(m-p)_t - (m^a - p^a)_t$]
4. Real Income 196
   (a) Logs of NSA and SA data [y_t and y_t^a],
   (b) The seasonal component [$y_t - y_t^a$]
5. Nominal Income 197
   (a) Logs of NSA and SA data [$(y+p)_t$ and $(y^a + p^a)_t$],
   (b) The seasonal component [$(y+p)_t - (y^a + p^a)_t$]
6. Inverse Velocity 197
   (a) Logs of NSA and SA data [$(m-p-y)_t$ and $(m^a - p^a - y^a)_t$],
   (b) The seasonal component [$(m-p-y)_t - (m^a - p^a - y^a)_t$]
7. The three-month local authority interest rate $R3_t$ (—), the learning-adjusted M_1 retail sight-deposit interest rate Rra_t (···), and the annual inflation *rate* $\Delta_4 p_t$ (---). 199

8. The NSA and SA disequilibrium measures $\beta' x_t$ (—) and $\beta' x_t^a$ (···) 205
9. Recursive least squares test statistics and estimates 208
(a) One-step residuals and $0 \pm 2\hat{\sigma}_t$
(b) Sequence of break-point Chow stats., scaled by their one-off 5% critical values
(c) Recursive estimates of the coefficient of $\Delta_3(m-p)_{t-1}/3$, with $\hat{\beta}_t \pm 2ese(\hat{\beta}_t)$
(d) Recursive estimates of the coefficient of $(\Delta p_t + \Delta p_{t-4})/2$, with $\hat{\beta}_t \pm 2ese(\hat{\beta}_t)$
(e) Recursive estimates of the coefficient of $\Delta^2 y_{t-2}$, with $\hat{\beta}_t \pm 2ese(\hat{\beta}_t)$
(f) Recursive estimates of the coefficient of $(R_t^* + R_{t-1}^* + R_{t-2}^*)/3$, with $\hat{\beta}_t \pm 2ese(\hat{\beta}_t)$
(g) Recursive estimates of the coefficient of $(m-p-y)_{t-1}$, with $\hat{\beta}_t \pm 2ese(\hat{\beta}_t)$
(h) Recursive estimates of the coefficient of the constant term, with $\hat{\beta}_t \pm 2ese(\hat{\beta}_t)$
(i) Recursive estimates of the coefficient of S_{1t}, with $\hat{\beta}_t \pm 2ese(\hat{\beta}_t)$
(j) Recursive estimates of the coefficient of S_{2t}, with $\hat{\beta}_t \pm 2ese(\hat{\beta}_t)$
(k) Recursive estimates of the coefficient of S_{3t}, with $\hat{\beta}_t \pm 2ese(\hat{\beta}_t)$
(ℓ) Actual and fitted values of $\Delta(m-p)_t$

Chapter 8.

1. Plot of Projected Second Cointegrating Vector among States 238
2. First Period Values and Predictions of them from the MPA and RBC Models (with suffixes MPA and RBC respectively) 249
(a) Consumption (C), (b) Investment (I), (c) Income (Y)
3. Second Period Values and Predictions of them from the MPA and RBC Models (with suffixes MPA and RBC respectively) 250
(a) Consumption (C), (b) Investment (I), (c) Income (Y)

Chapter 9.

1. $P_t - P_t'$ (—) versus $P_t - P_t''$ (---), $\alpha = -0.8$ 273
2. $P_t - P_t'$ (—) versus $P_t - P_t''$ (---), $\alpha = -4.5$ 273

Chapter 10.

1. Transition Probability Matrix 285
2. The EM algorithm 289
3. Y and S sequences 295
4. X sequence 296
5. Sequence of log-likelihood values 296
6. Actual and smoothed state sequences 297
   (a) Time-varying probabilities, MSE = 0.13
   (b) Constant probabilities, MSE = 0.27
7. Actual and fitted values 298
   (a) p^{00}, (b) p^{11}

List of Tables

Chapter 3.

1. Macroeconomic Variable Notation and Description 62
2. Best *Bayes Models* for Australian Macro Time Series 64
3. Forecasting Exercises for Australian Macroeconomic Time Series, 1988–91 83

Chapter 4.

1. Monte Carlo Results for Base Simulation 111
2. Monte Carlo Results with a Sample Size of 50 113
3. Monte Carlo Results with a High ECM Error Variance compared to the Base Case 114
4. Monte Carlo Results with a Low ECM Error Variance compared to the Base Case 115
5. Monte Carlo Results with y not in the Common Trends 116
6. Monte Carlo Results with Highly Autocorrelated Equilibrium Error ($\rho = 0.9$) 118
7. Monte Carlo Results with Independent Moving Average Errors ($\theta = 0.5$) 119
8. Monte Carlo Results with Positively Interrelated Moving-Average Errors 120
9. Monte Carlo Results with Negatively Interrelated Moving Average Errors 121
10. Monte Carlo Results with Contemporaneously Positively Correlated Errors 122
11. Monte Carlo Results with Contemporaneously 'Negatively' Correlated Errors 123
12. Estimation of CV with low variance in a 2-dimensional cointegrating space 124
13. Estimation of a cointegrating vector where the 'regressors' are cointegrated amongst themselves 126

14. Median Biases with a subset of the regressors cointegrated amongst themselves 127

Chapter 5.

1. Asymptotic Critical Values 148
2. Powers of Tests 150

Chapter 6.

1. HEGY tests of quarterly series 170

Chapter 7.

1. Augmented Dickey-Fuller Test Statistics 198
2. Cointegration Analysis for Pairs of NSA and SA Series 200
3. A Cointegration Analysis of NSA Data: $\{m - p, \Delta p, y, R^*\}$ 202
4. A Cointegration Analysis of SA Data: $\{m^a - p^a, \Delta p^a, y^a, R^*\}$ 203
5. A General Autoregressive Distributed Lag for Nominal Money, Conditional on Prices, Incomes, and Interest Rates (NSA Data) 206

A1. Data Definitions and Sources 215
A2. The Four Primary Breaks in M_1 216
B1. The Unrestricted Error Correction Model for NSA Data 218
B2. *F*- and Related Statistics for the Sequential Reduction from the Fifth-order ADL Model in Table 5 (NSA Data) 219

Chapter 8.

1. Tests for Integration in Data using ADF(4) 237
2. Tests of Cointegrating Relations in (15a) and (15b) 240
3. Estimates for Multiplier-Accelerator Model 245
A. Parameter Values for the BER Model 253

Chapter 9.

1. Projection Results for H_0^f, Exact Equation Case 274
2. Projection Results for H_0^f, Random Walk Case 275
3. Projection Results for H_0^g, Exact Equation Case 276
4. Projection Results for H_0^g, Random Walk Case 277
5. First Order Autoregressions 278

Chapter 10.

1. Estimation Results 295

List of Contributors

Fabio Canova European University Institute

Michael P. Clements Institute of Economics and Statistics, Oxford University

Francis X. Diebold Department of Economics, University of Pennsylvania, Philadelphia

Steven N. Durlauf Department of Economics, University of Wisconsin, Madison

Neil R. Ericsson Federal Reserve Board, Washington

Mary Finn Federal Reserve Bank of Richmond, Richmond

Colin Hargreaves National Centre for Development Studies, Australian National University

David Harris Department of Econometrics, Monash University

David F. Hendry Nuffield College, Oxford University, Oxford

Mark A. Hooker Department of Economics, Dartmouth College

Svend Hylleberg Institute of Economics, Aarhus University, Aarhus

Brett Inder Department of Econometrics, Monash University

Joon-Haeng Lee Korea Securities Research Institute, Seoul

Adrian R. Pagan Economics Division, Research School of Social Sciences, Australian National University

Peter C. B. Phillips Cowles Foundation for Research in Economics, Yale University

Hong-Anh Tran O.E.C.D., Paris

Gretchen C. Weinbach Department of Economics, University of Pennsylvania, Philadelphia

Foreword

These papers were invited for a special symposium as part of the Australasian Economic Modelling Conference in 1992. In alphabetic order, the speakers were Francis Diebold, Steven Durlauf, Neil Ericsson, Colin Hargreaves, David Hendry, Svend Hylleberg, Adrian Pagan and Peter Phillips. The paper by David Harris and Brett Inder was a contributed paper that we all recognised as worth adding and fitted well with the other papers. Thus this is not a conference volume but rather a set of papers collected together on the particular topic, 'Nonstationary Time Series Analysis and Cointegration', matching the particular research interests of the authors.

This annual conference series has now been renamed the 'Asia-Pacific Economic Modelling Conference' and rather than publish the invited papers in book form they will in future appear in a new journal which we propose to call the 'Asia-Pacific Economic Review'. The first issue of this will be officially January 1995, but hopefully it will be available before then. The major topic of the first issue will be 'Savings and Investment'. The journal specifically aims to publish quantitative economic papers, ranging from quantitative descriptive analyses, to computer general equilibrium models to econometric analyses and modelling. Contributed papers are welcome and should be sent to Colin Hargreaves at the Australian National University.

Before these conferences, we always run a three-day training course related to the main topic of the conference and taught by some of the invited speakers. In 1992, David Hendry, Peter Phillips and Adrian Pagan taught the course, which was entitled 'Business Cycle Modelling, Macroeconometrics and Cointegration'. I thank them warmly for giving such a splendid course. The panel discussion at the end of the course was exceptional such that we all regretted that we had not recorded it. Thus we decided to record the panel discussion at the end of the conference and some comments from that are summarised in Chapter 1.

To date, these conferences have been near Cairns in North Queensland, near the Great Barrier Reef. I thank all the many invited speakers who have made this conference series such a great success. To have about 10 to 12 world renowned economists/econometricians come together in such a relaxed environment, concentrating on a particular research topic, has led to the most interesting and fruitful discussions, in which everyone was able to take part.

I would also like to thank Clive Granger and Grayham Mizon for reviewing these papers and accepting them in this fine series that they have established. This book has been formatted using Microsoft's Word for Windows plus Design Science's MathType. Thanks also go to Debbie Robertson and my wife and eldest daughter, Marian and Philippa Hargreaves, for their typing and editorial help.

COLIN HARGREAVES
Australian National University

1

Introduction

The analysis of nonstationary time series, unit roots and cointegration has developed dramatically over the last 12 years. The papers here variously describe new methods, they evaluate methods, they provide useful overviews and they show detailed implementations that are helpful to practitioners. However these papers show not only developments in methods of estimating long-run relations (for seasonally unadjusted as well as adjusted data, for instance), but also how the whole subject matter is broadening out to have a profound affect on econometric analysis in general. Michael Clements and David Hendry's discussion of economic forecasting is based around an integral understanding of integration and cointegration including an analysis of whether one should use differenced models. Fabio Canova, Mary Finn and Adrian Pagan's analysis of evaluating real business cycle models and Steven Durlauf and Mark Hooker's analysis of Cagan's hyperinflation model equally work from an understanding of the different time series properties to be found in nonstationary versus stationary data. The concepts of integration and cointegration are here to stay although some of the demands made on the data are very high as we shall discuss later.

The first paper by Michael Clements and David Hendry is a substantive and innovative contribution to analysing economic forecasting, looking at the various sources of error such as 'parameter non-constancy, estimation uncertainty, variable uncertainty, innovation uncertainty, and model misspecification', and trying to organise these into a coherent theory of economic forecasting. This approach appears to have great potential and certainly leads to a better appreciation of the use of intercept adjustments to improve forecasting. One could see this sort of approach being used also for a well-organised analysis of the many different types of error involved in the creation of the data itself and how balancing and other techniques can best be used by statistical agencies.

The second paper by Peter Phillips also deals with forecasting performance but more as a tool for evaluating different models. This paper provides an empirical application of his recent work with Werner Ploberger on building 'Bayes models' for economic time series. These papers were collected together for a special symposium in 1992 on nonstationary time series analysis and cointegration, following a training course taught by David Hendry, Adrian Pagan and Peter Phillips. At the end of the symposium, a panel discussion went over some of the issues raised by these papers and a summary of the discussion is given below.

Peter Phillips' method provides one way of dealing specifically with an issue raised in the panel discussion, that of separately determining lag length from cointegrating rank in the Johansen reduced rank regression methods. Here the one criterion is used to determine lag order, parameter constancy, and the presence of deterministic and stochastic trends. The adaptive method is a highly data-based approach that allows changing *numbers* of coefficients in a model as time passes (in addition to changing values) and it allows explosive roots and changing orders of integration. The coherence of a statistical method and its data orientation were also discussed fully at the panel discussion.

The next four papers are analyses and expositions of techniques of estimating cointegrating relations. In Chapter 3, Colin Hargreaves provides an overview of many of the extant estimation methods and then some simulation results showing their different properties. Testing for cointegration is the subject of the paper by David Harris and Brett Inder. It is sometimes argued that it would be better, when determining whether a series has unit roots or not, to test a null of stationarity rather than the ADF null of nonstationarity. Kwiatkowski *et al.* (1992) provide such a unit root test and David Harris and Brett Inder extend this approach to include explanatory variables in order to test the null of cointegration as opposed to the ADF style null of no cointegration. They provide critical values, a Monte Carlo comparison to ADF style tests and an application as well.

The following two papers look at questions of seasonal integration and cointegration. Svend Hylleberg provides an overview of methods of testing for seasonal unit roots and seasonal cointegration, showing results with interesting graphics on a few particular data series. He also discusses a definition of seasonality and likely future developments in the analysis of seasonality. Neil Ericsson, David Hendry and Hong-Anh Tran look specifically at a question often evaded in applied work, that is, the question of using seasonally adjusted and/or unadjusted data and they describe encompassing tests to compare models estimated with each type of data. The paper provides a detailed description of an application of David Hendry's general-to-specific modelling approach to the demand for money in the UK and the associated testing procedures.

Fabio Canova, Mary Finn and Adrian Pagan propose testing a model's implied cointegrating and dynamic restrictions on the VAR representation as a way of evaluating small linear models such as real business cycle models. While similar to the procedure of assessing a model by its power to explain certain stylised facts, it does provide a rational metric for assessing whether the stylised facts or restrictions are satisfied although this does naturally depend on the statistical model specification adopted. In their paper, Steven Durlauf and Mark Hooker show how it is possible to test for rational bubbles separately from testing the specification of the Cagan hyperinflation model. Both Canova *et al.* and Durlauf and Hooker find evidence of misspecification of their respective underlying models. The sudden inflationary bubbles studied by Durlauf and Hooker may appear sometimes like sudden switches in regime which, rather than being random, may relate to underlying economic fundamentals. In this way, Francis Diebold, Joon-Haeng Lee and Gretchen Weinbach extend the Markov switching model so that the transition probabilities are related to the underlying economic fundamentals instead of being constant. They provide an estimation algorithm and a simulation example.

PANEL DISCUSSION

The first issue, raised by Francis Diebold, was about how to determine the appropriate loss function to evaluate forecasts. He argued that mean square error was used for little more than mathematical convenience and this was rather like telling an economic agent what his utility function should be; one really ought to determine the relevant loss function for each particular objective.

While this was fine for a particular objective, David Hendry argued that in the macro context, there is rarely one objective; government ministers will continually change what the objective is, even in the course of a short discussion. However this is very different to a broker on the stock market buying and selling a bond where there is a very clear objective. One participant commented on the problem of having to reconcile different point estimates so that they were internally consistent as otherwise they were not presentable to a minister.

Peter Phillips felt that all this raised the general question of what is the best model, i.e. forecast performance of individual series or linear combinations and transformations as for example the balance of payments and imports and exports.

He felt that the best model need not be best in terms of mean square error on every single variable. There had to be a trade-off; better 'general explanation' may be at the cost of high error in one or two variables. Peter Phillips felt we needed systems criteria to be applied to the full vector of variables across the full forecast horizon. He added that we will always have a problem in comparing the large multi-equation multi-parameter models with large numbers of exogenous variables with the more simplistic VAR models whose purpose is very different.

David Hendry then responded to a participant asking about the practical considerations when choosing the lag length in a Johansen VAR. David Hendry commented that it was very dangerous to underestimate the lag length as then one's inferences would not be robust. Overestimation of the lag length will change the actual size of the test relative to the nominal size because of the introduction of so many nuisance parameters and he mentioned his own work on Bartlett-type ($T-k$) corrections which will correct the size but not however improve the power. It was not quite clear as to what the 'loss' of power is; you simply get the power you got.

There was quite a long discussion about the number of parameters estimated in Johansen VAR's compared to the number of observations available. Peter Phillips felt less optimistic about the value of Bartlett-type corrections as he felt that the Johansen procedure has a serious power problem like that of the ADF unit root test. The Johansen system is unrestricted with respect to all transient dynamics and thus, in a 5 variable system with say 5 lags leading to 125 parameters, even with 150 observations, you are still in a small sample situation and so lack power.

There was an unresolved discussion as to whether the relevant comparison of the number of parameters was to the number of observations or the number of observations times the number of variables. Peter Phillips argued that if you wanted to bring in the number of variables then you ought to bring in the parameters of the variance-covariance matrix as well. He stressed that reduced rank regression, like other optimal restricted regression techniques, involves estimation of the error covariance matrix. In contrast, unrestricted VAR and even the BVAR method of Litterman (1986) are really single regression techniques.

David Hendry emphasised the difficulty of estimating just one cointegrating vector in a four dimensional system, let alone two, if you have not many degrees of freedom left over. He felt that these procedures should be used in a confirmatory rather than exploratory sense. He expressed concern at people seeing how many cointegrating vectors they came up with and then trying to interpret them. He felt that you should have a good idea as to how many cointegrating vectors you anticipate and what they should look like and then check that your findings are consistent with this.

Peter Phillips agreed strongly with this confirmatory orientation and noted that one thing missing in the literature was LM tests for cointegration with which he was presently working in a semiparametric context. He also expressed concern over *t*-ratios. In his simulation studies with Loretan and with Hansen (Phillips and Hansen, 1990, and Phillips and Loretan, 1991), the performance of asymptotic inference procedures in systems of equations where you have large numbers of variables on the right-hand side was substantially affected by the extent to which you 'sweat' everything out of the error term. When you push your standard error down by throwing everything into the regression but the kitchen sink, so to speak, the distribution of the *t*-ratios compensates by becoming too dispersed. This distorts inference. By contrast, the FM procedure seems to perform quite well with regard to inference.

With respect to the lack of observations and power, Adrian Pagan commented that what the Johansen and Phillips-Hansen Modified estimators were all about was making the regressors exogenous. Whereas in the past we would just be worried about a contemporaneous correlation with the error term for which we might use 2SLS etc, these new estimators are concerned to account also for the lagged correlation of the regressors with the current error terms; the lags here could potentially be very long. Many of these correlations are likely to be quite spurious, given the poorly measured data that we have, and so Adrian Pagan wondered about both the value and feasibility of doing this, given only 50 to 100 observations, feeling that the Cowles Commission people such as Koopmans would have been amazed to hear that we could even think of doing this sort of thing. He questioned whether trying to estimate these long-run covariances was just asking too much of the data.

Peter Phillips commented here that Koopmans left econometrics because he ultimately believed that it was very important to deal with serial correlation and, with the small samples available, Koopmans felt that it was just an impossible task. However he added that we need to bear in mind the important consideration that we are now dealing with I(1) systems and I(1) regressors and nonstationarity. This causes the problem that the endogeneity, that needs to be flushed out of the system in order to correct the regressors so that they are properly exogenous, is locked in with the fact that there is serial dependence in the system.

The problem is that the correct way of adjusting the regressors on the right-hand side relies on the difficult problem of estimating the spectrum at a point. And furthermore, estimating the spectrum at the origin is particularly difficult as we do not have an infinite data set but only n observations and we are trying to estimate the lowest frequency of all.

Peter Phillips emphasised the need for extensive diagnostic testing particularly over the issue of whether to include a constant or trend. With I(1) regressors, one is very likely to assume there is a deterministic trend present in the regression, to avoid as the spurious regression effects and so that classical tests have power against trend stationarity. However, the mechanical inclusion of trends can be distortionary and certainly effects forecasting performance. Even constant terms are problematic as they are essentially drift terms when we have unit roots.

The question arose about how each estimation method arrives at its results, whether in a series of building blocks leading up to the final result or in a single shot. Many people had commented during the symposium that it was hard to see how the Johansen method obtained its results and why they varied so much as one changed the lag length or sample size. Some commented on how they found it difficult explaining the Johansen results to their superiors as few people understood eigen vectors or knew how to explain them simply. Everyone however could understand OLS, even if nothing more than in terms of fitting a line to a scattergram. Both Peter Phillips and Colin Hargreaves commented on how robust the OLS estimator can be under all sorts of different mis-specifications.

Adrian Pagan said that this was one reason why he liked the FM method as it starts from basic OLS and goes in stages, estimating the long-run variances and covariances next and then carrying out the modification. One can see whether the modification changes the results dramatically and, if so, look carefully at the long-run variance matrix and thereby check whether the change is due or not to some spuriously high correlations resulting from some particular glitch in the data, e.g. one caused by a strike. While involving quite a good deal of hard work tracing how the estimator came to its results, at least it was possible with the FM estimator while, in comparison, the Johansen method seemed like a black box. One may have many diagnostic tests to check the specification but one could never be sure what was the effect of each aspect of the specification.

Svend Hylleberg commented that this seemed to be contrary to the general-to-specific modelling approach. He felt that what you don't have in the model is in the residuals and so he felt you did not gain much by going in this step by step way. Adrian Pagan tried to distinguish between going step-wise in one's estimator and in one's model. He certainly believed in putting everything into the model that should be there, but did not believe in performing fancy adjustments using what was left over. There evidently was a problem in definitions here in that it is not clear that the FM estimator is not a specific to general estimator; one could describe the estimator as starting implicitly from a model where the long-run variances have no effect and then seeing whether they do. Instead of adding another variable with its slope parameter one is adding parameters related to the variances.

A problem with both methods, Johansen's Reduced Rank Regression and Phillips-Hansen's Fully Modified OLS, tends to be the lack of reporting of exactly how the estimates were obtained. Johansen's estimator does not directly tackle the issue of lag length. The transient dynamics are absorbed in coefficient matrices that are rarely printed out. Even when they are it is hard to comment on such a plethora of coefficients where there is no economic theory to explain the signs and often the majority are statistically insignificant. What criteria should be used to delete the insignificant ones (lag length versus absolute t-value) is not discussed. When a researcher calculates ADF statistics with other researchers, they often differ as to what lag to knock out next and even how far to go back in the first place. Two separate researchers will nearly always lead to different augmentation and hence different results, even though one cannot really differentiate between their separate final models in terms of error diagnostics.

To be fair to people reporting Johansen results, it is hard to know how one should report them. As just stated, there is no economic theory to explain the transient dynamics and so very little that one can say about them. At the same time, reporting them would take up much valuable space in an article. Many papers on long-run relations, e.g. on Australian money demand (Hargreaves, 1991, and Juselius and Hargreaves, 1992), report the eigenvectors and values from dozens of runs of the Johansen estimator; to have reported all the lag coefficient matrices and their standard errors for every model estimated as well would not have been practicable.

A not dissimilar thing occurs with the FM estimator. One researcher may use pre-whitening while another does not. One researcher may use a Parzen kernel while another uses a Bartlett or a Quadratic Spectral kernel. One may use automatic band-width selection (Andrews, 1991) while another nominates a specific integer value. One may assume that the regressors were I(1) with a linear time trend, while another assumes a quadratic time trend. Peter Phillips talked of using even more complex model selection procedures, now incorporated into his long-run variance estimation routines. Just as each researcher may clearly specify the lag length used in the Johansen estimator, they may clearly specify their choices here but it is very hard to know the effects of each choice and how they interact and determine the final estimates. There seems to be a belief that these decisions with the FM estimator do not make such a big difference as different lag lengths for the Johansen estimator.

One must acknowledge that estimating the long run is essentially a much more complicated procedure than OLS and yet because the results often end up with a single equation for the long run it looks much simpler. In fact, Engle and Granger's 1987 article gave the impression that it was that simple and that OLS could be used. The additional idea that economic theory was essentially only about the long run added to the emphasis on the long run. The Johansen estimator tends to lead one to thinking that there is something almost magical about the long run that is worth studying on its own. Many papers that use the Johansen estimator have often left the problem after estimating the long run, sometimes blithely giving lip-service to how one should now go on to making a full model. Peter Phillips emphasised the view of many that estimating the long run is only a stepping stone to a more sharply specified, economically interpretable, system of equations.

We are moving into a new phase in our understanding of these issues. Questions about the identification of the system are being raised but it seems unclear as to how specifying a separate system of long-run equations affects the identification of an economic system. For a while it seemed as if one could analyse the long run on its own and that economic policy should only aim at controlling the long run. However there is no causal direction in a cointegrating vector, only a co-relation over time. The question of control in nonstationary systems needs to be addressed carefully for the new econometric understanding to be of use to policy-makers.

REFERENCES

ANDREWS, D. W. K. (1991), 'Heteroskedasticity and Autocorrelation Consistent Covariance Matrix Estimation', *Econometrica*, 59, pp. 817–858.

ENGLE, R. F. and C. W. J. GRANGER (1987), 'Co-integration and error Correction: Representation, Estimation, and Testing', *Econometrica*, Vol. 55, No. 2, pp. 251–76.

HARGREAVES, C. P. (1991), Australian Long-Run Money Demand, mimeo.

HARGREAVES, C. P. (ed.) (1992), *Macroeconomic Modelling of the Long Run*, Edward Elgar Publishing Ltd, Aldershot, UK.

JORESKOG, K. G. and D. SORBOM (1981), *LISREL V: Analysis of linear structural relationships by maximum likelihood and least squares methods*, Research Report 81–8, Department of Statistics, Uppsala, Sweden.

JUSELIUS, K. and C. P. HARGREAVES (1992), 'Long-Run Relations in Australian Monetary Data', Chapter 10, pp. 249–86, in Hargreaves (1992).

KWIATKOWSKI, D., P. C. B. PHILLIPS, P. SCHMIDT and Y. SHIN (1992), 'Testing the Null Hypothesis of Stationarity Against the Alternative of a Unit Root: How Sure are We that Economic Time Series Have a Unit Root?', *Journal of Econometrics*, 54, pp. 159–78.

PHILLIPS, P. C. B. and B. E. HANSEN (1990), 'Statistical Inference in Instrumental Variables Regression with I(1) Processes', *Review of Economic Studies,* Vol. 57, pp. 99–125.

PHILLIPS, P. C. B. and M. LORETAN (1991), 'Estimating Long-Run Eonomic Equilibria', *Review of Economic Studies,* Vol. 59, pp. 407–36.

2

Towards a theory of economic forecasting

Michael P. Clements and David F. Hendry[*]

Some recent developments in the theory of economic forecasting using econometric systems are reviewed. Measures of forecast uncertainty for conditional and unconditional forecasts suggest a limit to forecastability and show the potential advantages of combining types of forecast to improve accuracy; but pooling forecasts from rival econometric models violates encompassing. Mean square forecast error comparisons are criticized in favour of generalized forecast error second moments: only the latter are invariant to linear transformations. We delineate five sources of forecast uncertainty: parameter non-constancy; estimation uncertainty; variable uncertainty; innovation uncertainty; and model misspecification. A theory of intercept corrections to mitigate such errors is discussed. Asymptotic forecast error variance formulae for nonstationary economic time series depend on the treatment of unit roots and cointegration but work well in finite samples, and show that forecast evaluation based on differenced data may fail to reveal inadequate models.

1. INTRODUCTION

We consider several aspects of the theory of economic forecasting based on econometric models. The term forecast is used to denote a statement about a future event or set of events; prediction is used to denote an implication of a model, so forecasts are a sub-class of predictions. We do not assume that the data generation process (DGP) is constant over time or that the model coincides with that DGP,

[*] Financial support from the U.K. Economic and Social Research Council under grant R000233447 is gratefully acknowledged by both authors. We are indebted to Neil Ericsson for helpful comments.

although for developing technical analyses we will often make both assumptions to simplify the algebra. We also focus on linear models for analytical tractability, see Mariano and Brown (1991) for references to the forecasting literature for nonlinear models. The discussion seeks to interlink a number of recent developments as follows.

Forecasts are rarely purely model-based, so other methods of forecasting come into play via the specification of non-zero intercept corrections over the forecast period. The role of intercept corrections arises naturally in Section 2 when considering conditional and unconditional forecasts for I(0) and I(1) processes. The former are conditional on some events occurring (e.g. no realignments of the European exchange rate mechanism). The analysis mainly deals with univariate processes to highlight the basic statistical calculations involved in forecasting. Forecast error and error variance calculations are outlined. The properties of these are important in appraising both the limits to forecastability and possible improvements in forecasting. Although model-based forecasts are conditional in nature, like most practical forecasts, we show that even with correct specification, such forecasts can sometimes be improved via intercept corrections which exploit unconditional forecasts.

Section 3 concerns forecasting with stationary systems of equations. We develop a taxonomy of sources of forecast uncertainty which distinguishes five major categories, namely parameter non-constancy, sampling variability, variable mismeasurement, error uncertainty, and model misspecification, with a number of sub-classes. None of the sources of forecast accuracy can be completely removed, but some can be partially mitigated. For example, measurement errors in the initial conditions entail that forecasts are conditioned on values of variables that may be poorly correlated with the true current state of the economy. Alternatively, parameter uncertainty and structural change may be a serious problem during particularly turbulent periods. Both initial condition and parameter uncertainties suggest that forecast accuracy may be improved via suitable intercept corrections. The potential impacts of forecasts, or expectations, on outcomes and the Lucas (1976) critique are also discussed.

Further rationales for intercept corrections are then discussed since forecasts from large-scale macroeconometric models are rarely based on the model alone but also reflect the skills of the models' proprietors. We draw on Hendry and Clements (1993), who provide an econometric theory of intercept corrections, to explain why such corrections typically improve forecast accuracy relative to model-based forecasts (see Turner, 1990). The forms of intercept corrections naturally depend on their rationale. Our theory does not seek to describe the reasoning of macroeconometric forecasters, but shows that the efficacy of their habitual practices can be understood.

The discussion of forecast comparisons in Section 4 centres on the limitations of the commonly used criterion of mean square forecast error (MSFE) for assessing forecast accuracy across models. MSFEs are criticized as a basis for model selection due to their non-invariance to linear transforms for a model class that is invariant. A generalized forecast error second moment (GFESM) measure that is invariant is described. MSFEs are also inadequate in that they fail to reflect non-constancy and non-encompassing, where the latter concerns whether the forecasts of one model can explain the forecast errors made by another. Each model has available to it the same information set as its rivals, so that if a rival model explains an additional feature, the model under scrutiny could be amended to do so also.

Any need to pool forecasts from distinct models can be taken as evidence of model misspecification. However, pooling between complementary sources may yield gains, since the previous strictures are not applicable to different types of information. For example, survey data can be a useful adjunct within models (rather than a substitute for econometric systems): the importance for forecast accuracy of the reliability of the initial conditions is made apparent below, and survey information may provide an alternative reading on the current state of the economy.

Since many economic time series need to be treated as nonstationary, probably I(1) processes, we develop representations for cointegrated systems in Section 5. That investigators may choose to analyse I(0) transformations of the variables highlights the concerns expressed in Section 4 about the lack of invariance of MSFE based measures to such transformations. Section 6 presents asymptotic formulae for h-step ahead MSFE matrices, derived by Clements and Hendry (1992), which reveal the lack of invariance of such measures to linear transformations. Further, they show how the assessment of forecast accuracy (based on MSFEs) over models depends upon the particular transformation of the variables chosen for forecasting. If an MSFE measure is used to evaluate the accuracy of various forecasts of the changes in an I(1) variable, then there is little chance of forecast inaccuracy revealing poor models (e.g. ones which omit cointegrating relations). Vector Autoregressive Representations (VARs) are often expressed in changes and hence omit error-correction feedbacks. However, evaluating VARs in differences by their ability to predict the changes of a series will have low power to detect misspecification, whereas the asymptotic formulae derived for the GFESM measure suggest that there is an excellent chance of uncovering such misspecified models. Conversely, VARs in levels often retain integrated components of variables and so could predict poorly.

In Section 7, we discuss the outcome of a Monte Carlo study investigating the implications for forecast accuracy of estimating the parameters of the econometric model, rather than knowing them *a priori*. The findings also help judge the usefulness of the asymptotic results as a guide to small sample outcomes. In particular, we investigate the impact of under-specifying the cointegrating rank versus retaining 'spurious' levels combinations. The Monte Carlo is based on a bivariate system, and we conjecture that this should favour unrestricted systems (such as VARs in levels) relative to multivariate settings.

Section 8 records conclusions and recommendations for forecasting practice.

Issues of model size and aggregation are not investigated as we have no useful points to make. Intuition suggests that for a restricted budget allocation, the larger the model beyond the minimum size needed to forecast the main aggregates of interest, the poorer its equations, and hence the poorer the forecasts. Historical track records are not discussed either, for the same reason; for the UK, the ESRC Macromodelling Bureau have investigated this matter in detail (see e.g. Wallis *et al.* 1986, 1987 and Wallis and Whitley, 1991). Also, seasonality is not considered: for recent results concerning I(0) and I(1) data respectively, see Hylleberg (1992) and Ericsson, Hendry and Tran (1992).

Finally, there are many methods of forecasting, which although conceptually distinct, in practice draw on each other. Most macroeconometric model-based forecasts make use of several methods, according them complementary rather than adversarial roles. Possible methods range from guessing; oracles; 'expert systems' (see Artis *et al.*, 1992); extrapolating recent tendencies; leading indicators (see Coen *et al.* (1969) and Stock and Watson, 1989); surveys of plans and anticipations by consumers and business; time-series models (see Kalman (1960), Box and Jenkins (1970), Nelson (1972) and Harvey and Shephard, 1992), including VARs (see Doan, Litterman and Sims (1984) and Todd (1990); or Hendry and Mizon (1993) and Clements and Mizon (1991) for a different view); as well as econometric systems. In using intercept corrections, for example, the forecaster is engaged in many of these methods of forecasting.

Irrespective of the method selected, three ingredients underpin forecasting:

(i) that there are regularities to capture;

(ii) that such regularities are informative about the future; and:

(iii) that the selected method captures them.

The first two are properties (or otherwise) of the economic system, common to all forecasting methods, whereas the third depends specifically on the method proposed. Economies appear to exhibit some constant features which can be embodied in appropriate methods; but not all features are constant. To be useful, therefore, macroeconometric models must:

(a) be well specified, which requires their congruence, so that the models fully embody the available data information;

(b) dominate alternatives, which is their encompassing ability, namely how well they account for the results obtained by rival explanations;

(c) remain constant even when economic policies change, which is their invariance to regime shifts and structural change;

(d) be accurately and precisely estimated, to minimize the impact of estimation uncertainty.

In so far as econometric models can forecast the future, they do so on the basis of available past and present information, and hence are formalized and tested leading-indicator systems.

2. FORECASTING IN UNIVARIATE PROCESSES

We begin by reviewing the theory of forecasting for stationary and integrated variables when the model and the DGP coincide and the latter is constant. We define forecasts, forecast errors, forecast error variances, and make explicit the distinction between conditional and unconditional forecasts. We then show how a combination of both may deliver a smaller forecast error variance than that of either alone, and how this trick can be incorporated to improve model-based forecasts. A univariate first-order autoregressive model (denoted AR(1)) will suffice to illustrate the main issues:

$$w_t = \mu + \psi w_{t-1} + v_t \ . \tag{1}$$

For simplicity, the disturbance term $\{v_t\}$ is taken to be an independent normal random variable with zero mean and constant variance σ^2, denoted by $\mathrm{IN}(0,\sigma^2)$. Let $\mathbf{x}_t = (1:w_t)'$ and $\boldsymbol{\theta} = (\mu:\psi)'$.

2.1 Stationarity

In the stationary case, the unconditional mean and variance exist:

$$E[w_t] = \frac{\mu}{1-\psi} \qquad \text{and} \qquad V[w_t] = \frac{\sigma^2}{1-\psi^2} \tag{2}$$

where E[·] and V[·] denote expectation and variance respectively. Given the information available at time T, for known parameters, the conditional mean is the minimum mean square error forecast of w in the next period, denoted $\widetilde{w}_{T+1}$:

$$\mathrm{E}\left[w_{T+1} | w_T\right] = \widetilde{w}_{T+1} = \mu + \psi w_T \tag{3}$$

with forecast error: $e_{T+1} = w_{T+1} - \widetilde{w}_{T+1} = v_{T+1}$. Hence $\mathrm{V}[w_{T+1}|w_T] = \mathrm{V}[e_{T+1}|w_T] = \sigma^2$ so the error variance equals the conditional variance.

In practice, μ and ψ must be estimated and the forecast error for estimated parameter values $\hat{\mu}$ and $\hat{\psi}$ becomes:

$$\hat{e}_{T+1} = w_{T+1} - \hat{w}_{T+1} = v_{T+1} + (\mu - \hat{\mu}) + (\psi - \hat{\psi}) w_T = v_{T+1} + \left(\theta - \hat{\theta}\right)' \mathbf{x}_T .$$

The variance of the forecast error is (ignoring a slight dependence of $\hat{\theta}$ on $\mathbf{x}_T$):

$$\mathrm{V}\left[\hat{e}_{T+1} | w_T\right] = \sigma^2 + \mathbf{x}_T' \mathrm{V}\left[\hat{\theta}\right] \mathbf{x}_T, \tag{4}$$

which thus depends upon the variance σ^2 of the innovation process, the variances of the estimated parameters $\mathrm{V}[\hat{\theta}]$, and the initial conditions. The conditional variance is less than the unconditional variance when the parameters are known (so $\mathrm{V}[\hat{\theta}] = \mathbf{0}$), but otherwise could be larger or smaller. In (4) (ignoring the first observation):

$$\begin{aligned} \mathrm{V}\left[\hat{\theta}\right] = \mathrm{V}\begin{bmatrix} \hat{\mu} \\ \hat{\psi} \end{bmatrix} &= \sigma^2 \mathrm{E}\begin{bmatrix} T & \sum_{t=1}^{T} w_{t-1} \\ \sum_{t=1}^{T} w_{t-1} & \sum_{t=1}^{T} w_{t-1}^2 \end{bmatrix}^{-1} \\ &\cong T^{-1}\begin{bmatrix} \sigma^2 + \mu^2 (1+\psi)(1-\psi)^{-1} & -\mu(1+\psi) \\ -\mu(1+\psi) & (1-\psi)^2 \end{bmatrix}. \end{aligned} \tag{5}$$

The last approximation is valid asymptotically and is used below.

Conditional on information available at the beginning of the forecast period, for known parameters, the h-step ahead forecast is obtained by backward iteration of $w_{T+h} = \mu + \psi w_{T+h-1}$, so denoting it by $\widetilde{w}_{T+h}$:

$$\widetilde{w}_{T+h} = \frac{\mu\left(1-\psi^h\right)}{1-\psi} + \psi^h w_T. \tag{6}$$

This is the conditional mean of w_{T+h} given w_T. In the stationary case, ψ^h tends to zero as h increases, and the forecast converges to the unconditional mean $\mu/(1-\psi)$. In this sense, stationary processes are well behaved. The actual value of w_{T+h} from (1) is:

$$w_{T+h} = \frac{\mu\left(1-\psi^h\right)}{1-\psi} + \psi^h w_T + \sum_{i=0}^{h-1} \psi^i v_{T+h-i}. \tag{7}$$

Thus, the multi-period forecast error variance for known parameters is:

$$\mathrm{V}\left[e_{T+h}\right] \equiv \mathrm{V}\left[w_{T+h} - \widetilde{w}_{T+h}\right] = \frac{\sigma^2\left(1-\psi^{2h}\right)}{1-\psi^2}. \tag{8}$$

Then, $\mathrm{V}[e_{T+h}]$ is also $\mathrm{V}[w_{T+h}|w_T]$, and from (8), it increases with h, converging to the unconditional variance of the process. We call the forecast horizon h_α at which the conditional variance is within α% of the unconditional variance the 'horizon of no information', and this establishes the limit to forecastability beyond the unconditional mean. For different purposes, such as forecasting GNP growth or the change in stock prices, α might be 10% or 1% of the unconditional variance.

When parameters are not known but have to be estimated, (6) becomes:

$$\hat{w}_{T+h} = \frac{\hat{\mu}\left(1-\hat{\psi}^h\right)}{1-\hat{\psi}} + \hat{\psi}^h w_T, \tag{9}$$

and hence:

$$\hat{e}_{T+h} = \sum_{i=0}^{h-1}\left(\mu\psi^i - \hat{\mu}\hat{\psi}^i\right) + \left(\psi^h - \hat{\psi}^h\right)w_T + \sum_{i=0}^{h-1}\psi^i v_{T+h-i}. \tag{10}$$

Thus, assuming that $\mathrm{E}[\hat{e}_{T+h}]$ is approximately zero[1]:

$$\begin{aligned}\mathrm{V}\left[\hat{e}_{T+h}|w_T\right] = {} & \frac{\sigma^2\left(1-\psi^{2h}\right)}{1-\psi^2} + \mathrm{E}\left[\left\{\sum_{i=0}^{h-1}\left(\mu\psi^i - \hat{\mu}\hat{\psi}^i\right)^2\right\}\right] + \mathrm{V}\left[\psi^h - \hat{\psi}^h\right]w_T^2 \\ & + \; 2\mathrm{E}\left[\left\{\sum_{i=0}^{h-1}\left(\mu\psi^i - \hat{\mu}\hat{\psi}^i\right)\right\}\left(\psi^h - \hat{\psi}^h\right)\right]w_T.\end{aligned} \tag{11}$$

[1] For symmetric error distributions this is a good approximation as can be shown using antithetic variates even though the parameter estimators are biased (see e.g. Hendry and Trivedi, 1972).

The four components correspond to the innovation variance accumulation, the intercept variance, the autoregressive parameter variance, and their covariance, but in awkward combinations of sums of powers. In the appendix, we approximate (11) by:

$$\mathrm{V}\left[\hat{e}_{T+h}|w_T\right] = \frac{\sigma^2\left(1-\psi^{2h}\right)}{1-\psi^2} + \mathbf{d}'\mathrm{V}\left[\hat{\theta}\right]\mathbf{d}, \tag{12}$$

where

$$\mathbf{d}' = \left(\frac{\left(1-\psi^h\right)}{1-\psi} : \left\{\frac{\mu\left[1-h\psi^{h-1}(1-\psi)\right]}{(1-\psi)^2} + h\psi^{h-1}w_T\right\}\right).$$

Given (5), analytical and numerical approximations can be calculated, but the former are unenlightening when $\mu \neq 0$.

2.2 Nonstationarity

The class of nonstationary processes considered contains stochastic trends, integrated of order one, denoted I(1). Many of the essential points arise when $\psi = 1$ in (1):

$$w_t = w_{t-1} + \mu + v_t \quad \text{or} \quad \Delta w_t = \mu + v_t \quad \text{where} \quad v_t \sim \mathrm{IN}\left(0, \sigma^2\right). \tag{13}$$

This class of equation is called difference stationary, since its first difference is a stationary process under the stated assumptions. In reality, economic time series may be integrated to higher orders, and nonstationarity may emanate from regime shifts and structural breaks, which may be better treated as deterministic in nature.[2] Nevertheless, we will investigate (13). Solving:

$$w_t = w_0 + \mu t + \sum_{i=1}^{t} v_i, \tag{14}$$

and hence past shocks persist indefinitely. Now, $\mathrm{E}[w_t] = w_0 + \psi t$, which trends over time, in contrast to the constant unconditional mean in the stationary case. However, the conditional mean of w_{T+1} given w_T, remains the same form as in (3), namely $\mathrm{E}[w_{T+1}|w_T] = \mu + w_T$.

[2] See the debate about whether US GNP has a unit root (e.g. Campbell and Mankiw, 1987, and references therein). Distinguishing between stochastic and deterministic non-stationarity is difficult in practice, not least because the debate turns on whether 'non-stationarity of economic time series is produced by the cumulation of permanent shocks at each observation point or by few large shocks occurring infrequently', Reichlin (1989, p.231).

The conditional variance of w_t in (13) for known parameters is $\mathsf{V}[w_t|w_{t-1}] = \sigma^2$. However, the unconditional variance is $\mathsf{V}[w_t] = \sigma^2 t$, which increases linearly over time, making distant forecasts of the level unreliable. The h-step ahead forecast for known parameters, conditional on information available at time T is:

$$\widetilde{w}_{T+h} = \mu + w_{T+h-1} = \mu h + w_T. \tag{15}$$

Thus, the forecast is of a change in the variable from the initial condition with a local trend. The conditional multi-period forecast error is:

$$e_{T+h} = w_{T+h} - \widetilde{w}_{T+h} = \mu h + w_T + \sum_{i=0}^{h-1} v_{T+h-i} - (\mu h + w_T) = \sum_{i=0}^{h-1} v_{T+h-i} \tag{16}$$

This has a cumulative error, with a variance that increases at $O(h)$ in the forecast horizon h:

$$\mathsf{V}[e_{T+h}] = h\sigma^2 \tag{17}$$

in contrast to the $O(1)$ stationary forecast error variance in (8). Conversely, from (13), changes cannot be forecast more accurately than their unconditional mean μ.

Stationary combinations of nonstationary variables (e.g. due to cointegration or differencing) will have multi-period forecast error variances that converge to the unconditional variance of the process (as in (8)): thus, Section 2.1 reapplies for forecasting Δw_t. However, the forecast error variances of the constituent series will increase with the forecast horizon (as in (17)). Engle and Yoo (1987) demonstrate these propositions using the vector moving average representation of a system of cointegrated variables (see *inter alia* Engle and Granger, 1987).

The h-step forecast error for levels using estimated parameter values is:

$$\begin{aligned}\hat{e}_{T+h} &= w_{T+h} - \hat{w}_{T+h} = \mu h + w_T + \sum_{i=0}^{h-1} v_{T+h-i} - \left(\hat{\mu} h + \hat{\psi}^h w_T\right) \\ &= (\mu - \hat{\mu})h + \left(1 - \hat{\psi}^h\right) w_T + \sum_{i=0}^{h-1} v_{T+h-i}.\end{aligned} \tag{18}$$

As before, we neglect the coefficient biases, treating such conditional forecasts as unbiased. When ψ is correctly imposed as unity, the forecast variance trends as the forecast horizon increases, as in (17), namely:

$$\mathsf{V}[e_{T+h}] = h\left(\sigma^2 + h\mathsf{V}[\hat{\mu}]\right) \cong h\sigma^2 (1 + h/T). \tag{19}$$

Generally, for estimated parameters in I(1) processes, the variance of the forecast error is hard to derive due to the non-standard nature of the distribution. However, the limiting distribution is normal when estimating (13) unrestrictedly for non-zero μ (see West, 1988): the estimate of ψ converges at a rate of $T^{3/2}$ (so its variance can be neglected) whereas $V[\hat{\mu}] = 4\sigma^2/T$, emphasizing the importance of accurately estimating the local trend.

2.3 Combining conditional and unconditional forecasts

Forecasts from macroeconomic models are conditional forecasts given the initial values of the variables as well as any projected future time paths of non-modelled variables. From the asymptotic formulae derived by Schmidt (1977) and Baillie (1979*a*, 1979*b*) for multi-step mean squared forecast errors, Chong and Hendry (1986) show that the conditional forecast error variances for a stationary process can exceed the unconditional variance of the process. Hendry and Clements (1993) develop their suggestion that a weighted average of the conditional and unconditional predictions may have a smaller variance than either alone. We draw on those analyses to provide a potential justification for intercept corrections and derive the form of intercept correction that would be entailed. We assume stationarity, so the h-step ahead conditional MSFE_h is given by (11), and suppress the intercept μ for simplicity to obtain:

$$\mathrm{E}\left[\hat{e}^2_{T+h}\middle|w_T\right] = \text{MSFE}_h = \frac{\sigma^2\left(1-\psi^{2h}\right)}{1-\psi^2} + \mathrm{E}\left[\left(\psi^h - \hat{\psi}^h\right)^2\right]w^2_T. \tag{20}$$

The first term is the contribution of future disturbances, and the second is parameter uncertainty, which we evaluate using the asymptotic formula in Baillie (1979*a*, equation 1.6, p.676) to give:

$$\text{MSFE}_h = \frac{\sigma^2\left(1-\psi^{2h}\right)}{1-\psi^2} + T^{-1}w^2_T h^2 \psi^{2(h-1)}\left(1-\psi^2\right). \tag{21}$$

Details of this derivation are provided by Ericsson and Marquez (1989, p.5), and as a special case of the appendix. The first term in (21) is monotonically increasing in h, approaching the unconditional variance of w_t asymptotically, but the term $h^2\psi^{2(h-1)}$ has a maximum at $h = -1/\log \psi$, and so is not monotonic. Hence for certain values of the parameters, the value of the expression in (21) may exceed the unconditional variance given by $V[w_t]$ in (2).

Next, consider the behaviour of a composite predictor formed by linearly combining the conditional predictor $\hat{w}$ with $\overline{w}$, where the latter does not depend on time and is a function of the information set at time T; for example, $\overline{w}$ could be the mean of a long sample of historical data. Intuitively, w_t can be thought of as a growth rate so that $\overline{w}$ is a sample estimate of the average growth rate of the series, but we treat it as having zero variance. In our example, the w_t process has a zero mean, so we set $\overline{w} = 0$. The composite predictor is:

$$\widetilde{w}_{T+h} = \alpha\overline{w} + (1-\alpha)\hat{w}_{T+h} = (1-\alpha)\hat{w}_{T+h}, \tag{22}$$

where $0 \leq \alpha \leq 1$, and the associated h-step ahead composite forecast error is:

$$\widetilde{e}_{T+h} \equiv w_{T+h} - \widetilde{w}_{T+h} = \left(w_{T+h} - \hat{w}_{T+h}\right) + \alpha\hat{w}_{T+h} = \alpha\overline{e}_{T+h} + (1-\alpha)\hat{e}_{T+h}, \tag{23}$$

where $\hat{e}_{T+h}$ is defined following (9) and $\overline{e}_{T+h} \equiv w_{T+h} - \overline{w} = w_{T+h}$. As (22) reveals, the benefit (if any) is due to 'shrinkage' and we are presently investigating more general shrinkage methods (such as Stein-James) applied to multi-period forecasts (see e.g. Judge and Bock, 1978).

Again treating the bias as negligible, the conditional MSFE for the composite predictor is:

$$\begin{aligned} \mathrm{E}\left[\widetilde{e}_{T+h}^2 \middle| w_T\right] = \alpha^2 \mathrm{E}\left[\overline{e}_{T+h}^2 \middle| w_T\right] + (1-\alpha)^2 \mathrm{E}\left[\hat{e}_{T+h}^2 \middle| w_T\right] \\ +2\alpha(1-\alpha)\mathrm{E}\left[\overline{e}_{T+h}\hat{e}_{T+h} \middle| w_T\right] \end{aligned} \tag{24}$$

which after some rearrangement and using (21) becomes:

$$\mathrm{E}\left[\widetilde{e}_{T+h}^2 \middle| w_T\right] = \alpha^2\psi^{2h}w_T^2 + \frac{\sigma^2\left(1-\psi^{2h}\right)}{1-\psi^2} + \frac{(1-\alpha)^2}{T}w_T^2 h^2 \psi^{2(h-1)}\left(1-\psi^2\right). \tag{25}$$

Minimizing (25) with respect to α yields:

$$\alpha^* = \left[1 + \frac{T\psi^2}{h^2\left(1-\psi^2\right)}\right]^{-1} \tag{26}$$

which does not depend on w_T. When $\alpha^* \neq 0$, the overall MSFE may be reduced by using the composite predictor defined in (23) (compare (21) and (25)). However, α^* will be close to zero unless h is large and T is small, so the gains may not be large and may be lost by estimating the long-run mean. From (21), the conditional variance will typically be less than the unconditional initially, and then rise above it as h increases, suggesting increasing α from 0 to 1 as h increases (or from zero rising to a maximum and then back to zero, since the conditional and unconditional forecast error variances converge). From (23) this implies an intercept correction that adjusts the model prediction towards zero by an increasing amount as the number of steps ahead increases, and then tails off to leave the 'pure' model forecasts.

3. FORECASTING IN STATIONARY SYSTEMS

3.1 Open systems

In any practical forecasting exercise, there will be a number of additional complicating factors to the simple univariate forecasting schemes with known parameters discussed in Section 2, including the following:

i forecasting is usually undertaken from a system;
ii there are non-modelled variables which must be separately forecast;
iii the system is usually nonlinear and involves longer lagged reactions than just one period;
iv parameters are unknown and have to be estimated;
v the initial conditions are subject to error and revision;
vi the model may not characterize the data to within an innovation error;
vii parameters may change in the forecast period; and
viii the error process distribution may change over time (see e.g. Engle, 1982).

Many of these problems can be classified within a general taxonomy of the factors that contribute to forecast error. We establish the general framework before discussing ways in which the individual sources of forecast error can be ameliorated. In some cases, the appropriate course of action suggested by our analysis can be implemented by specifying non-zero equation errors over the forecast period, namely intercept corrections, following Hendry and Clements (1993).

Consider a structural linear stochastic econometric model of k endogenous variables $\mathbf{y}_t$:

$$\mathbf{B}_0\mathbf{y}_t + \mathbf{B}_1\mathbf{y}_{t-1} + \cdots + \mathbf{B}_r\mathbf{y}_{t-r} = \mathbf{C}_0\mathbf{z}_t + \mathbf{C}_1\mathbf{z}_{t-1} + \cdots + \mathbf{C}_s\mathbf{z}_{t-s} + \varepsilon_t \quad (27)$$

where $\varepsilon_t \sim \mathsf{IN}(\mathbf{0},\Sigma)$. The $\{\mathbf{B}_i\}$ are the $k \times k$ coefficient matrices of the k vector of endogenous variables lagged i periods, and the $\{\mathbf{C}_i\}$ are the $k \times n$ coefficient matrices on the non-modelled and policy variables $\mathbf{z}_t$. Assuming that $\mathbf{B}_0$ is nonsingular, the reduced form is:

$$\begin{aligned}\mathbf{y}_t &= \mathbf{B}_0^{-1}\mathbf{C}_0\mathbf{z}_t + \cdots + \mathbf{B}_0^{-1}\mathbf{C}_s\mathbf{z}_{t-s} - \mathbf{B}_0^{-1}\mathbf{B}_1\mathbf{y}_{t-1} - \cdots - \mathbf{B}_0^{-1}\mathbf{B}_r\mathbf{y}_{t-r} + \mathbf{B}_0^{-1}\varepsilon_t \\ &= \Gamma_0\mathbf{z}_t + \cdots + \Gamma_s\mathbf{z}_{t-s} + \Phi_1\mathbf{y}_{t-1} + \cdots + \Phi_r\mathbf{y}_{t-r} + \eta_t.\end{aligned} \quad (28)$$

Writing the system in companion form, one lag is fully general, and hence we consider:

$$\mathbf{A}_0\mathbf{y}_t + \mathbf{A}_1\mathbf{y}_{t-1} = \mathbf{C}\mathbf{z}_t + \varepsilon_t \tag{29}$$

with reduced form:

$$\mathbf{y}_t = \Gamma\mathbf{z}_t + \Phi\mathbf{y}_{t-1} + \eta_t \tag{30}$$

where $\eta_t \sim \mathrm{IN}(\mathbf{0},\Omega)$. If the autoregressive component of the model is dynamically stable, $\Phi^t \to \mathbf{0}$ as $t \to \infty$, and by backward substitution:

$$\mathbf{y}_t = \sum_{i=0}^{\infty} \Phi^i \Gamma \mathbf{z}_{t-i} + \sum_{i=0}^{\infty} \Phi^i \eta_{t-i}, \tag{31}$$

so that $\mathbf{y}_t$ is the cumulation of all past $\mathbf{z}$'s and disturbances, with declining weights.

Let $\mathbf{w}_t' = (\mathbf{z}_t' : \mathbf{y}_{t-1}')$ and $\Pi = (\Gamma:\Phi)$, so that the system (30) becomes:

$$\mathbf{y}_t = \Pi\mathbf{w}_t + \eta_t. \tag{32}$$

One-step forecasts for time $T+i$ made at $T+i-1$ with known data but estimated parameters — as used in testing parameter constancy — are:

$$\hat{\mathbf{y}}_{T+i} = \hat{\Pi}\mathbf{w}_{T+i}, \tag{33}$$

with forecast errors $\mathbf{y}_{T+i} - \hat{\mathbf{y}}_{T+i} = \hat{\mathbf{e}}_{T+i}$, where:

$$\hat{\mathbf{e}}_{T+i} = \left(\Pi - \hat{\Pi}\right)\mathbf{w}_{T+i} + \eta_{T+i} = \left(\mathbf{I} \otimes \mathbf{w}_{T+i}'\right)\left(\Pi - \hat{\Pi}\right)^{v} + \eta_{T+i}, \tag{34}$$

when $(\cdot)^v$ denotes column vectoring and $\otimes$ is the Kronecker product, defined by $\mathbf{A} \otimes \mathbf{B} = \{b_{ij}\mathbf{A}\}$. If $\hat{\Pi}$ is unbiased for Π, to a first approximation, $\mathrm{E}[\hat{\mathbf{e}}_{T+i}] = \mathbf{0}$ and:

$$\mathrm{V}[\hat{\mathbf{e}}_{T+i}] = \Omega + (\mathbf{I} \otimes \mathbf{w}_{T+i}')\mathrm{V}\left[\hat{\Pi}^v\right](\mathbf{I} \otimes \mathbf{w}_{T+i}) = \Theta_i, \tag{35}$$

so that (see Calzolari (1981) and Chong and Hendry, 1986):

$$\hat{\mathbf{e}}_{T+i} \underset{app}{\sim} \mathrm{IN}\left(\mathbf{0},\Theta_i\right) \tag{36}$$

where Θ_i reflects both parameter and innovation error uncertainty. Ex post, for model evaluation over h periods, on the null of correct specification and parameter constancy:

$$\sum_{i=1}^{h} \hat{\mathbf{e}}_{T+i}' \hat{\Theta}_i^{-1} \hat{\mathbf{e}}_{T+i} \underset{app}{\sim} \chi^2(kh). \tag{37}$$

Next, we derive the vector analogues of the h-step ahead conditional forecast errors and error variances for univariate processes given in Section 2 by (8), (11), (17) and (19). To begin with, we assume that parameters are constant, and their values, initial conditions and future $\mathbf{z}$'s are known and correct. The *1*-step ahead forecasts for time $T+j$ made at time $T+j-1$ are, from (30):

$$\widetilde{\mathbf{y}}_{T+j} = \mathbf{\Phi}\widetilde{\mathbf{y}}_{T+j-1} + \mathbf{\Gamma}\mathbf{z}_{T+j} \tag{38}$$

and the *1*-step ahead forecast errors are defined recursively by:

$$\widetilde{\eta}_{T+j} = \mathbf{y}_{T+j} - \widetilde{\mathbf{y}}_{T+j} = \mathbf{\Phi}\left(\mathbf{y}_{T+j-1} - \widetilde{\mathbf{y}}_{T+j-1}\right) + \eta_{T+j} = \mathbf{\Phi}\widetilde{\eta}_{T+j-1} + \eta_{T+j} \tag{39}$$

so that by backward substitution:

$$\widetilde{\eta}_{T+j} = \sum_{i=0}^{j-1} \mathbf{\Phi}^i \eta_{T+j-i} + \mathbf{\Phi}^j \widetilde{\eta}_T = \sum_{i=0}^{j-1} \mathbf{\Phi}^i \eta_{T+j-i}, \tag{40}$$

since $\widetilde{\mathbf{y}}_T = \mathbf{y}_T$ which implies that $\widetilde{\eta}_T = \mathbf{0}$. Hence the *j*-step ahead conditional forecast error can be expressed as a weighted average of all previous *1*-step ahead errors (which are just the future realizations of the underlying error process, given our assumptions). Some of these errors will cumulate indefinitely when $\mathbf{\Phi}$ has a root of unity.

The forecast error variance from (40) is given by:

$$\mathrm{E}\left[\widetilde{\eta}_{T+j}\widetilde{\eta}'_{T+j}\middle|\mathbf{y}_T\right] = \mathrm{E}\left[\sum_{i=0}^{j-1}\sum_{k=0}^{j-1} \mathbf{\Phi}^i \eta_{T+j-i}\eta'_{T+j-k}\mathbf{\Phi}^{k\prime}\right] = \sum_{i=0}^{j-1} \mathbf{\Phi}^i \mathbf{\Omega} \mathbf{\Phi}^{i\prime}. \tag{41}$$

The forecast standard errors increase with the horizon ahead, but converge to the variance of $\mathbf{y}_t$, given current and past $\mathbf{z}_t$, when all the latent roots of $\mathbf{\Phi}$ lie inside the unit circle, and otherwise diverge indefinitely. Given stationarity, the variance of $\mathbf{y}_t$ can be derived from (31) as:

$$\mathrm{E}\left[(\mathbf{y}_T - \mathrm{E}[\mathbf{y}_T])(\mathbf{y}_T - \mathrm{E}[\mathbf{y}_T])'\right] = \sum_{i=0}^{\infty} \mathbf{\Phi}^i \mathbf{\Omega} \mathbf{\Phi}^{i\prime}. \tag{42}$$

When parameters are estimated but the model coincides with the constant DGP, a generalization of (41) to account for parameter uncertainty is possible (see Schmidt (1977), Spitzer and Baillie (1983), Chong and Hendry (1986) and Calzolari, 1987).

3.2 Closed systems

We now consider dynamic forecasts from a closed system where parameters are subject to change. The closed assumption keeps the algebra tractable. Although all non-deterministic variables must be forecast in operational usage of models, we appreciate that autoregressive schemes may not well approximate the rules used by modellers to pre-predict non-modelled variables. All the variables are denoted by $\mathbf{w}_t$ as before, and we write the closed system as:

$$\mathbf{w}_t = \Upsilon \mathbf{w}_{t-1} + \nu_t \tag{43}$$

where $\nu_t \sim \mathsf{IN}(\mathbf{0},\Omega)$. Let the parameter estimates be $\hat{\Upsilon}$ where $\text{plim}_{T\to\infty}\hat{\Upsilon} = \Upsilon_p \neq \Upsilon$ (in general) due to model misspecification. Then, commencing from initial conditions $\hat{\mathbf{w}}_T$ which may differ from the true values ($\hat{\mathbf{w}}_T \neq \mathbf{w}_T$), j-step ahead forecasts for j=1,...,h are:

$$\hat{\mathbf{w}}_{T+j} = \hat{\Upsilon}\,\hat{\mathbf{w}}_{T+j-1} = \hat{\Upsilon}^j\,\hat{\mathbf{w}}_T. \tag{44}$$

We allow for the possibility that the system does not remain the same in the forecast period, so that in fact Υ changes to Υ_j^* (j = 1,...,h) and the mean and variance of the error also change:

$$\begin{aligned}\mathbf{w}_{T+j} &= \Upsilon_j^*\,\mathbf{w}_{T+j-1} + \nu_{T+j} = \nu_{T+j} + \Upsilon_j^*\,\nu_{T+j-1} + \Upsilon_j^*\,\Upsilon_{j-1}^*\,\mathbf{w}_{T+j-2} \\ &= \sum_{i=0}^{j-1}\Upsilon_{j-i+1}^{**}\,\nu_{T+j-i} + \Upsilon_0^{**}\mathbf{w}_T,\end{aligned} \tag{45}$$

where $\nu_{T+j} \sim \mathsf{D}(\mu_j,\Omega_j)$, and by convention $\Upsilon_{j+1}^* = \mathbf{I}$, when:

$$\Upsilon_{j-i+1}^{**} = \prod_{k=0}^{j-1}\Upsilon_{j-k}^*. \tag{46}$$

Letting $\hat{\nu}_{T+j} = \mathbf{w}_{T+j} - \hat{\mathbf{w}}_{T+j}$, the j-step ahead forecast error can then be decomposed into the following five components:

$$\begin{aligned}\hat{\nu}_{T+j} &= \sum_{i=0}^{j-1}\Upsilon_{j-i+1}^{**}\,\nu_{T+j-i} + \Upsilon_0^{**}\mathbf{w}_T - \hat{\Upsilon}^j\,\hat{\mathbf{w}}_T \\ &\cong \sum_{i=0}^{j-1}\Upsilon_{j-i+1}^{**}\,\nu_{T+j-i} + \left(\Upsilon_0^{**} - \Upsilon^j\right)\mathbf{w}_T + \left(\Upsilon^j - \Upsilon_p^j\right)\mathbf{w}_T \\ &\quad + \left(\Upsilon_p^j - \hat{\Upsilon}^j\right)\mathbf{w}_T + \Upsilon_p^j\left(\mathbf{w}_T - \hat{\mathbf{w}}_T\right),\end{aligned} \tag{47}$$

where the estimated parameter multiplying the error in the initial conditions has been replaced by the sample period plim (dropping a term of $O_p(T^{-1})$). The five components of forecast error are due to error accumulation; parameter non-constancy; model misspecification (i.e. inconsistent parameter estimates); sampling variability; and mismeasured initial conditions. The middle three terms can be combined to form the composite deviation $(\Upsilon_0^{**} - \hat{\Upsilon}^j)\mathbf{w}_T$, but (47) will prove useful.

In the case of open systems like (30), with off-line forecasts denoted by $\breve{\mathbf{z}}_{T+j}$, the *1*-step ahead error in forecasting the value of $\mathbf{y}_{T+j}$, $\hat{\eta}_{T+j} = \mathbf{y}_{T+j} - \hat{\mathbf{y}}_{T+j}$ is:

$$\begin{aligned}\hat{\eta}_{T+j} &= \Gamma_j^* \mathbf{z}_{T+j} - \hat{\Gamma}\breve{\mathbf{z}}_{T+j} + \Phi_j^* \mathbf{y}_{T+j-1} - \hat{\Phi}\hat{\mathbf{y}}_{T+j-1} + \eta_{T+j} \\ &= \left(\Gamma_j^* - \Gamma\right)\mathbf{z}_{T+j} + \left(\Gamma - \Gamma_p\right)\mathbf{z}_{T+j} + \left(\Gamma_p - \hat{\Gamma}\right)\mathbf{z}_{T+j} + \hat{\Gamma}\left(\mathbf{z}_{T+j} - \breve{\mathbf{z}}_{T+j}\right) \\ &\quad + \left(\Phi_j^* - \Phi\right)\mathbf{y}_{T+j-1} + \left(\Phi - \Phi_p\right)\mathbf{y}_{T+j-1} + \left(\Phi_p - \hat{\Phi}\right)\mathbf{y}_{T+j-1} \\ &\quad + \hat{\Phi}\left(\mathbf{y}_{T+j-1} - \hat{\mathbf{y}}_{T+j-1}\right) + \eta_{T+j}.\end{aligned} \tag{48}$$

There is an *h*-step ahead expression for (48) analogous to (47), but it adds little insight.

Returning to the closed system, from (45) and the first line of (47), the expectation of the *j*-step forecast error given the observed initial condition $\hat{\mathbf{w}}_T$ is:

$$\mathrm{E}\left[\hat{\boldsymbol{\nu}}_{T+j} \middle| \hat{\mathbf{w}}_T\right] \cong \sum_{i=0}^{j-1} \Upsilon_{j-i+1}^{**} \boldsymbol{\mu}_{T+j-i} + \Upsilon_0^{**} \mathrm{E}\left[\mathbf{w}_T \middle| \hat{\mathbf{w}}_T\right] - \Upsilon_p^j \hat{\mathbf{w}}_T \tag{49}$$

assuming that $\mathrm{E}[\hat{\Upsilon}] \cong \Upsilon_p$. Thus, forecasts will be biased unless the future disturbances have zero means, the parameters remain constant, their estimates are essentially unbiased,[3] and the initial conditions are correctly imputed. The result in (49) is in stark contrast to the implications of a theory which assumes that the model and DGP coincide and that the latter is constant, in which case essentially unbiased forecasts will be obtained. Subtracting (49) from (47) yields:

$$\begin{aligned}\hat{\boldsymbol{\nu}}_{T+j} - \mathrm{E}\left[\hat{\boldsymbol{\nu}}_{T+j} \middle| \hat{\mathbf{w}}_T\right] &= \sum_{i=0}^{j-1} \Upsilon_{j-i+1}^{**} \left(\boldsymbol{\nu}_{T+j-i} - \boldsymbol{\mu}_{T+j-i}\right) \\ &\quad + \Upsilon_0^{**}\left(\mathbf{w}_T - \mathrm{E}\left[\mathbf{w}_T \middle| \hat{\mathbf{w}}_T\right]\right) - \left(\hat{\Upsilon}^j - \Upsilon_p^j\right)\hat{\mathbf{w}}_T.\end{aligned} \tag{50}$$

We consider $\mathrm{E}[\mathbf{w}_T|\hat{\mathbf{w}}_T]$ in Section 3.4.4 below. Assuming the components in (50) are independent, the formula for the forecast error variance becomes (letting $\boldsymbol{\nu}^*_{T+j-i} = \boldsymbol{\nu}_{T+j-i} - \boldsymbol{\mu}_{T+j-i}$):

$$\begin{aligned}\mathrm{V}\left[\hat{\boldsymbol{\nu}}_{T+j} \middle| \hat{\mathbf{w}}_T\right] &= \mathrm{E}\left[\sum_{i=0}^{j-1}\sum_{k=0}^{j-1} \Upsilon_{j-i+1}^{**} \boldsymbol{\nu}^*_{T+j-i} \boldsymbol{\nu}^{*\prime}_{T+j-k} \Upsilon_{j-k+1}^{**\prime}\right] \\ &\quad + \Upsilon_0^{**} \mathrm{V}\left[\mathbf{w}_T \middle| \hat{\mathbf{w}}_T\right] \Upsilon_0^{**\prime} + (\mathbf{I} \otimes \hat{\mathbf{w}}_T') \mathrm{V}\left[\hat{\Upsilon}_j^v\right] (\mathbf{I} \otimes \hat{\mathbf{w}}_T).\end{aligned} \tag{51}$$

[3] Here we intend that the bias in the *j*th power remains negligible relative to other sources of forecast bias, and do not asume that powers of unbiased estimators are unbiased.

When the forecast shocks are serially independent, the first term in (51) reduces to:

$$\sum_{i=0}^{j-1} \Upsilon^{**}_{j-i+1} \Omega_{j-i} \Upsilon^{**\prime}_{j-i+1} . \tag{52}$$

The result in (51) contrasts with the form of forecast error variance that an investigator would compute *ex ante*, assuming parameter constancy. This is given by (41) (plus the final term in (51) when Υ is estimated).

3.3 Empirical models

Econometrics would not be viable if it required the assumption that the model was the DGP, but it is possible to develop a theory of congruent models which do not need to be facsimiles of the DGP. An empirical model is the data population counterpart of the postulated theory specification, and arises as a reduction of the DGP (see Hendry, 1987, 1993). Such a model is congruent if it matches the data evidence in all measurable respects — which does not require that it is either complete or correct. Key elements of congruency include that the empirical model's error is a homoscedastic innovation against the available information; that the conditioning variables are weakly exogenous for the parameters of the model; and that those parameters are constant. In addition, an encompassing model will be undominated in its class. It is possible to design models with all of these characteristics which thereby exhaust the available information and in essence represent the DGP up to an innovation error which cannot be distinguished from the actual DGP error on the given data. In the context of forecasting, therefore, Υ_p and Υ can be conflated in congruent models, (52) applies within (51), and Ω is the 'smallest' that can be obtained given available information. However, even if the model coincided with the DGP in-sample on this basis, parameters could alter over forecast periods, so we continue to allow for that possibility in the next section.

3.4 Sources of forecast uncertainty

In both (47) for the closed system and (48) for the open system we can delineate five main sources of forecast error, namely (using (47) as the illustration):

I parameter change: $(\Upsilon^*_j - \Upsilon) \neq \mathbf{0}$;
II model misspecification: $(\Upsilon - \Upsilon_p) \neq \mathbf{0}$, or $\{\nu_t\}$ not being an innovation;
III estimation uncertainty: $\mathsf{V}[(\Upsilon_p - \hat{\Upsilon})^{\nu}] \neq \mathbf{0}$;
IV variable mismeasurement: $(\mathbf{w}_T - \hat{\mathbf{w}}_T) \neq \mathbf{0}$; or in the open system:
  a initial condition uncertainty: $(\mathbf{y}_T - \hat{\mathbf{y}}_T) \neq \mathbf{0}$;

b cumulative mistakes in forecasting endogenous variables: $(\mathbf{y}_{T+j-1} - \hat{\mathbf{y}}_{T+j-1}) \neq \mathbf{0}$;

c non-modelled variable uncertainty: $(\mathbf{z}_{T+j} - \breve{\mathbf{z}}_{T+j}) \neq \mathbf{0}$ for $j = 0,...,h$;

d incorrect categorization of some variables as exogenous;

e lack of invariance to policy changes in exogenous variables;

V error accumulation due to $\sum_{i=o}^{j-1} \Upsilon^{**}_{j-i+1} \mathbf{v}^{*}_{T+j-i}$.

We comment in turn on the possibility of reducing the associated sources of uncertainty given that models are not used in an unadulterated manner by their proprietors. Since adjustments to model-based forecasts generally result in value-added in terms of improved forecast accuracy (see, for example, Wallis *et al.* (1986, Table 4.8) and Wallis *et al.* (1987, Figs. 4.3, 4.4)), we discuss the sources of forecast errors in relation to intercept corrections to ascertain why it is that such adjustments might be expected to work. In particular, we do not assume that forecasts are conditional expectations given the DGP.

For instance, forecasters sometimes set models 'back on track', so that under- (over-) prediction of an equation in the past leads to a commensurate adjustment in the future. Suppose that the model is not misspecified, and that the forecaster decides to add in the residual of the current period to the next period's forecast for *1*-step ahead forecasts. We illustrate with a scalar process, using the DGP given by (1) when $\mu = 0$, so that the 'adjusted' forecast for period $T+1$ based on period T information is:

$$\hat{w}^{*}_{T+1} = \hat{\psi} w_T + \hat{e}_T \tag{53}$$

with a forecast error given by:

$$\widetilde{e}_{T+1} = w_{T+1} - \hat{w}^{*}_{T+1} = (\psi - \hat{\psi}) w_T + v_{T+1} - \hat{e}_T, \tag{54}$$

where the unadjusted forecast and its forecast error $\hat{e}_{T+1}$ are as before, namely:

$$\hat{w}_{T+1} = \hat{\psi} w_T, \tag{55}$$

and:

$$\hat{e}_{T+1} = w_{T+1} - \hat{w}_{T+1} = (\psi - \hat{\psi}) w_T + v_{T+1}. \tag{56}$$

From (54), therefore:

$$\widetilde{e}_{T+1} = (\psi - \hat{\psi}) w_T + (v_{T+1} - v_T) + (v_T - \hat{e}_T), \tag{57}$$

and since from (56) lagged one period:

$$(v_T - \hat{e}_T) = -(\psi - \hat{\psi}) w_{T-1}, \tag{58}$$

then:

$$\widetilde{e}_{T+1} = (\psi - \hat{\psi})\Delta w_T + \Delta v_{T+1} = \Delta \hat{e}_{T+1}. \tag{59}$$

From (59), the relationship between the two forecast errors shows that setting forecasts back on track has induced the difference in the original forecast error. Comparing the second moments of the two possible forecast errors we find that:

$$\mathrm{V}[\widetilde{e}_{T+1}] = \mathrm{V}[\Delta \hat{e}_{T+1}] = \mathrm{V}[\hat{e}_{T+1}] + \mathrm{V}[\hat{e}_T] - 2\,\mathrm{Cov}[\hat{e}_{T+1}, \hat{e}_T], \tag{60}$$

where Cov[·] denotes a covariance. Assuming stationarity, so that $\mathrm{V}[\hat{e}_j] = \mathrm{V}[\hat{e}_t]$ $\forall\, j,t$:

$$\mathrm{V}[\widetilde{e}_{T+1}] = 2\mathrm{V}[\hat{e}_t] - 2\rho_e \mathrm{V}[\hat{e}_t] = 2(1 - \rho_e)\mathrm{V}[\hat{e}_t], \tag{61}$$

where ρ_e is the correlation coefficient between adjacent (ordinary) forecast errors, $\{\hat{e}_t\}$. Hence, setting the forecast on track reduces the expected squared forecast error if $\rho_e > ½$. In terms of the parameters of the DGP and the estimator of the econometric model, ρ_e can be written as:

$$\rho_e = \psi \Big/ \left(1 + \frac{1 - \psi^2}{\mathrm{V}[\hat{\psi}]}\right). \tag{62}$$

For a given value of $\psi > ½$, $\rho_e > ½$ will only hold for a sufficiently imprecise estimator of ψ. A necessary condition for this form of residual adjustment to work is that ψ is positive. This parallels the requirement that the estimated residuals are at least positively autocorrelated.

By itself, this rationale for an intercept correction is not overly interesting. However, particular forms of structural breaks can generate positive autocorrelation in post-estimation sample residuals, so the present form of correction may prove effective in such circumstances. We now consider the five main sources of forecast error in turn, abstracting from the presence of other sources to highlight salient effects.

3.4.1 Parameter change

This can take many forms, and is probably the main cause of serious forecasting errors when models are used operationally. Model-based forecasts implicitly assume that Γ, Φ and the zero mean of the error distribution remain constant, when in fact they may alter to Γ_j, Φ_j and μ_j, say, where the subscript denotes the lead over time T.

When models are not congruent specifications of the economic mechanism, changes in one part of a system can induce apparent parameter change in other equations (see Hendry, 1979). Simulation evidence in Favero and Hendry (1992) suggests that this can produce important predictive failures. Conversely, when all variables are weakly stationary, even badly misspecified models can predict as accurately as anticipated.

We consider the consequences for forecast errors of a structural break in ψ in the scalar process (1) from the sample period value to a value ψ^* at $t = T+1,...,T+h$, and examine the efficacy of different strategies for robustifying a model's forecasts when the timing and form of the break are not known. For simplicity, we abstract from estimation uncertainty and an intercept, and maintain an unaltered error distribution over the entire period. For $i = 1,...,h$:

$$w_{T+i} = \psi^* w_{T+i-1} + v_{T+i} = \psi^{*i} w_T + \sum_{j=0}^{i-1} \psi^{*i} v_{T+i-j}. \tag{63}$$

The forecaster, however, uses:

$$\hat{w}_{T+i} = \psi^i w_T, \tag{64}$$

which results in the sequence of forecast errors given by:

$$\hat{v}^*_{T+i} = w_{T+i} - \hat{w}_{T+i} = \left(\psi^{*i} - \psi^i\right) w_T + \sum_{j=0}^{i-1} \psi^{*j} v_{T+i-j}. \tag{65}$$

The average sequence of conditional forecast errors is:

$$E\left[\hat{v}^*_{T+i} | w_T\right] = \left(\psi^{*i} - \psi^i\right) w_T. \tag{66}$$

Thus, the sequence of forecast errors will, on average: (*i*) all have the same sign, positive if $\psi^* > \psi$ and negative if the opposite strict inequality holds; (*ii*) diminish to zero as $h \to \infty$, since the conditional forecasts approach the unconditional forecasts (of zero, given that it is a zero mean stationary process, marking the limit to forecastability); but (*iii*) typically will increase initially. The most useful implication is (*i*), which implies that on average the forecast errors will be positively autocorrelated. This is distinct from the usual autocorrelation due to multi-step forecasting and will be observable *ex post* in the *1*-step forecasts over the break period.

Suppose now that the structural break has taken place prior to the forecast period, between periods $T-1$ and T, say, but that this is unbeknown to the forecaster. Again abstracting from considerations of parameter uncertainty, but allowing for a non-zero intercept:

$$w_{T+i} = \gamma^* + \psi^* w_{T+i-1} + v_{T+i}, \tag{67}$$

for $i = 0,...,h$, noting the difference from (63) in the timing of the break. The timing is important, because the forecaster now observes the period T residual given by:

$$e_T^* = w_T - \hat{w}_T = \gamma^* - \gamma + (\psi^* - \psi)w_{T-1} + v_T, \tag{68}$$

since $\hat{w}_T = \gamma + \psi w_{T-1}$. To forecast period T+1, there are two options: (*i*) use $\hat{w}_{T+1} = \gamma + \psi w_T$ so the period T error is ignored with an associated forecast error of:

$$e_{T+1}^* = \gamma^* - \gamma + (\psi^* - \psi)w_T + v_{T+1}, \tag{69}$$

or, (*ii*) set the forecast back on track using $\tilde{w}_{T+1} = \hat{w}_{T+1} + e_T^*$, which produces the forecast error:

$$\begin{aligned} \tilde{e}_{T+1} &= \gamma^* - \gamma + (\psi^* - \psi)w_T + v_{T+1} - e_T^* \\ &= (\psi^* - \psi)\Delta w_T + \Delta v_{T+1} \\ &= \Delta e_{T+1}^* \end{aligned} \tag{70}$$

where the second equality follows from substituting for e_T^* from (68) (see (53)–(62) for details).

As earlier, the strategy of adding back the previous period's residual will prove successful if the correlation between the errors exceeds ½ (for instance, if $(\gamma^* - \gamma)$ is large). Now, the relevant correlation is between e_{T+1}^* and e_T^*, whereas before, $\{e_t\}$ was assumed stationary in the absence of model misspecification or changes in the DGP, and thus had a constant correlation. Equation (70) shows that some forms of intercept correction can offer partial protection against structural breaks of unknown form that have occurred but the model proprietor is unaware of that information at the time of forecasting. If no break has occurred, the cost of using (70) will be an increase in the forecast error variance unless $\rho_e > ½$ in (62).

To forecast period T+2, compare two alternatives to the forecast from the (misspecified) model which is:

$$\hat{w}_{T+2} = \gamma + \psi\gamma + \psi^2 w_T. \tag{71}$$

The first method uses $\tilde{w}_{T+2}$, defined by:

$$\tilde{w}_{T+2} = \gamma + \psi\tilde{w}_{T+1} = \gamma + \psi\hat{w}_{T+1} + \psi e_T^* = \hat{w}_{T+2} + \psi e_T^* \tag{72}$$

whereas the second uses $\vec{w}_{T+2}$:

$$\vec{w} = \hat{w}_{T+2} + e_T^* \tag{73}$$

In the first alternative, the (false) model is used to generate a *2*-step ahead forecast using the adjusted *1*-step ahead forecast. As can be seen from (72), this amounts to adjusting the (false) conditional expectation by a fraction ψ of e_T^*. The second strategy adjusts the conditional expectation by the full extent of the error. For these two predictors the forecast errors are:

$$\widetilde{e}_{T+2} = e_{T+2}^* - \psi e_T^* = \Delta_2 e_{T+2}^* + (1-\psi) e_T^* \tag{74}$$

and:

$$\ddot{e}_{T+2} = e_{T+2}^* - e_T^* = \Delta_2 e_{T+2}^* \tag{75}$$

where e_{T+2}^* denotes the *2*-step ahead forecast error from using the (incorrect) conditional expectation (cf. (68)). From (66) we deduced that the structural break would induce positive autocorrelation in the sequence $\{e_{T+i}^*\}$, which also have a non-zero mean: this holds *a fortiori* if an intercept change is allowed. For sufficient positive autocorrelation (74) and (75) will on average be smaller than the forecast error from the conditional expectation based on the false assumption of no break.

3.4.2 Model misspecification

While this is a pervasive problem, from (47) and (51) the main impacts on forecast accuracy concern inconsistent parameter estimates and non-innovation errors. The latter are susceptible to design and if the model is a congruent representation of the DGP, then the former plays an intermediate role in (47) and serves as the central tendency for the parameter estimation variance in (51). Otherwise, the main cost is too large a value of the system's error variance matrix.

3.4.3 Estimation uncertainty

Estimation uncertainty in the form of $(\Upsilon_p - \hat{\Upsilon})$, or $(\Gamma_p - \hat{\Gamma})$ and $(\Phi_p - \hat{\Phi})$, is dependent on the size and information content of the sample, the quality of the model, the innovation error variance, the choice of the parameterization, and the selection of the estimator. The degree of integration of the data and whether unit roots are imposed or implicitly estimated will also interact with the above factors. We report the results of research into the impact of estimation uncertainty on forecast accuracy in Section 7, after recording asymptotic forecast error variances for VARs in levels and differences in Section 6. In general, many of the problems just noted are amenable to amelioration by efficient specification, modelling and estimation strategies (see, e.g., Banerjee *et al.* (1993) on the implications of nonstationarity for modelling practice).

When parameter values are uncertain, we showed above that the h-step forecast confidence intervals do not always increase monotonically as the horizon of the forecast increases, but may first expand then contract. Since the forecast variance must eventually equal the unconditional data variance for weakly stationary processes, conditional dynamic forecasts can have larger variances than unconditional. The calculations in Hendry (1986) suggest that this can occur within the policy relevant forecast horizon which would allow scope for pooling forecasts (e.g. between unconditional and conditional predictors as discussed in Section 2.3). Alternatively, there is scope for imposing parameter values rather than estimating them, since forecast error variances will fall when restrictions are nearly satisfied by the data: parsimonious yet valid parameterizations can pay dividends here (see Clements and Hendry, 1993*b*).

3.4.4 Variable uncertainty

Incorrectly measured initial values. Badly measured initial values may impart an important source of error to short-term forecasts. The frequency with which preliminary data are subsequently revised suggests that forecasts may often be conditioned on data measured with a substantial degree of error. Indeed, Wallis (1989, p.32) attributes an important role to measurement-error related issues in justifying intercept corrections: 'the continuing presence of data discrepancies and delays is one reason why there remains a role for informed judgement in forecasting'. Several possibilities suggest themselves for improving current practice. First, the econometric model could be used to predict the initial conditions for the current forecast, given the better established initial values from a few periods earlier. Below, we establish a role for model predictions of initial conditions in the presence of measurement errors, and show how this can be incorporated as an intercept correction. Second, there may be extraneous information available, such as survey information which can help confirm or amend other measurements on the initial state of the economy.

The following analysis is again based on Hendry and Clements (1993). We assume that only the latest observation is measured with error, and set the estimated parameters of the econometric model to their true values. The DGP is (1) with $\mu = 0$:

$$w_t = \psi w_{t-1} + v_t \quad \text{for} \quad t=1,\ldots,T+h, \tag{76}$$

where $v_t \sim \mathrm{IN}(0,\sigma_v^2)$ and $|\psi| < 1$. The observed series w_T^* for the Tth period is the actual series w_T measured with error ω_T:

$$w_T^* = w_T + \omega_T, \tag{77}$$

where $\omega_T \sim \mathrm{IN}(0, \sigma_\omega^2)$ and $\mathrm{E}[w_T \omega_T] = 0$.

To derive the implications of initial condition uncertainty for intercept corrections, we use the literature on 'signal extraction' popularized in e.g. macroeconomics by Lucas (1973) in his 'islands model'. The problem is to derive the optimal predictor of w_{T+h} given information available at period T, when the period T observation w_T^* is error ridden. Thus, the information set (denoted by $\mathcal{I}_T^*$) contains the values of w_s for $s < T$; knowledge of the model given by (76); and w_T^*. The optimal predictor is given by $\mathrm{E}[w_{T+h}|\mathcal{I}_T^*]$, which must be inferior to the predictor in the absence of measurement error, namely $\mathrm{E}[w_{T+h}|\mathcal{I}_T]$, where $\mathcal{I}_T$ replaces w_T^* by w_T.

From (76) by backward substitution:

$$w_{T+h} = \psi^h w_T + \sum_{j=0}^{h-1} \psi^j v_{T+h-j}, \tag{78}$$

where the expectation of the last term is zero for all finite forecast horizons. Consequently, we view the forecasting task as a '2-stage' procedure: first, obtain the best estimate of w_T, which we denote by $\widetilde{w}_T$, and then forecast w_{T+h} from $\psi^h \widetilde{w}_T$. Conceptually, we split the information set into two; (*i*) the model given by (76), and (*ii*) an information set which comprises the past values of observed data, and the relationship between w_T^* and w_T given by (77). We consider the optimal predictor of w_T given each of these in turn, and then combine the two predictors in an optimal fashion by choosing weights that minimize the squared prediction error (cf. Nelson, 1972, for example).[4] Hendry and Clements (1993) show that the composite predictor has the usual form:

$$\mathrm{E}\left[w_T \middle| \mathcal{I}_T^*\right] = (1-\lambda)\hat{w}_T + \lambda \psi w_{T-1}, \tag{79}$$

where λ weights the two component predictors, and $\hat{w}_T = k_T w_T^*$ where k_T depends on the relative variability of the measurement error component to the underlying uncertainty in the DGP. Minimizing the squared prediction error from (79) yields a value of λ, say λ^*, so that the optimal predictor becomes:

$$\widetilde{w}_T = \left(1-\lambda^*\right)\hat{w}_T + \lambda^* \psi w_{T-1} = w_T^* - \left(1-\lambda^*\right)(1-k_T)w_T^* - \lambda^* e_T^* \tag{80}$$

where e_T^* is the period T model error:

---

[4] Survey based information could be incorporated in a similar way.

$$e_T^* = w_T^* - \psi w_{T-1}. \tag{81}$$

This formula has a straightforward interpretation since λ^* varies positively with the variability of the measurement error. Thus, the noisier the measurement error, the greater the reliance on the model-based predictor and in the limit, as the measurement error variance becomes infinitely large, $\lambda^* \to 1$ and the predictor is ψw_{T-1}. Thus, the best predictor of the initial condition becomes the *1*-step ahead conditional expectation, based on the model (76) at T–1, completely ignoring the period T observation. The converse holds if measurement error variance goes to zero.

The second stage forecast of w_{T+h} using the first-stage estimate yields:

$$\widetilde{w}_{T+h} = \psi^h \widetilde{w}_T = \psi^h w_T^* - \psi^h\left[(1-\lambda^*)(1-k_T)w_T^* - \lambda^* e_T^*\right]. \tag{82}$$

When predicting using $\psi^h w_T^*$ rather than $\psi^h w_T$, the required intercept correction is given by the second term in (82). For example, when $k_T \approx 1$, the last term is approximately $-\lambda^* \psi^h e_T^*$, which diminishes in h under the assumption that w_t is stationary.

Cumulative errors due to successively forecasting endogenous variables. These cannot be overcome in principle and arise in both open and closed dynamic systems due to error accumulation from the endogenous variables — see Section 3.4.5 for further details.

Projections of non-modelled variables. In open systems, off-line projections of non-modelled variables have to be made, and in generating future paths for these variables the forecaster will be engaged in the sorts of activities noted in Section 1. It is conceivable that mistakes in extrapolating non-modelled variables may be compensated for by other 'corrections' (see Wallis and Whitley, 1991). Non-modelled variable uncertainty can only be reduced by devoting greater resources to modelling the exogenous and policy variables; having a better feel for what the future will bring forth; and taking account of extraneous information on their likely evolution. Survey information has a potentially useful role in improving off-line forecasting.

Feedbacks onto non-modelled variables. Another form of variable-based forecast error is hidden in (48) which is based on the legitimacy of treating the **z** variables as strongly exogenous (see Engle *et al.* 1983). Whereas weak exogeneity is sufficient at the estimation stage to justify ignoring the marginal process generating the **z**s, to support the generation of multi-step ahead forecasts from equations with conditioning variables such as (30) we require the absence of lagged feedback from the **y**s on to the **z**s. Such feedbacks violate conditioning on future non-modelled variables, but are testable from sample information. The absence of feedback is the condition that the **y**s do not Granger-cause the **z**s (see Granger, 1969). Intuitively, one may expect misleading forecasts to result in the worst instances of this form of misspecification: for example, suppose wages are forecast assuming a fixed future trajectory for prices.

Invariance to policy changes. To support policy analysis, when the **z**s are policy instruments, we require the super exogeneity of the **z**s for the parameters of the forecasting system, and not just their weak exogeneity: this issue is analysed in Engle and Hendry (1993) and Favero and Hendry (1992). A failure of super exogeneity may arise if agents alter the way they form expectations (the 'critique' in Lucas, 1976), or if regime shifts occur out of sample. The impact of forecasts on outcomes (the bandwagon effect) is only a problem to the extent that it is capricious and cannot be modelled as part of the forecasting device. Many models explicitly incorporate terms to capture agents' expectations concerning future values of variables, so that in principle at least this effect can be controlled. Models can then be solved to ensure model-consistent expectations such that the expectations of the future values of the variables coincide with the values for those variables generated by the model. Although the 'rationality' of expectations is a contentious issue, the expectations variant of the Lucas critique does not of itself condemn the forecasting enterprise to failure. In any specific instance, the critique is testable, and in general may lack force (Favero and Hendry, 1992). In other cases, potential effects from regime shifts can be determined ex ante: see Hendry and Ericsson (1991) for a model of financial innovation which could have been implemented in part prior to the change having its effect. In general, if anticipated future changes in policy are reasonably correlated with past episodes, then previous *ex post* errors may suggest a pattern for future intercept corrections. Nevertheless, if it is desired to use a model across policy regimes, either the relevant parameters must be invariant, or the effect of the policy change must be incorporated.

3.4.5 Error accumulation

The inherent uncertainty in the DGP places a limit on the predictability of nonstationary dynamic mechanisms, although no empirical model is likely to attain such a bound. For I(0) transforms, such an error source is bounded by the unconditional variance, but for the level of the economy, the variance is unbounded. Put simply, anything is possible in the indefinite future: imagine the level and composition of GNP today had the Industrial Revolution not occurred, or if a nuclear war had. Fortunately, only a few periods ahead are usually necessary for economic stabilization policy (e.g. 4–8 quarters), and only medium-term averages are needed for growth and investment strategies such as education and infrastructure.

In practice many models have errors which are not even innovations against their own information sets, and hence they can be improved by better modelling (e.g. by not imposing 'theoretically plausible' but data-rejected constraints); better methodology (e.g. general-to-simple); and higher quality data banks of information. The innovation error in a model is not a given, but is derived from the design of the model, so congruent, encompassing models should out-perform others on this aspect, especially against models with autocorrelated residuals.

Asymmetric error distributions potentially induce biased forecasts. Positive inflation shocks might be more likely than negative, so models which assumed mean-zero forecast errors would on average underpredict inflation. Also, error variance changes (e.g. heteroscedasticity) will lead to increased average squared errors, which will be reflected in many measures of forecast inaccuracy.

4. FORECAST COMPARISONS

Forecasts from large models reflect the judgements of their proprietors as well as the models' properties, so historical forecast track records may reveal little about model validity. When evaluating forecasts, the difficulties are compounded as follows by basing forecast comparisons on mean square forecast errors (MSFEs).

4.1 Mean square forecast error

Despite the fact that MSFEs, or their square roots, are commonly used for forecast comparisons across alternative models or methods, they have potentially serious drawbacks. Earlier criticisms of MSFE measures have pointed out that minimum MSFE is neither necessary nor sufficient for a model to have constant parameters, to provide accurate forecasts or to encompass rival models, all of which are serious deficiencies (see e.g. Ericsson, 1992). However, Clements and Hendry (1993*a*) criticize MSFE criteria for their lack of invariance to transformations for which the forecasting model is invariant. They show that scalar MSFEs are only unambiguous for assessing forecast accuracy in the univariate case and then only for l-step ahead forecasts.

In systems, the complete MSFE matrix (the second moment matrix of forecast errors) is not invariant, and the trace of the MSFE matrix (TMSFE) (one of the most commonly applied measures) certainly is not: MSFEs constitute an inadequate basis for method or model selection because they are not invariant to nonsingular, scale-preserving linear transforms, while linear models are invariant to such transforms. Consequently, for multi-step forecasts in systems of equations, a specific model or method can have the minimum MSFE (MMSFE) for one linear function of predicted variables, but this does not imply MMSFE for another linear function. Such non-invariance entails that rankings across models and/or methods based on MSFE measures can vary by choosing alternative yet isomorphic representations of a given process (e.g. levels, differences, or cointegrating vectors). This problem is especially serious in Monte Carlo studies of forecasting methods, where misleading outcomes can result due to the arbitrary choice of which linear transformation of the data to examine. However, it is an equally pertinent issue empirically, and would appear to vitiate the use of MSFE comparisons in many applications.

The intuition is straightforward. Suppose we attempt to compare the performance of the multi-step forecast of the level of a variable $\mathbf{x}$ and the forecast of its first difference $\Delta\mathbf{x}$. We assume these are each optimal h-step ahead predictors of $\mathbf{x}_{T+h}$ and $\Delta\mathbf{x}_{T+h}$, in that they are the conditional expectations and therefore MMSFE predictors. That they cannot be compared follows because the optimal predictors and associated MSFEs differ (see Granger and Newbold, 1977, p. 285). A less obvious point is that rankings between forecasts on MSFE-based criteria are not invariant to linear transformations of the variable to be forecast. So, model A may out perform model B in MMSFE terms when predicting $\mathbf{x}_{T+h}$, but the situation may be reversed when both models predict $\Delta\mathbf{x}_{T+h}$.

More formally, the following results concerning the (in)variance of various MSFE measures are established by Clements and Hendry (1993*a*). The MSFE measures that we consider are the TMSFE, the determinant of the MSFE matrix, $|\mathbf{V}_h|$, and a measure that relates to the complete MSFE matrix, which we term MSFEM. One model dominates another on MSFEM if $\mathbf{d}'\mathbf{V}_h\mathbf{d} < \mathbf{d}'\mathbf{V}_h^*\mathbf{d}$ for all non-zero $\mathbf{d}$, where $\mathbf{V}_h$ is the h-step ahead MSFE matrix for that model, and $\mathbf{V}_h^*$ is the same quantity for a rival model.

The class of transformations considered is that under which the following model is invariant:

$$\Phi\mathbf{s}_t = \mathbf{u}_t \tag{83}$$

where $\mathbf{u}_t \sim \mathrm{ID}(\mathbf{0},\Omega)$, when $\mathbf{s}_t' = (\mathbf{x}_t' : \mathbf{z}_t')$, $\mathbf{x}_t$ are the N variables to be forecast and $\mathbf{z}_t$

are k available predetermined variables (perhaps just $\mathbf{x}_{t-1}$) with $\mathbf{\Phi} = (\mathbf{I}:-\mathbf{B})$ say. The parameters are $(\mathbf{B} : \mathbf{\Omega})$, where $\mathbf{\Omega}$ is symmetric positive semi-definite. Then the likelihood and generalized variance of the system in equation (83) are invariant under scale-preserving, nonsingular transformations of the form:

$$\mathbf{M\Phi P}^{-1}\mathbf{Ps}_t = \mathbf{Mu}_t \quad \text{or} \quad \mathbf{\Phi}^*\mathbf{s}_t^* = \mathbf{u}_t^* \tag{84}$$

so $\mathbf{u}_t^* \sim \text{ID}(\mathbf{0},\mathbf{M\Omega M}')$, where $\mathbf{M}$ and $\mathbf{P}$ are nonsingular $N \times N$ and $(k+N) \times (k+N)$ matrices respectively, such that $|\mathbf{M}| = 1$ and $\mathbf{P}$ is block upper triangular such that $\mathbf{\Phi}^* = \mathbf{M\Phi P}^{-1} = (\mathbf{I} : -\mathbf{B}^*)$ and $\mathbf{s}_t^* = \mathbf{Ps}_t$.

Forecasts and forecast confidence intervals for the original system transformed to $\mathbf{x}_t^*$ will be identical to those made initially from the transformed system. No restrictions are imposed by these transforms, so the systems given by equations (83) and (84) are isomorphic. Transformations in the class shown in equation (84) are regularly undertaken in applied work, and include as special cases, differences, cointegrating combinations, substitution of identities, and differentials *inter alia.*

Consider the transform from $\mathbf{x}_t$ to $\mathbf{x}_t^*$ in (84) when $|\mathbf{M}| = 1$. The h-step ahead forecast errors resulting from $\mathbf{x}_t$ and $\mathbf{x}_t^*$ are denoted by $\mathbf{u}_t$ and $\mathbf{e}_t$ respectively. Then, for transformations using $\mathbf{M}$ only:

(i) TMSFE for $\mathbf{e}_t$ is not in general equivalent to that for $\mathbf{u}_t$: $tr(\mathbf{MV}_h\mathbf{M}') \neq tr(\mathbf{V}_h)$;
(ii) $|\mathbf{V}_h|$ is invariant when $|\mathbf{M}| = 1$: $|\mathbf{MV}_h\mathbf{M}'| = |\mathbf{V}_h|$;
(iii) MSFEM is invariant:

$$\mathbf{d}_1'\mathbf{V}_h\mathbf{d}_1 < \mathbf{d}_1'\mathbf{V}_h^*\mathbf{d}_1 \quad \forall\ \mathbf{d}_1 \neq \mathbf{0}$$

implies that

$$\mathbf{d}'\mathbf{MV}_h\mathbf{M}'\mathbf{d} < \mathbf{d}'\mathbf{MV}_h^*\mathbf{M}'\mathbf{d} \quad \forall\ \mathbf{d} \neq \mathbf{0}.$$

Thus, providing a common basis is used, contemporaneous transformations of the variables will not induce a switch in ranking.

For transformations using $\mathbf{P}$, as happens with the transformation from $\mathbf{x}_t$ to $\Delta\mathbf{x}_t$, these results continue to hold with the following exceptions:

(i) TMSFE is valid for *1*-step forecasts;
(ii) $|\mathbf{V}_h|$ is not invariant except for $h = 1$, when the invariance follows from the conditional nature of the expectations operator as with (i).
(iii) MSFEM is not invariant for $h > 1$.

More generally for $h > 1$, the covariance terms between the different step ahead forecast errors have to be taken into account, and are altered by $\mathbf{P}$-type transforms.

This suggests using the complete (i.e. stacked) forecast error second moment matrix as a measure, or for an invariant scalar criterion, its determinant, which is the generalized forecast error second moment, denoted GFESM.

4.2 Generalized forecast error second moment matrix

The GFESM for an h-step ahead forecast is formed by stacking the forecast errors from *1* to h step ahead forecasts as in equation (85):

$$|\Phi_h| = |\mathrm{E}[\breve{\mathbf{e}}\breve{\mathbf{e}}']| \tag{85}$$

where $\breve{\mathbf{e}}' = [\mathbf{e}'_{T+1}, \mathbf{e}'_{T+2}, \ldots, \mathbf{e}'_{T+h-1}, \mathbf{e}'_{T+h}]$. Transforming the data by $\mathbf{M}$ where $|\mathbf{M}| = 1$ must leave the criterion in (85) unaffected; transforming by $\mathbf{P}$ leaves the error process unaffected (see (84)) so again (85) is invariant.

The MSFEM criterion can be generalized to apply to the pooled or stacked forecast error second moment matrix $\mathbf{\Phi}_h$. In that case, the model denoted by ~ dominates ^ if $\hat{\mathbf{\Phi}}_h - \widetilde{\mathbf{\Phi}}_h \succ 0$, that is, if the difference between the two estimates of the stacked forecast error second moment matrix is positive semi-definite. It follows immediately that MSFEM dominance on the stacked MSFE matrix is sufficient but not necessary for GFESM dominance, that is:

$$\hat{\mathbf{\Phi}}_h - \widetilde{\mathbf{\Phi}}_h \succ 0 \quad \text{implies} \quad |\hat{\mathbf{\Phi}}_h| \geq |\widetilde{\mathbf{\Phi}}_h|,$$

since $\hat{\mathbf{\Phi}}_h$ and $\widetilde{\mathbf{\Phi}}_h$ are positive definite (see, e.g. Dhrymes (1984) proposition 66 p.77). Thus GFESM dominance is seen to be a weaker condition than MSFEM applied to the stacked MSFE matrix. It can also be related to the statistical literature on the predictive likelihood measure (on which, see e.g. Bjørnstad (1990) for a survey).

4.3 Pooling of forecasts and forecast encompassing

Tests for forecast encompassing concern whether the *1*-step forecasts of one model can explain the forecast errors made by another. Due to the autocorrelation between successive step ahead forecast errors, this concept can not be made operational for multi-step ahead forecasts. Tests of forecast encompassing were proposed as a feasible way to evaluate large-scale econometric models by Chong and Hendry (1986) who establish that any need to pool forecasts from disparate sources indicates model misspecification. Mizon (1984), Mizon and Richard (1986) and Hendry and Richard (1989) provide general discussions of encompassing; Diebold (1989) offers an attempted reconciliation of the pooling and encompassing approaches; and Lu and Mizon (1991) and Ericsson (1992) analyse forecast encompassing in detail.

Generally if $\hat{\mathbf{x}}_t$ denotes the vector of *1*-step forecasts from model 1 of $\mathbf{x}_t$, and $\tilde{\mathbf{x}}_t$ the corresponding forecasts from model 2, then the pooled forecast is:

$$\overline{\mathbf{x}}_t = \hat{\mathbf{x}}_t + \Gamma(\tilde{\mathbf{x}}_t - \hat{\mathbf{x}}_t). \tag{86}$$

If Γ is neither a zero nor unit matrix, then both models are valuable in forecasting $\mathbf{x}_t$ but neither is sufficient. However, the linear combination of forecasts in (86) generally is not a sensible way to handle information: two models which individually fail to capture the salient features of the data are unlikely to combine on a systematic basis to produce good forecasts and it seems preferable to sort out the models.[5]

Forecast encompassing is formally equivalent to Nelson's (1972) procedure for computing 'conditional efficiency' (see Granger and Newbold (1973), as Nelson only applied the procedure to within-sample predictions, rather than to explicitly evaluate forecasts). A forecast is said to be conditionally efficient if the variance of the forecast error from a combination of that forecast and a rival forecast is not less than that of the original forecast alone. We concur with Granger and Newbold (1977) that 'the conditional efficiency criterion (is) of great potential value in actual forecast evaluation' (p.283), but have less sympathy with the combining of sample period fitted values from different models in an attempt to improve forecasts. Forecast encompassing was intended to evaluate the forecasts from large-scale macroeconometric models for which standard methods are not operational, and is not in general a good way to evaluate models when parametric encompassing tests are applicable. However, as an application of the encompassing principle, the concept of forecast encompassing is more soundly based than the other *ad hoc* evaluation criteria which Granger and Newbold (1973) show are generally inadequate or even misleading.

Forecast encompassing is invariant for both **P** and **M** type transforms, unlike TMSFEs which for *1*-step ahead forecasts only have this property for **P** transforms (see Section 4.1). Consider testing for forecast encompassing by fitting:

$$\mathbf{x}_t - \hat{\mathbf{x}}_t = \Gamma(\tilde{\mathbf{x}}_t - \hat{\mathbf{x}}_t) + \zeta_t. \tag{87}$$

An advantage of the formulation in (87) is that it remains balanced[6] under relatively weak conditions when $\mathbf{x}_t$ is integrated of order one. When $\Gamma = \mathbf{0}$, the first

[5] Contrast Nelson (1972, p. 914), 'if the bum on the street corner offers free tips to the decision maker on his way to the office, these will be incorporated in composite predictions if they result in any reduction in expected loss'. One can gauge the effect on within-sample predictions of such tips, but their casual nature does not instil confidence in their reliability in the future.

[6] Balanced in the sense that both sides of the equation will have the same order of integration, that is, they will be stationary. We are grateful to Stephen Hall for drawing our attention to this.

system forecast encompasses the second, but it fails to do so if $\mathbf{\Gamma} \neq \mathbf{0}$. A switch from $\mathbf{x}_t$ to $\Delta\mathbf{x}_t$ leaves the test unaffected:

$$\Delta\mathbf{x}_t - \Delta\hat{\mathbf{x}}_t = \mathbf{\Gamma}(\Delta\tilde{\mathbf{x}}_t - \Delta\hat{\mathbf{x}}_t) + \zeta_t, \tag{88}$$

since the $\mathbf{x}_{t-1}$ cancel. Further, suppose that forecast encompassing holds for a set of variables in a linear system, so $\mathbf{\Gamma} = \mathbf{0}$ in (87) or (88). Then forecast encompassing must hold for any linear combination of that set of variables with the usual *caveat* about the possibility of conflicts from finite sample tests at fixed significance levels. Consequently, despite its limitations, if forecast encompassing can be established for all relevant variables, that will ensure population MSFE dominance for all linear combinations such that $|\mathbf{M}| = 1$. Forecast encompassing is sufficient to ensure *1*-step GFESM dominance, but the model which performs best on $|\hat{\mathbf{\Phi}}|$ need not be able to encompass all the contending models. Ericsson (1992) shows that in scalar processes, *1*-step MMSFE is not sufficient to ensure forecast encompassing although it is necessary, so that encompassing ensures MMSFE. Further, he shows that MMSFE is neither necessary nor sufficient to ensure parameter constancy and his forecast parameter encompassing test is required to avoid conflicts. Ericsson (1992) and Lu and Mizon (1991) also discuss the relative merits of, and relations between, forecast error encompassing test statistics and parameter constancy tests.

5. SYSTEMS FOR NONSTATIONARY DATA

Whether economic time series are integrated of order unity (denoted I(1)) or are stationary has important implications for their forecastability. We established in Section 2 that I(1) variables can only be forecast with increasingly wide confidence intervals, whereas stationary, cointegrating linear combinations of such variables have finite confidence intervals as the horizon grows.

In Sections 6 and 7 we report asymptotic forecast error variance formulae and a Monte Carlo study of these (both from Clements and Hendry, 1992); the formulae shed light on the relative merits of different models and/or estimation methods when the number and location of unit roots in the system is unknown. As a precursor, this section briefly reviews cointegration in systems of equations.

Cointegration implies reduced-rank type restrictions on VARs for integrated variables (see, *inter alia*, Engle and Granger (1987) and Johansen, 1988). Our interest is in the implications for forecast uncertainty of using various transformations of the variables and of imposing cointegrating restrictions and unit roots at the estimation stage. We categorize three commonly used model specifications in terms of the number of unit roots and cointegrating restrictions imposed or implicitly estimated in each case.

Let $\mathbf{x}_t$ be an $N \times 1$ vector of time series variables, generated by the first-order dynamic linear system:

$$\mathbf{x}_t = \mathbf{A}\mathbf{x}_{t-1} + \psi\mathbf{D}_t + \nu_t \tag{89}$$

where $\mathbf{v}_t \sim \mathrm{IN}(\mathbf{0},\Omega)$, for $t = 1,2,...,T$. In (89), $\mathbf{D}_t$ contains deterministic components, and the initial value $\mathbf{x}_0$ is fixed. Equation (89) can be reparameterized as:

$$\Delta\mathbf{x}_t = \Pi\mathbf{x}_{t-1} + \psi\mathbf{D}_t + \nu_t, \tag{90}$$

where $\Pi = (\mathbf{A}-\mathbf{I}_n)$. Since r^{th}-order lag systems can be stacked into a first-order form, there is no loss of generality in analyzing the system (90). Ignoring deterministic nonstationarities, the rank n of Π determines how many independent linear combinations of the variables x_{it} are stationary. We assume that none of the roots of $|\mathbf{I}-\lambda\mathbf{A}| = 0$ lie inside the unit circle, so that explosive variables are excluded. Then, if we preclude I(2) variables (see Johansen, 1992), three cases arise: $n = N$, so that all N variables in $\mathbf{x}_t$ are I(0); $n = 0$, so that there are no linear combinations which are stationary; and finally, $0 < n < N$, so there are n cointegrated linear combinations of $\mathbf{x}_t$. For the intermediate case, we can write $\Pi = \alpha\beta'$ where α and β are $N \times n$ of rank n, and $\beta'\mathbf{x}_t$ are the cointegrated combinations of $\mathbf{x}_t$.

A more convenient form of (89) is given by the following example of a **P**-transform. Reformulate the system by partitioning $\mathbf{x}_t$ into $(\mathbf{x}'_{at}:\mathbf{x}'_{bt})$ where $\beta'\mathbf{x}_t$ and $\Delta\mathbf{x}_{bt}$ are I(0) by construction. Let $\mathbf{w}'_t = (\mathbf{x}'_t\beta:\Delta\mathbf{x}'_{bt}) = (\mathbf{w}'_{at}:\mathbf{w}'_{bt})'$ and $\alpha' = (\alpha'_a:\alpha'_b)$ which is $(n \times n : n \times (N-n))$, then:

$$\mathbf{w}_t = \mathbf{C}\mathbf{w}_{t-1} + \psi_0\mathbf{D}_t + \varepsilon_t \tag{91}$$

where $\varepsilon_t \sim \mathrm{IN}(\mathbf{0},\Sigma)$. In (91), $\psi_0 = (\psi'\beta:\psi'_b)' = \mathbf{Q}\psi$ where $\psi_b = \mathbf{J}'\psi$ when $\mathbf{J}' = (\mathbf{0}:\mathbf{I})$ and:

$$\mathbf{Q} = \begin{bmatrix} \beta' \\ \mathbf{J}' \end{bmatrix}, \quad \mathbf{C} = \begin{bmatrix} (\mathbf{I}+\beta'\alpha) & 0 \\ \alpha_b & 0 \end{bmatrix} = \begin{bmatrix} \mu & 0 \\ \alpha_b & 0 \end{bmatrix},$$

and:

$$\Sigma = \begin{bmatrix} \beta'\Omega\beta & \beta'\Omega\mathbf{J} \\ \mathbf{J}'\Omega\beta & \mathbf{J}'\Omega\mathbf{J} \end{bmatrix}. \tag{92}$$

The three estimation methods to be analysed in Sections 6 and 7 can be categorized in terms of the number of common trends (Stock and Watson, 1988) or unit roots, $N-n$, and cointegrating restrictions, n, imposed by the estimation method. The unrestricted VAR estimator (UV) estimates Π as a full-rank matrix so that neither unit roots nor cointegrating restrictions are imposed. The polar case is the estimator of the VAR in differences (DV), for which Π is set to the null matrix so that N unit roots are imposed. The intermediate case is illustrated by the maximum likelihood (ML) estimator of Johansen (1988) or alternatively the two-step procedure of Engle and Granger (1987).

The theory of inference in cointegrated systems has been developed by Phillips and his co-authors (see e.g. Phillips, 1991) and Johansen (1988), amongst others: also see Engle and Yoo (1991) for a good exposition with examples. The formulae in the next section abstract from issues of parameter estimation uncertainty.

6. ASYMPTOTIC FORECAST ERROR VARIANCES

In this section, we note the asymptotic formulae of Clements and Hendry (1992) which demonstrate how the assessment of forecast accuracy (based on MSFEs) over models depends upon the transformation of the variables being forecast. Specifically, we consider two 'rival' models: an essentially correctly specified model, in that valid cointegrating relationships are modelled, and the DV model which omits levels terms. If forecast accuracy is based on the MSFE criterion (see Section 4) and models are assessed in terms of their ability to predict the differences in the process, then it will become apparent that there is little to choose between the correctlyspecified and DV models.

The asymptotic formulae for h-step ahead forecast uncertainty provide a first-order approximation for empirical implementation and make Monte Carlo outcomes more interpretable: the usefulness of the asymptotic results as a guide to small sample outcomes is evaluated in Section 7. Conceptually, the asymptotic formulae for known parameters set the estimators of the parameters and covariance matrix of the disturbances of the econometric models equal to their population parameters in the DGP. The results in this section make use of the representations of a cointegrated system derived in Section 5. Full details of the following formulae and derivations are given in Clements and Hendry (1992).

There are three main types of forecast error second moment that could be calculated. If we denote a forecast error moment by $\mathsf{M}[\cdot]$, then in principle we could calculate the following quantities: $\mathsf{M}[\mathbf{e}_{T+h}]$, $\mathsf{M}[\mathbf{e}_{T+h}|\mathbf{x}_T]$ or $\mathsf{M}[\mathbf{e}_{T+h}|\mathbf{X}_T^1]$. The first of these corresponds to the Monte Carlo in Clements and Hendry (1992), where $\mathbf{X}_{T+h}^1$ is resampled on each replication and results are averaged across replications; the last involves conditional forecasting after parameter estimation, and corresponds to having one set $\mathbf{X}_T^1$ with sets of independent forecasts generated from the last data point. The middle is a hybrid, albeit one which is regularly used for analytically deriving forecast error variance formulae, where taking expectations over $\mathbf{x}_T$ will yield the first. To illuminate the Monte Carlo, we consider unconditional moments of the form $\mathsf{M}[\mathbf{e}_{T+h}]$.

For the correctly specified model, ignoring parameter estimation uncertainty, we can show that the h-step ahead forecast error variance matrix, for predicting $\mathbf{x}_{T+h}$, for the model given by (89) is:

$$\mathsf{V}\left[\mathbf{e}_{x,T+h}\right]=\mathsf{V}\left[\mathbf{e}_{x,T+h}|\mathbf{x}_T\right]=\sum_{s=0}^{h-1}\mathbf{A}^s\mathbf{\Omega}\mathbf{A}^{s\prime} \tag{93}$$

showing the equivalence between the conditional and unconditional second moments. A similar expression holds for the $\mathbf{P}$-transform of the model given by (91) for $\mathbf{w}_t$:

$$\mathsf{V}\left[\mathbf{e}_{w,T+h}\right]=\mathsf{V}\left[\mathbf{e}_{w,T+h}|\mathbf{x}_T\right]=\sum_{s=0}^{h-1}\mathbf{C}^s\mathbf{\Sigma}\mathbf{C}^{s\prime} \tag{94}$$

Since $\mathbf{A} = (\mathbf{I} + \alpha\beta')$, it has roots equal to unity, so the TMSFE for $\mathbf{x}$ from (93) is $O(h)$, while the TMSFE for $\mathbf{w}$ from (94) is $O(1)$ in h, reflecting the fact that $\mathbf{x}_t \sim \mathsf{I}(1)$ but $\mathbf{w}_t \sim \mathsf{I}(0)$ (cf. Engle and Yoo (1987) and Lütkepohl, 1991). If instead we consider assessing the accuracy of predictions of $\Delta\mathbf{x}_{T+h}$ we obtain the variance formula for $h > 1$:

$$\mathsf{V}\left[\mathbf{e}_{\Delta x,T+h}\right]\equiv\mathsf{V}\left[\mathbf{e}_{\Delta x,T+h}|\mathbf{x}_T\right]=\mathbf{\Omega}+\sum_{s=0}^{h-2}\mathbf{A}^s\alpha\beta'\mathbf{\Omega}\beta\alpha'\mathbf{A}^{s\prime} \tag{95}$$

where the second term is absent for $h = 1$. In fact, the MSFE in (95) is $O(1)$ despite the terms in $\mathbf{A}^s$, since $\mathbf{A}^s\alpha \equiv \alpha\mu^s \to \mathbf{0}$ as $s \to \infty$ since $\mu = \mathbf{I} + \beta'\alpha$ from (92).

Formulae (93), (94) and (95) are the relevant asymptotic formulae for the h-step ahead MSFE matrix for a correctly specified model, that is, when the model coincides with the DGP. In contrast, the DV model is misspecified when $n \neq 0$, and so we need to derive the appropriate asymptotic variance expressions for the DV model in terms of $\mathbf{x}$, $\mathbf{w}$ and $\Delta\mathbf{x}$. We find that the equivalence of the conditional and unconditional forecast moments no longer holds.

Define the forecasts from the DV model for $\Delta\mathbf{x}$ by setting $\Delta\mathbf{x}_{T+h}$ equal to the (population) growth rate, denoted $\mathbf{K}\kappa$, (where κ is a vector of ones) so when $\text{rank}(\beta'\mathbf{d}) = n$

$$\Delta\mathbf{x}_{T+h}=\mathbf{K}\kappa \quad \text{where} \quad \mathbf{K}=\mathbf{I}-\alpha(\beta'\alpha)^{-1}\beta', \tag{96}$$

in which case the h-step ahead forecast error conditional second moment is still given by (95), while $M[\mathbf{e}_{T+h}]$ is:

$$\begin{aligned} V\left[\widetilde{\mathbf{e}}_{\Delta x,T+h}\right] &= \alpha\mu^{h-1}V\left[\mathbf{w}_{at}\right]\mu^{h-1\prime}\alpha' + \Omega + \sum_{s=0}^{h-2}\alpha\mu^{s}\beta'\Omega\beta\mu^{s\prime}\alpha' \\ &= \Omega + \alpha V\left[\mathbf{w}_{at}\right]\alpha', \end{aligned} \tag{97}$$

where the last term on the first line is present only for $h > 1$ (see Clements and Hendry, 1992). The difference between the correctly specified model and DV model formulae for the unconditional h-step ahead MSFE is (97) minus (95), namely:

$$\alpha\mu^{h-1}V\left[\mathbf{w}_{at}\right]\mu^{h-1\prime}\alpha' \tag{98}$$

which goes to zero in h. Hence, for MSFE measures of forecast accuracy for predicting $\Delta\mathbf{x}$ there is no gain to using the correctly specified model as against the DV at anything but the shortest forecast horizons. Conversely, forecast evaluation for moderate horizons h based on $\Delta\mathbf{x}$ will lack power to discriminate between rival specifications.

The analogous expressions for forecast accuracy for the DV model for $\mathbf{x}$ indicate that $M[\mathbf{e}_{T+h}|\mathbf{x}_T]$ is again equivalent to the correctly specified model case (93), but that $M[\mathbf{e}_{T+h}]$ is:

$$V\left[\widetilde{\mathbf{e}}_{x,T+h}\right] = \sum_{r=0}^{h-1}\sum_{q=0}^{h-1}\alpha\mu^{r}V\left[\mathbf{w}_{at}\right]\mu^{q\prime}\alpha' + \sum_{s=0}^{h-1}\mathbf{A}^{s}\Omega\mathbf{A}^{s\prime} \tag{99}$$

Comparing this expression to (93), we can see that the unconditional variance will always exceed the conditional since the first term is a positive definite matrix.

The expressions for forecast variances of $\mathbf{w}$ in the DV model can be found by suitably combining the expressions for $\mathbf{x}$ and $\Delta\mathbf{x}$ from (97) and (99).

Figure 1 illustrates the extent of the relative losses from using the misspecified DV model when forecast accuracy is assessed in terms of the three transformations of the variables. The figure compares the asymptotic TMSFEs for the DV model with those from the correctly specified model, for a particular set of parameter values. The line denoting the ratio of the DV model to the correctly specified model for $\Delta\mathbf{x}$ rapidly approaches unity as h increases, so the TMSFEs of the correctly and incorrectly specified models differ by a term which goes to zero in h. This is why forecast evaluation in differences may lack power to reveal model misspecification.

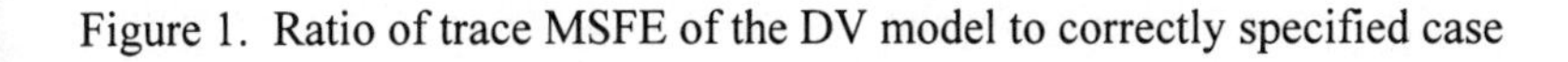

Figure 1. Ratio of trace MSFE of the DV model to correctly specified case

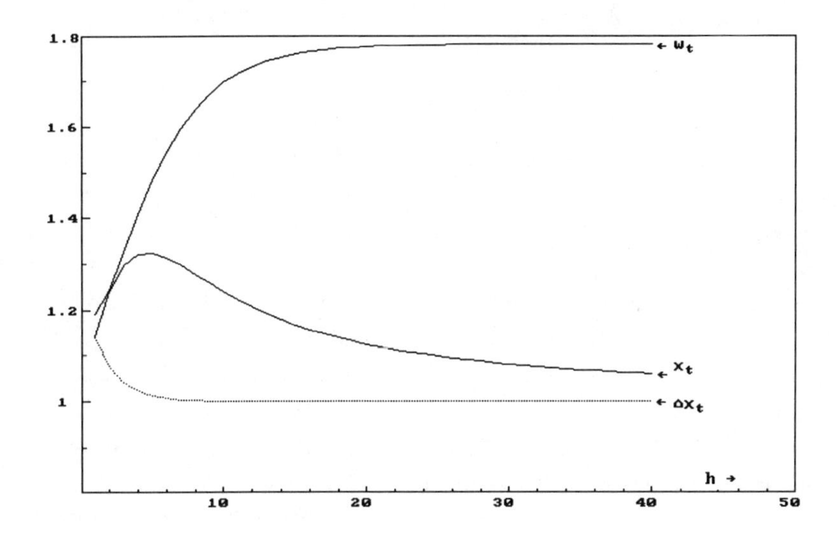

In terms of $\mathbf{x}$, the relative losses from using the DV model initially increase then decrease in the forecast horizon, but do not disappear. The relative loss to using the DV model is largest for $\mathbf{w}$, and the ratio converges in h (here to an 80% increase in TMSFE).

Finally, we consider the asymptotic behaviour of the GFESM. In Section 4 we established its invariance to the transformation on which forecast accuracy is assessed, whether parameters are known or have to be estimated from the data. For the correctly specified model, we can establish the relationship between the measure for the *1*-step ahead and the *h*-step ahead forecasts. For a *2*-step forecast, since the covariance $\mathsf{E}[\mathbf{e}_{x,T+1}\mathbf{e}'_{x,T+2}] = \Omega\mathbf{A}'$, the complete second moment matrix is:

$$|\Phi_2| = \begin{vmatrix} \Omega & \Omega\mathbf{A}' \\ \mathbf{A}\Omega & \Omega + \mathbf{A}\Omega\mathbf{A}' \end{vmatrix} = |\Omega| \cdot \left|(\Omega + \mathbf{A}\Omega\mathbf{A}') - (\mathbf{A}\Omega)\Omega^{-1}(\Omega\mathbf{A}')\right| = |\Omega|^2 \quad (100)$$

This recursion continues so that $|\Phi_h|^{1/h} = |\Omega|$ for *h*-step ahead forecasts. Thus, for a correct specification with known parameters, the generalized second moment of the *1*-step forecast error determines the ranking for all h. This suggests using $|\Phi_h|^{1/h}$ for forecast comparisons, which will not affect the qualitative outcome but will stabilize the measure for large h. When the model is misspecified, as in the case of the DV model, there may be additional information from further step ahead forecasts. The GFESM criterion remains invariant to the class of transformations considered above, but the asymptotic formula for the DV model no longer collapses as in (100) to a simple function of the *1*-step ahead value.

7. A MONTE CARLO STUDY

A Monte Carlo was carried out by Clements and Hendry (1992) to assess the relative merits of different models and estimation methods in a bivariate I(1) cointegrated system, when the number and location of unit roots in the system was unknown to the investigator. The importance of parameter uncertainty was highlighted by comparing the simulation outcomes with control variate estimates and the asymptotic formulae presented in Section 6.

Other recent Monte Carlo studies of forecast accuracy include Engle and Yoo (1987), Brandner and Kunst (1990) and Chambers (1991). However, all three are based on MSFE measures and are subject to the criticisms of non-invariance outlined in Section 4. Clements and Hendry (1992) demonstrate the force of the critique for the Engle and Yoo (1987) study, and then base subsequent assessments of forecast accuracy on the GFESM measure. Their calculations were carried out over a range of experimental designs and the results summarized for each estimation method with response surfaces that related the deviation of the Monte Carlo estimate from the control variate value (based on assuming no parameter uncertainty, to control for the variation in the random numbers) to the design parameters: the autoregressive parameters, and the covariance structure of the random numbers (proxying the system disturbance terms).

Specifically, they were able to show that Engle and Yoo's result of improved relative forecast accuracy from imposing long-run constraints by using the Engle and Granger (1987) two-step procedure (EG), compared to an unrestricted vector autoregression, is not robust to isomorphic transformations of the model. For an estimation sample size of 100, EG registers gains relative to UV on the TMSFE criterion for $\mathbf{x}$, but the situation is reversed for $\mathbf{w}$. In terms of the GFESM criterion, EG shows only small gains relative to UV, while the relative inferiority of DV is apparent.

Results for a smaller estimation sample of 50 favour EG relative to UV on the GFESM criterion, and also ML relative to UV, indicating the importance of imposing valid restrictions when there are fewer degrees of freedom.

In terms of the invariant GFESM measure, the results reveal little benefit from imposing reduced rank cointegration restrictions unless the sample size is small. Although care is required in generalizing from a bivariate system (since the number of cointegration restrictions increases more than proportionately to the dimension of the system), we expect our claim — that using too few cointegration vectors imposes greater costs in forecast inaccuracy than retaining non-cointegrating terms — to apply to larger systems. Intuitively, the former excludes the relationships which keep differentials between the levels of variables stationary, whereas the latter has parameter estimates on I(1) variables which converge at $O_p(T)$ (or faster) as against the growth in forecast error variance of $O(h)$ for h-step ahead forecasts. Hence the conjecture that ML should dominate UV in larger systems is expected to hold because the number of restrictions implied by cointegration increases with the size of the system, rather than because non-cointegrating I(1) levels terms are being excluded. Conversely, imprecise parameter estimates act in the opposite direction.

Finally, the asymptotic second moment formulae appear to be a reasonably good guide to the small sample (estimation size of 100) results of the Monte Carlo, and the Monte Carlo control variables are highly correlated with the Monte Carlo estimates when parameters are estimated.

8. SUMMARY AND CONCLUSIONS

Of the forecasting methods noted in Section 1, macroeconometric systems seem to offer the only sustainable basis for forecast judgements which also consolidate accumulating empirical and theoretical knowledge. The nonstationarity apparent in many economic time series has important implications for forecasting just as it does for econometric modelling and inference. Nonstationary time series may be cointegrated, so we presented formulae for forecast error variances for cointegrated systems and reported on the accuracy of these formulae in finite samples based on Monte Carlo evidence. The sensible practice of comparing forecasts in terms of differenced variables reduces discrimination of good models from bad, because after a few steps ahead, their forecast error variances converge. Other evaluation methods are required for selecting policy-relevant models. Including too many cointegrating relations seems less detrimental than omitting relevant ones; the converse finding could have posed a conflict for policy analysis.

Econometricians often analyse transformations of the original variables, especially in nonstationary processes. This has important implications for evaluating forecast accuracy since linear models are invariant under nonsingular, scale-preserving linear transformations, but mean square error measures are not. Although alternative rankings can occur when parameters are estimated, this correctly reflects the potential benefits of parsimony (see Clements and Hendry,

1993*b*). In any case, selecting the model with the minimum MSFE is an inadequate strategy since the model need not have constant parameters, nor encompass the forecast errors of competing models.

There appears to be a relatively short forecast horizon over which econometric models can yield marked gains relative to unconditional forecasts. This problem is exacerbated by the potential non-monotonicity of forecast variances as the horizon increases. We reconsidered the benefits of pooling complementary sources of forecast information, as against the weaknesses inherent in pooling forecasts based on competing models, which are implicit encompassing tests. We evaluated the role of intercept corrections, drawing on a recent taxonomy to demonstrate that specific forms of such corrections can make forecasts more robust to model misspecification and to structural change. This finding is consistent with the empirical evidence that intercept corrections improve model-based forecasts.

A comprehensive list of sources of forecast error was developed, and a positive aspect of our paper was to reveal where econometrics could contribute value added to the forecasting process in a world where models are mimics rather than facsimiles of the data generation process, where that process is not constant over time, and where investigators disagree about model choice.

APPENDIX

We use three approximations to obtain a more tractable expression following Schmidt (1977), Baillie (1979*b*) and Chong and Hendry (1986). Let:

$$\hat{\psi} = \psi + \delta \tag{101}$$

where δ is $O_p(1/\sqrt{T})$ so that powers of δ are asymptotically negligible, then:

$$\hat{\psi}^h = (\psi + \delta)^h \cong \psi^h + h\delta\psi^{h-1} = \psi^h + h\psi^{h-1}(\hat{\psi} - \psi). \tag{102}$$

Consequently:

$$\mathrm{V}\left[\hat{\psi}^h - \psi^h\right] \cong \mathrm{V}\left[h\psi^{h-1}(\hat{\psi} - \psi)\right] = h^2\psi^{2(h-1)}\mathrm{V}[\hat{\psi}]. \tag{103}$$

This result also follows from the usual formula for a nonlinear estimation function:

$$\mathrm{V}\left[\hat{\psi}^h - \psi^h\right] \cong \frac{\partial\psi^h}{\partial\psi}\mathrm{V}[\hat{\psi}]\frac{\partial\psi^h}{\partial\psi}. \tag{104}$$

Next:

$$\mu\psi^j - \hat{\mu}\hat{\psi}^j = \psi^j(\mu - \hat{\mu}) + \mu(\psi^j - \hat{\psi}^j) - (\mu - \hat{\mu})(\psi^j - \hat{\psi}^j), \tag{105}$$

where the final term is negligible relative to the first two, and hence from (102):

$$\begin{aligned}\sum_{j=0}^{h-1}\left(\mu\psi^j - \hat{\mu}\hat{\psi}^j\right) &\cong (\mu - \hat{\mu})\sum_{j=0}^{h-1}\psi^j + \mu(\psi - \hat{\psi})\sum_{j=1}^{h-1} j\psi^{j-1} \\ &= (\mu - \hat{\mu})\frac{1-\psi^h}{1-\psi} + (\psi - \hat{\psi})\frac{\mu\left(1 - h\psi^{h-1}(1-\psi) - \psi^h\right)}{(1-\psi)^2} \\ &\cong (\mu - \hat{\mu})\frac{1-\psi^h}{1-\psi} + (\psi - \hat{\psi})\frac{\mu\left(1 - h\psi^{h-1}(1-\psi)\right)}{(1-\psi)^2} \\ &= \left(\theta - \hat{\theta}\right)' \mathbf{b},\end{aligned} \tag{106}$$

(say), where:

$$\mathbf{b}' = \left(\frac{1-\psi^h}{1-\psi} : \frac{\mu\left(1 - h\psi^{h-1}(1-\psi)\right)}{(1-\psi)^2}\right)$$

Then:

$$\mathrm{E}\left[\left\{\sum_{j=0}^{h-1}\left(\mu\psi^j - \hat{\mu}\hat{\psi}^j\right)\right\}^2\right] = \mathbf{b}'\mathrm{V}\left[\hat{\theta}\right]\mathbf{b}. \tag{107}$$

Finally, using (102) and (106):

$$\begin{aligned}\mathrm{E}\left[\left\{\sum_{j=0}^{h-1}\left(\mu\psi^j - \hat{\mu}\hat{\psi}^j\right)\right\}\left(\psi^h - \hat{\psi}^h\right)\right] w_T &= h\psi^{h-1}\mathbf{b}'\mathrm{E}\left[\left(\theta - \hat{\theta}\right)(\psi - \hat{\psi})\right] w_T \\ &= h\psi^{h-1}\mathbf{b}'\mathrm{V}\left[\hat{\theta}\right]\mathbf{s}w_T,\end{aligned} \tag{108}$$

where $\mathbf{s}' = (0 : 1)$. Letting $h\psi^{h-1}w_T\mathbf{s} = \mathbf{c}$ and $\mathbf{d} = \mathbf{b} + \mathbf{c}$, (11) becomes:

$$\mathrm{V}\left[\hat{e}_{T+h} \middle| w_T\right] = \sigma^2\frac{1-\psi^{2h}}{1-\psi^2} + \mathbf{d}'\mathrm{V}\left[\hat{\theta}\right]\mathbf{d}, \tag{109}$$

where:

$$\mathrm{V}\left[\hat{\theta}\right] = T^{-1}\begin{bmatrix}\sigma^2 + \mu^2\frac{1+\psi}{1-\psi} & -\mu(1+\psi) \\ -\mu(1+\psi) & 1-\psi^2\end{bmatrix}, \tag{110}$$

from which our analytical approximations are calculated.

REFERENCES

ARTIS, M., S. MOSS and ORMEROD, P. (1992), 'A Smart Automated Macroeconometric Forecasting System', unpublished paper, University of Manchester.

BAILLIE, R. T. (1979*a*), 'Asymptotic prediction mean squared error for vector autoregressive models', *Biometrika*, 66, 3, 675–8.

—— (1979*b*), 'The asymptotic mean squared error of multistep prediction from the regression model with autoregressive errors', *Journal of the American Statistical Association*, 74, 175–84.

BANERJEE, A., J. J. DOLADO, J. W. GALBRAITH and D. F. HENDRY (1993), *Co-integration, Error Correction, and the Econometric Analysis of Non-Stationary Data*, Oxford University Press.

BJØRNSTAD, J. F. (1990), 'Predictive Likelihood: A Review', *Statistical Science*, 5, 242–65.

BOX, G. E. P. and G. M. JENKINS (1970), *Time Series Analysis Forecasting and Control*, San Francisco: Holden-Day.

BRANDNER, P. and R. S. KUNST (1990), 'Forecasting Vector Autoregressions — The Influence of Cointegration. A Monte Carlo Study', Research Memorandum, 265, Institute for Advanced Studies, Vienna.

CALZOLARI, G. (1981), 'A note on the variance of ex post forecasts in econometric models', *Econometrica*, 49, 1593–6.

—— (1987), 'Forecast Variance in Dynamic Simulation of Simultaneous Equations Models', *Econometrica*, 55, 6, 1473–6.

CAMPBELL, J. Y. and N. G. MANKIW (1987), 'Are output fluctuations transitory?' *Quarterly Journal of Economics*, 102, 857–80.

CHAMBERS, M. J. (1991), 'A Note on Forecasting in Co-Integrated Systems', unpublished paper, Department of Economics, University of Essex.

CHONG, Y. Y. and D. F. HENDRY (1986), 'Econometric Evaluation of Linear Macro-Economic Models', *Review of Economic Studies*, LIII, 671–90.

CLEMENTS, M. P. and D. F. HENDRY (1992), 'Forecasting in Cointegrated Systems', Institute of Economics and Statistics Discussion Paper No. 139, Oxford.

—— (1993*a*), 'On the limitations of comparing mean squared forecast errors', *Journal of Forecasting*, forthcoming.

—— (1993*b*), 'Model Selection when Forecasting', unpublished paper, Institute of Economics and Statistics, Oxford.

CLEMENTS, M. P. and G. E. MIZON (1991), 'Empirical analysis of macroeconomic time series: VAR and structural models', *European Economic Review*, 35, 887–32.

COEN, P. G., E. D. GOMME and M. G. KENDALL (1969), 'Lagged Relationships in Economic Forecasting', *Journal of the Royal Statistical Society*, A, 132, 133–63.

DHRYMES, P. J. (1984), *Mathematics for Econometrics*, 2nd edition, Springer-Verlag, New York.

DICKEY, D. A. and W. A. FULLER (1979), 'Distribution of the Estimators for Autoregressive Time Series with a Unit Root', *Journal of American Statistical Association*, 74, 427–31.

DIEBOLD, F. X. (1989), 'Forecast combination and encompassing: Reconciling two divergent literatures', *International Journal of Forecasting*, 5, 589–92.

DOAN, T., R. LITTERMAN and C. A. SIMS (1984), 'Forecasting and conditional projection using realistic prior distributions', *Econometric Reviews*, 3, 1–100.

ENGLE, R. F. (1982), 'Autoregressive Conditional Heteroskedasticity with Estimates of the Variance of United Kingdom Inflation', *Econometrica*, 50, 4, 987–1007.

ENGLE, R. F. and C. W. J. GRANGER (1987), 'Cointegration, and Error-Correction: Representation, Estimation and Testing', *Econometrica*, 55, 251–76.

ENGLE, R. F. and D. F. HENDRY (1993), 'Testing Super Exogeneity and Invariance in Regression Models', *Journal of Econometrics*, 56, 119–39.

ENGLE, R. F., D. F. HENDRY and J.-F. RICHARD (1983), 'Exogeneity', *Econometrica*, 51, 277–304.

ENGLE, R. F. and B. S. YOO (1987), 'Forecasting and Testing in Cointegrated Systems', *Journal of Econometrics*, 35, 143–59.

ENGLE, R. F. and B. S. YOO (1991), 'Cointegrated Economic Time Series: An Overview with New Results' in R. F. ENGLE and C. W. J. GRANGER (eds.), *Long-Run Economic Relationships*, Oxford University Press, 237–66.

ERICSSON, N. R. (1992), 'Parameter Constancy, Mean Square Forecast Errors, and Measuring Forecast Performance: An Exposition, Extensions, and Illustration', *Journal of Policy Modeling*, 14, 4, 465–98.

ERICSSON, N. R. and J. R. MARQUEZ (1989), 'Exact and Approximate Multi-Period Mean-Square Forecast Errors for Dynamic Econometric Models', International Finance Discussion Paper No. 348, Federal Reserve Board, Washington D.C.

ERICSSON, N. R., D. F. HENDRY and H.-A. TRAN (1992), 'Cointegration, Seasonality, Encompassing and the Demand for Money in the United Kingdom', International Finance Discussion paper, Federal Reserve Board,Washington D.C.

FAVERO, C. and D. F. HENDRY (1992), 'Testing the Lucas Critique: A Review', *Econometric Reviews*, 11, 265–306.

GRANGER, C. W. J. (1969), 'Investigating Causal Relations by Econometric Models and Cross-Spectral Methods', *Econometrica*, 37, 224–38.

GRANGER, C. W. J. and P. NEWBOLD (1973), 'Some comments on the evaluation of economic forecasts', *Applied Economics*, 5, 35–47.

—— (1977), *Forecasting Economic Time Series*, New York: Academic Press. (Second Edition, 1986.)

HARVEY, A. and N. SHEPHARD (1992), 'Structural Time Series Models', in G. S. Maddala, C. R. Rao and H. D. Vinod (eds.), *Handbook of Statistics*, Vol. 11: Econometrics, North-Holland, Amsterdam.

HENDRY, D. F. (1979), 'Predictive Failure and Econometric Modelling in Macroeconomics: The Transactions Demand for Money', Ch. 9 in P. Ormerod (ed.), *Modelling the Economy*, London: Heinemann Educational Books.

—— (1986), 'The Role of Prediction in Evaluating Econometric Models', *Proceedings of the Royal Society*, A407, 25–33.

—— (1987), 'Econometric Methodology: A Personal Perspective', Chapter 10 in T. Bewley (ed.), *Advances in Econometrics*, Cambridge: Cambridge University Press.

—— (1993), *Dynamic Econometrics*, forthcoming: Oxford University Press.

HENDRY, D. F. and M. P. CLEMENTS (1993), 'On a Theory of Intercept Corrections in Macro-Economic Forecasting', in S. Holly (ed.), *Money, Inflation and Employment: Essays in Honour of Sir James Ball*, forthcoming: Edward Elgar.

HENDRY, D. F. and N. R. ERICSSON (1991), 'Modeling the Demand for Narrow Money in the United Kingdom and the United States', *European Economic Review*, 35, 833–81.

HENDRY, D. F. and G. E. MIZON (1993), 'Evaluating Dynamic Econometric Models by Encompassing the VAR', in P. C. B. Phillips (ed.), *Models, Methods and Applications of Econometrics*, Oxford: Basil Blackwell, 272–300.

HENDRY, D. F. and J.-F. RICHARD (1989), 'Recent Developments in the Theory of Encompassing', in B. Cornet and H. Tulkens (ed.), *Contributions to Operations Research and Econometrics. The Twentieth Anniversary of CORE*. Cambridge, Mass.: MIT Press, 393–440.

HENDRY, D. F. and P. K. TRIVEDI (1972), 'Maximum Likelihood Estimation of Difference Equations with Moving Average Errors: A Simulation Study', *Review of Economic Studies*, 39, 117–45.

HYLLEBERG, S. (ed.) (1992), *Modelling Seasonality*, Oxford University Press.

JOHANSEN, S. (1988), 'Statistical Analysis of Cointegration Vectors', *Journal of Economic Dynamics and Control*, 12, 231–54.

JOHANSEN, S. (1992), 'A representation of vector autoregressive processes integrated of order 2', *Econometric Theory*, 8, 188–202.

JOHANSEN, S. and K. JUSELIUS (1990), 'Maximum Likelihood Estimation and Inference on Cointegration — With Applications to the Demand for Money', *Oxford Bulletin of Economics and Statistics*, 52, 169–210.

JUDGE, G. G. and M. E. BOCK (1978), *The Statistical Implications of Pre-Test and Stein-Rule Estimators in Econometrics*. Amsterdam: North-Holland.

KALMAN, R. E. (1960), 'A New Approach to Linear Filtering and Prediction Problems', *Journal of Basic Engineering*, 82, 35–45.

LU, M. and G. E. MIZON (1991), 'Forecast Encompassing and Model Evaluation', in P. Hackl and A. H. Westlund (eds), *Economic Structural Change, Analysis and Forecasting*. Berlin: Springer-Verlag, 213–318.

LUCAS, R. E. (1973), 'Some International Evidence on Output-Inflation Tradeoffs', *American Economic Review*, 63, 326–34.

—— (1976), 'Econometric Policy Evaluation: A Critique', in K. Brunner and A. H. Meltzer (eds.), *The Phillips Curve and Labor Markets*, 19–46, Amsterdam: North-Holland.

LÜTKEPOHL, H. (1991), *Introduction to Multiple Time Series Analysis*, Springer-Verlag.

MARIANO, R. S. and B. W. BROWN (1991), 'Stochastic-Simulation Tests of Nonlinear Econometric Models', in L. R. KLEIN (ed.), *Comparative Performance of U.S. Econometric Models*, Oxford University Press, 250–59.

MIZON, G. E. (1984), 'The Encompassing Approach in Econometrics', in D. F. Hendry and K. F. Wallis (eds), *Econometrics and Quantitative Economics*, Oxford: Basil Blackwell, 135–72.

MIZON, G. E. and J.-F. RICHARD (1986), 'The Encompassing Principle and its Application to Non-Nested Hypothesis Tests', *Econometrica*, 54, 657–78.

NELSON, C. (1972), 'The Prediction Performance of the FRB-MIT-PENN Model of the US Economy', *American Economic Review*, 62, 902–17.

PHILLIPS, P. C. B. (1991), 'Optimal inference in cointegrated systems', *Econometrica*, 59, 2, 283–306.

REICHLIN, L. (1989), 'Structural change and unit root econometrics', *Economic Letters*, 31, 231–3.

SCHMIDT, P. (1977), 'Some small sample evidence on the distribution of dynamic simulation forecasts', *Econometrica*, 45, 97–1005.

SPITZER, J. J. and R. T. BAILLIE (1983), 'Small Sample Properties of Predictions from the Regression Model with Autoregressive errors', *Journal of the American Statistical Association*, 258–63.

STOCK, J. H. and M. W. WATSON (1988), 'Testing for Common Trends', *Journal of the American Statistical Association*, 88, 404, 1097–107.

—— (1989), 'New Indexes of Coincident and Leading Economic Indicators', *NBER Macro-Economic Annual*, 351–409.

TODD, R. M. (1990),'Improving economic forecasts with Bayesian vector autoregression', chapter 10 in C. W. J. Granger (ed.), *Modelling Economic Series. Readings in Econometric Methodology*, Oxford: Oxford University Press, 214–34.

TURNER, D. S. (1990), 'The Role of Judgement in Macroeconomic Forecasting', *Journal of Forecasting*, 9, 315–45.

WALLIS, K. F. (1989), 'Macroeconomic forecasting: a survey', *Economic Journal*, 99, 28–61.

WALLIS, K. F. and J. D. WHITLEY (1991), 'Sources of Error in Forecasts and Expectations: U.K. Economic Models 1984–8', *Journal of Forecasting*, 10, 231–53.

WALLIS, K. F., M. J. ANDREWS, P. G. FISHER, J. A. LONGBOTTOM and J. D. WHITLEY (1986), *Models of the UK Economy: A Third Review by the ESRC Macroeconomic Modelling Bureau*, Oxford: Oxford University Press.

WALLIS, K. F., P. G. FISHER, J. A LONGBOTTOM, D. S. TURNER and J. D. WHITLEY (1987), *Models of the UK Economy: A Fourth Review by the ESRC Macroeconomic Modelling Bureau*, Oxford: Oxford University Press.

WEST, K. D. (1988), 'Asymptotic Normality when Regressors have a Unit Root', *Econometrica*, 56, 1397–418.

3

Bayes models and forecasts of Australian macroeconomic time series

Peter C.B. Phillips[*]

This paper provides an empirical implementation of some recent work by the author and Werner Ploberger on the development of *Bayes models* for time series. The methods offer a new data-based approach to model selection, to hypothesis testing and to forecast evaluation in the analysis of time series. A particular advantage of the approach is that modelling issues such as lag order, parameter constancy, and the presence of deterministic and stochastic trends all come within the compass of the same statistical methodology, as do the evaluation of forecasts from competing models. The paper shows how to build parsimonious empirical *Bayes models* using the new approach and applies the methodology to some Australian macroeconomic data. *Bayes models* are constructed for thirteen quarterly Australian macroeconomic time series over the period 1959(3)–1987(4). These models are compared with certain fixed format models (like an AR(4) + linear trend) in terms of their forecasting performance over the period 1988(1)–1991(4). The *Bayes models* are found to be superior to these forecasting exercises for two of the thirteen series, while at the same time being more parsimonious in form.

1. INTRODUCTION

Not all econometric models are designed as instruments for forecasting Nevertheless, the capacity of one model to forecast adequately in comparison with competing models is an important element in the evaluation of its overall

[*] All of the computations reported in this paper were performed by the author on a 486-33 PC using programs written in GAUSS-386i (Version 2.2). The author thanks Sam Ouliaris for supplying the data, the NSF for research support under Grant No. SES 9122142, and Glena Ames for her skill and effort in keyboarding the manuscript.

performance. Indeed, many of the procedures that are presently used to appraise a model's performance involve summary statistics that depend in one way or another on the model's within-sample and outside-sample tracking behaviour. Thus, in spite of the multiplicity of objectives in econometric modelling, one common characteristic is the attempt each model makes to explain the data, or certain subsets of the data conditionally on other data. This attempted explanation often leads directly, but sometimes indirectly, to a model's 'probability distribution of the data'. Again, this may be a conditional distribution, and the statistical procedures that are employed may mean that only certain characteristics of the distribution rather than the full distribution are modelled. However, this common element of econometric modelling provides a basis by which different models can be compared. Thus, one model's explanation of the data can be compared with that of another model in terms of their implied 'probability distributions of the data'. In a similar way, one model's predictions can be compared with those of a competing model in terms of the respective 'probability distributions of the prediction errors'.

These ideas underlie some recent work by the author (1992) and by the author and Werner Ploberger (1991, 1992) on the development of *Bayes models* for time series. *Bayes models* are essentially location models conditional on the data that is available to the latest observation. In these models, the location estimate or systematic part of the model is nonlinear and time varying even when the underlying 'true model' is linear in parameters and variables like an autoregression. The location estimate is a predictor given by the current best estimate, using prior information and the available data, of the value of the dependent variable in the next period. The predictor is calculated as the conditional mean of the dependent variable given data to the latest available observation. Here the conditional expectation is taken with respect to the probability measure of the data implied by the given model and the prior distribution of the parameters. We call this measure the *Bayes model* measure. As more data accumulates, this *Bayes model* measure of the data becomes independent of the prior and is therefore 'objective' in the well defined sense that it ultimately depends only on the form of the model and the observed data. Since the Bayes measure is distinct for different models, it may be used as the basis for comparing competing models in terms of their implied 'probability distributions or predictive distributions of the data', as discussed in the last paragraph.

This paper implements the above ideas in an empirical study of Australian macroeconomic data. Our purpose is to seek out the best *Bayes models* in a certain generic class of time series models for each data set and then evaluate the adequacy of these models against certain fixed format competitor models in terms of one-period ahead forecasts. The methodology is based on earlier work in Phillips and Ploberger (1991, 1992) and Phillips (1992), which will be briefly reviewed in section 2 of the paper.

2. MODEL AND FORECAST EVALUATION USING THE PIC CRITERION

Our set up is the single equation stochastic linear regression model

$$y_t = \beta' x_t + \varepsilon_t, \quad (t = 1, 2, \ldots) \tag{1}$$

where the dependent variable y_t and the error ε_t are real valued stochastic processes on a probability space (Ω, F, P). Accompanying y_t is the filtration $F_t \subset F (t = 0, 1, 2, \ldots)$ to which both y_t and ε_t are adapted. Usually it is convenient to think of F_t as the σ-field generated by $\{\varepsilon_t, \varepsilon_{t-1}, \ldots\}$ and in the cases we consider this will always be appropriate. The regressors x_t $(k \times 1)$ in (1) are defined on the same space and are assumed to have the property that x_t is F_{t-1}-measurable. The errors ε_t satisfy $E(\varepsilon_t | F_{t-1}) = 0$, so that the conditional mean function in (1) is correctly specified under the probability measure P.

An example of (1) that is frequently empirically relevant is the 'ARMA (p,q) + trend(r)' model. This model can be written in difference format as

$$\Delta y_t = h y_{t-1} + \sum_{i=1}^{p-1} \varphi_i \Delta y_{t-i} + \sum_{j=1}^{q} \psi_j \varepsilon_{t-j} + \sum_{k=0}^{r} \delta_k t^k + \varepsilon_t \tag{2}$$

which is especially convenient because it accommodates an autoregressive unit root under the simple restriction $h = 0$. We call the parameter $a = 1+h$ the 'long run autoregressive coefficient' since this parameter is instrumental in determining the shape of the spectrum of y_t at the origin — see Phillips (1991) for elaboration on this point.

In (2) there are $k = p + q + r + 1$ parameters. When $q > 0$, some of the regressors, *viz.* the ε_{t-j}, are not observed. Recursive techniques are then required, either to construct the likelihood as in the use of the Kalman filter, or in repeated linear regressions that involve the construction of estimates of the lagged errors ε_{t-j} as in the Hannan and Rissanen (1982, 1983) recursion. When $q = 0$ in (2) the model is an 'AR(p) + trend (r)'. When $r = -1$ there is no intercept in the model, when $r = 0$ there is a fitted intercept, and when $r = 1$ there is a fitted linear trend. These are the specialisations of (2) that are of primary interest in empirical applications.

The order parameters p, q and r in (2) are not known in practical applications and the model is in any event best regarded as just an approximate generating mechanism. Various methodologies for dealing with this complication are available. Those that concern us here are based on formal statistical order selection methods such as the commonly used criteria AIC and BIC. These criteria and their statistical properties in stationary systems are discussed in detail in the recent book by Hannan and Deistler (1988). When the system is potentially nonstationary as in (2) with $h = 0$ the properties of these criteria are less well understood although they have been studied, notably by Paulsen (1984), Tsay (1984) and Pötscher (1989).

Our approach to the order selection problem is based on the analysis in Phillips and Ploberger (1992) and is closely related to the principle underlying the BIC criterion of Schwarz (1978), viz. to select the model with the highest *a posteriori* probability. This approach has a compelling advantage over AIC and BIC in that it naturally accommodates models of nonstationary time series and has generally superior sampling performance (see Phillips and Ploberger (1992) for simulation evidence on this point). The probability measure used to determine our criterion is the measure associated with the *Bayes model* corresponding to (1). This model is formally derived for the case of Gaussian errors $\varepsilon_t = iid\, \mathrm{N}(0, \sigma^2)$ in (1) and has the form

$$y_t = \hat{\beta}'_{t-1} x_t + v_t, \quad where \quad v_t|_{F_{t-1}} \equiv \mathrm{N}(0, f_t) \tag{3}$$

with

$$f_t = \sigma^2 \{1 + x_t' A_{t-1}^{-1} x_t\}, \quad A_t = \sum_1^t x_s x_s' \tag{4}$$

and where $\hat{\beta}_{t-1}$ is the least squares estimate of β based on information in F_{t-1}.

The Phillips and Ploberger analysis shows that, under a uniform prior on β and a Gaussian likelihood the passage via Bayes rule to the posterior density of β implies the replacement of the model (1) by the time varying parameter model (3). We therefore call (3) the *Bayes model* corresponding to (1). Note that the systematic part of (3), $\hat{\beta}'_{t-1} x_t$, is the best estimate or predictor of the location of y_t given information in F_{t-1}. This location estimate is identical to the maximum likelihood estimate of the best predictor of the next period observation, i.e. it is precisely the predictor we would use in classical inference. Thus, the *Bayes model* is identical to the classical model that is actually used to make predictions (in place of (1)). From this perspective there is no difference between the Bayesian and classical approaches. However, we can go further in our approach and find the probability measure associated with the *Bayes model* (3). This is a forward looking measure that can be described by its conditional density given F_{t-1}. This density is given by the Radon Nikodym (RN) derivative of the measure at t (say Q_t) with respect to the measure at $t-1$ (Q_{t-1}), i.e.

$$
\begin{aligned}
dQ_t/dQ_{t-1} &= pdf_Q(y_t|F_{t-1}) \\
&= (2\pi f_t)^{-1/2}\exp\{-(1/2f_t)v_t^2\} \equiv N(0,f_t), \quad t=k+1,\ k+2,\ \ldots\ .
\end{aligned}
\tag{5}
$$

We use the notation $\mathrm{pdf}_Q(\cdot)$ here to signify that this is the density corresponding to Q-measure. Note that it is defined as soon as there are enough observations in a trajectory to estimate the k-vector β. Thus, (3), (4) and (5) are defined for $t \geq k+1$. The measure Q_t that appears in (5) is called the *Bayes model* measure, i.e. the measure corresponding to the *Bayes model* (3). This measure is σ-finite and, as shown in Phillips and Ploberger (1992), can also be defined in terms of the following RN derivative

$$
dQ_t/dP_t = \left|(1/\sigma^2)A_t\right|^{-1/2}\exp\{(1/2\sigma^2)\hat{\beta}_t' A_t \hat{\beta}_t\},
\tag{6}
$$

which is taken with respect to the reference measure P_t for the model (1) in which $\beta = 0$ (i.e. the probability measure of the $N(0, \sigma^2 I_t)$ distribution).

Associated with every *Bayes model* of the form (3) is a σ-finite measure Q_t. Different models of the same data may be compared in terms of the *Bayes model* measures that are associated with them. The natural mechanism for making such comparisons is the likelihood ratio. Suppose, for example, that we have two models of the form given in (3), one with k parameters and the other with $K \geq k$ parameters. Indexing the variables in (3) by the number of parameters we now have two *Bayes models* of the data: one with k parameters that we write as

$$
\mathbf{H}(Q_n^k): y_{n+1} = \hat{\beta}_n(k)' x_{n+1}(k) + v_{n+1}(k);
$$

and the second, more complex model with K parameters

$$
\mathbf{H}(Q_n^K): y_{n+1} = \hat{\beta}_n(K)' x_{n+1}(K) + v_{n+1}(K).
$$

The likelihood ratio of the measures associated with $\mathrm{H}(Q_n^k)$ and $\mathrm{H}(Q_n^K)$ is given by the RN derivative dQ_n^k/dQ_n^K. This quantity can be calculated by taking the ratio of the RN derivatives that define Q_n^k and Q_n^K in terms of the reference measure P_n, i.e. the ratio of the corresponding expressions given by (6) for each model. Thus,

$$dQ_n^k / dQ_n^K = \left(dQ_n^k / dP_n\right) / \left(dQ_n^K / dP_n\right)$$
$$= \left|\tfrac{1}{\sigma^2} A_n(k)\right|^{-\frac{1}{2}} \left|\tfrac{1}{\sigma^2} A_n(K)\right|^{-\frac{1}{2}} \exp\left\{\left(\tfrac{1}{2\sigma^2}\right)\left[\hat{\beta}_n(k)' A_n(k) \hat{\beta}_n(k) - \hat{\beta}_n(K)' A_n(K) \hat{\beta}_n(K)\right]\right\}. \quad (7)$$

This likelihood ratio measures the support in the data for the more restrictive model H(Q_n^k) against that of the more complex model H(Q_n^K). When we assign equal prior odds to the two competing models our decision criterion is to accept H(Q_n^k) in favour of H(Q_n^K) when $dQ_n^k / dQ_n^K > 1$.

Since σ^2 in (7) is usually unknown we must supply an estimate of this scale parameter before the criterion can be used in practice. Phillips and Ploberger (1992) suggests the use of $\hat{\sigma}_K^2$, the least squares estimate of σ^2 from the more complex model H(Q_n^K). Our order estimator is then given by

$$\hat{k} = \text{argmin}_k \text{PIC}_k \quad (8)$$

where

$$\text{PIC}_k = \left(dQ_n^K / dQ_n^k\right)\left(\hat{\sigma}_K^2\right). \quad (9)$$

Observe that $\hat{k}$ maximises $1/\text{PIC}_k = dQ_n^k / dQ_n^K\,(\hat{\sigma}_K^2)$ and thereby selects the model most favoured over H(Q_n^K) according to the density of the data.

An alternative form of the PIC criterion (9) that is given in Phillips and Ploberger (1992) is based on the predictive densities of the competing Bayes models, i.e. H(Q_n^k) and H(Q_n^K). By comparing the densities for these models over the same subsample of data, say $n > K$, we have

$$\text{PICF}_k = \frac{dQ_n^k}{dQ_n^K}\left(\hat{\sigma}_K^2\right)\Big| F_K = \prod_{K+1}^{n} \left(\hat{f}_t^K / \hat{f}_t^k\right)^{\frac{1}{2}} \exp\left\{\sum_{K+1}^{n}\left[v_t(K)^2 / 2\hat{f}_t^K - v_t(k)^2 / 2\hat{f}_t^k\right]\right\} \quad (10)$$

where

$$\hat{f}_t^k = \hat{\sigma}_K^2\left(1 + x_t(k)' A_{t-1}(k)^{-1} x_t(k)\right), \quad \hat{f}_t^K = \hat{\sigma}_K^2\left(1 + x_t(K)' A_{t-1}(K)^{-1} x_t(K)\right);$$

$$v_t(k) = y_t - \hat{\beta}_{t-1}(k)' x_t(k), \qquad v_t(K) = y_t - \hat{\beta}_{t-1}(K)' x_t(K).$$

Note that (10) differs from (9) only to the extent that in (10) the density ratio is conditional on data in F_K. In effect, the initialisation in (9) is at $t = 0$ on the field F_0, whereas in (10) the initialisation is at $t = K$ on the field F_K where there is enough sample data to estimate both models H(Q_t^k) and H(Q_t^K). Both criteria PIC and PICF are scale invariant because of the presence of the error variance estimate $\hat{\sigma}_K^2$ in their definitions. Note, however, that PICF is invariant to linear transformations of the regressors $x_t(k)$ and $x_t(K)$ in the two models H(Q_t^k) and

$H(Q_t^K)$, whereas PIC is not. Use of PIC therefore presumes that there is some natural form of the regressor variables, as there is for example in AR and ARMA models with deterministic time trends. This is the class of model that will be used in the application of our methods that is reported below.

As indicated earlier, our approach is related to the principle underlying the BIC criterion, which leads to the order estimator

$$\text{argmin}_k\left[\text{BIC}_k = \ln\hat{\sigma}_k^2 + k\ln(n)/n\right].$$

When the data are stationary and ergodic it is easily shown that our criterion PIC is asymptotically equivalent to BIC (see Phillips and Ploberger, 1992). However, when the data are nonstationary the criterion PIC imposes a greater penalty than BIC on the presence of additional nonstationary regressors. The simulations in Phillips and Ploberger (1991) show that PIC generally outperforms BIC as an order estimator criterion for both stationary and nonstationary data, at least in Gaussian models.

The PIC and PICF criteria are also related to the MDL and PMDL criteria of Rissanen (1986, 1987*a*, 1987*b*), *viz*.

$$\text{argmin}_k\left[\text{MDL}_k = \ln\hat{\sigma}_k^2 + \ln|A_n(k)|/n\right], \ \text{argmin}_k\left[\text{PMDL}_k = \sum_{t=K+1}^{n}\{\ln\hat{\sigma}_{kt} + \hat{e}_{t+1}^2/\hat{\sigma}_{kt}^2\}\right],$$

(see e.g. Mills and Prasad, 1992, for these formulae). Clearly, the MDL criterion is closely related to PIC in that the penalty term $\ln|A_n(k)|/n$ involves the data rather than simply a parameter count as in the BIC criterion penalty $k\ \ln(n)/n$. Note, however, that the PIC penalty involves the term $|A_n(k)/\hat{\sigma}_K^2|$ and is therefore scale invariant. In the Rissanen predictive criterion PMDL, $\hat{e}_{t+1}$ and $\hat{\sigma}_{kt}^2$ are defined by

$$\hat{e}_{t+1}^2 = \sum_{j=0}^{t}\left(y_{j+1} - x_{j+1}(k)'\hat{\beta}_t\right)^2,$$

and

$$\hat{\sigma}_{kt}^2 = \sum_{j=1}^{t}\left(y_j - x_j(k)'\hat{\beta}_t\right)^2 \Big/ t.$$

Writing

$$\exp\{-(½)\text{PMDL}_k\} = \prod_{t=k}^{n-1}(1/\hat{\sigma}_{kt})\exp\{-\hat{e}_{t+1}^2/2\hat{\sigma}_{kt}^2\},$$

we see that PMDL is related in form to our PICF. The criteria differ, however, because (i) PICF employs the recursive one step ahead squared forecast errors $v_t^2(k) = (y_t - x_t(k)'\beta_{t-1})^2$ rather than the sum of squared predictive errors $\hat{e}_{t+1}^2$ as in PMDL; and (ii) PICF employs the forecast error variance $\hat{f}_t^k$ rather than the error variance estimator $\hat{\sigma}_{kt}^2$ as in PMDL.

The PICF criterion (10) has a very interesting interpretation as a form of encompassing test statistic. For, if $dQ_n^k/dQ_n^K(\hat{\sigma}_K^2)|_{F_K} > 1$ the evidence in the sample suggests that the density for the model with k parameters exceeds the density of the model with K parameters when both are evaluated at the sample data. This is equivalent to saying that the model with k parameters encompasses the model with K parameters in terms of their respective probability densities. Thus, when $dQ_n^k/dQ_n^K(\hat{\sigma}_K^2)|_{F_K} > 1$, the *Bayes model* $\mathrm{H}(Q_n^k)$ encompasses the *Bayes model* $\mathrm{H}(Q_n^K)$ in terms of the probability distribution of the sample data over the period $t = K+1, \ldots, n$. This might be called distributional encompassing for $t \in [K+1, n]$.

Obvious extensions of this principle apply for subperiods of the overall sample. Moreover, the principle can be extended to evaluate the forecasts from competing models. For instance, $\mathrm{H}(Q_n^k)$ and $\mathrm{H}(Q_n^K)$ can be compared in terms of their respective performance in one-period ahead forecasts over the period $t = n+1, \ldots, N$. The *Bayes model* forecast encompassing test statistic for this period is

$$dQ_n^k/dQ_n^K\left(\hat{\sigma}(K)^2\right)\Big|_{F_n} = \prod_{t=n+1}^{N}\left(g_t^K/g_t^k\right)^{1/2}\exp\left\{-\left(1/2\hat{\sigma}_t^2(K)g_t^k\right)v_t(k)^2 + \left(1/2\hat{\sigma}_t^2(K)g_t^K\right)v_t(K)^2\right\}. \tag{11}$$

Note that in this formulation the variance estimate $\hat{\sigma}_t^2(K)$ evolves recursively over the forecast period. Again, $\mathrm{H}(Q_n^k)$ encompasses $\mathrm{H}(Q_n^K)$ in terms of forecast performance over the period $[n+1, N]$ when $dQ_n^k/dQ_n^K(\hat{\sigma}_K^2)|_{F_K} > 1$.

Bayes models like $\mathrm{H}(Q_n^k)$ may be permitted to evolve in a natural way as more observations become available. Thus, period by period we may employ the PIC criterion (8) to select the appropriate value, $\hat{k}_t$, of k for the sample data up to observation t–1, prior to making the one-period ahead forecast of the value of y_t. This leads to an evolving sequence of best *Bayes models*

$$\mathbf{H}\left(Q_t^B\right): y_t = \hat{\beta}_{t-1}\left(\hat{k}_{t-1}\right)' x_t\left(\hat{k}_{t-1}\right) + v_t\left(\hat{k}_{t-1}\right)$$

which are determined recursively using the PIC criterion (8) period after period. It is then possible to compare the best *Bayes model* sequence $\mathrm{H}(Q_t^B)$ with a fixed format *Bayes model* sequence $\mathrm{H}(Q_t^F)$ that employs a fixed number of parameters (F). The comparison can be made in terms of their respective predictive densities over a forecast horizon such as $t \in [n+1, N]$. In this case the forecast-encompassing test statistic is

$$dQ_N^B / dQ_N^F \left(\hat{\sigma}^2(\hat{k})\right)\Big|_{F_n}$$
$$= \prod_{t=n+1}^{N} \left(g_t^F / g_t^{\hat{k}_{t-1}}\right)^{½} \exp\left\{-\left(1/2\hat{\sigma}_t^2(\hat{k}_{t-1})g_t^{\hat{k}_{t-1}}\right)v_t(\hat{k}_{t-1})^2 + \left(1/2\hat{\sigma}_t^2(\hat{k}_{t-1})g_t^F\right)v_t(F)^2\right\}. \quad (12)$$

We would favour the best *Bayes model* sequence $\{H(Q_t^B)\}_{n+1}^N$ over the sequence of fixed format models $\{H(Q_t^F)\}_{n+1}^N$ if

$$dQ_n^B / dQ_N^F \left(\hat{\sigma}^2(\hat{k})\right)\Big|_{F_n} > 1, \quad (13)$$

that is, if the sequence $H(Q_t^B)$ generates forecasts over $t = n+1, ..., N$ that encompass the forecasts of the fixed format sequence of models $H(Q_t^F)$. Note that in (12) and (13) we use recursive estimates of the error variance from the best *Bayes model* sequence, since these are consistent for σ^2 when (1) is the actual generating mechanism.

The most important property of $H(Q_t^B)$ is that this sequence of models adapt to the data. When fewer parameters are needed to model the data the sequence will respond by eliminating unnecessary parameters. When more are needed, the sequence adapts by enlarging the model, either by adding more lags or by adding deterministic trend polynomial regressors, as appropriate. Since the PIC criterion can also be used to test for the presence of a unit autoregressive root the best *Bayes model* sequence $H(Q_t^B)$ can also be designed to include unit roots whenever these are supported by the data.

In general, we may expect $H(Q_t^B)$ to have fewer parameters than $H(Q_t^F)$, especially when *F* incorporates a linear trend and several lags. Reasonable choices for $H(Q_t^F)$ depend on the time interval of observation. Thus, for annual data a fixed format model of the type 'AR(3) + linear trend' may seem a sensible baseline competitor. For seasonally adjusted quarterly data an 'AR(4) + linear trend' may be reasonable and for monthly data one might choose models with longer lag lengths including an 'AR(12) + linear trend' as a baseline competitor. Some of these alternatives will be used in the empirical work that follows.

3. AUSTRALIAN MACROECONOMIC DATA

The data we use are quarterly and monthly Australian macroeconomic time series. The quarterly series cover the period 1959(3)–1991(4) and the monthly series cover the periods 1959(1)–1991(12) and 1967(7)–1991(12). All variables except interest rates and stock prices are seasonably adjusted. Table 1 gives details of the thirteen series that we use and the variable notation that we employ.

Table 1: Macroeconomic Variable Notation and Description

Variable	Description	Frequency	Sample Period	Forecast Period
C	Aggregate private final consumption expenditure ($m; sa)	Quarterly	1959(3)–1987(4)	1988(1)–1991(4)
RC	Aggregate real private final consumption exp. ($m average 1984/5 prices; sa)	"	"	"
GDP	Gross domestic product ($m; sa)	"	"	"
RGDP	Real gross domestic product ($m average 1984/5 prices; sa)	"	"	"
PGDP	Implicit price deflator for GDP (sa)	"	"	"
CPI	Consumer price index (1981 = 100; sa)	"	"	"
U	Unemployment rate (%; sa)	"	"	"
WR	Wage rate: Average earnings of non-farm wage and salary earners ($/week)	"	"	"
RWR	Real wage rate (= WR/PGDP)	"	"	"
SP500	Australian share price index: all ordinaries (31 December 1979 = 500)	monthly	1959(1)–1987(12)	1988(1)–1991(12)
Int1	Money market 13 week Treasury Notes (% pa, yield)	"	1969(7)–1987(12)	"
Int2	Capital market 2 year Treasury Bonds (% pa, yield)	"	"	"
Int3	Capital market 10 year Treasury Bonds (% pa, yield)	"	"	"

All of the series except interest rates are lagged. Interest rates are taken in levels (% pa) and reciprocals of levels. The latter transformation (i.e. $x \to 1/x$) is variance stabilising and reduces the volatility in the series that tends to occur at higher interest rate levels. The reciprocal transformation was found to work well for US bond yields in Phillips (1992) and is therefore used again here. All of the series are graphed in Figures 1(a)–13(a). Figure 11'(a), for instance, shows the short term interest rate, Int1, over the period 1969(7)–1991(12). The increased volatility in this series at higher levels of Int1 is apparent. Figure 11(a) graphs the series in reciprocals, i.e. 1/Int1, over the same period. The effects of stabilising the volatility in this case are quite clear from the two figures. This feature of interest rate data is less apparent for the intermediate rate, Int2, and the long term rate, Int3. However, the transformation is used for both these series as well and the graphs are shown in Figures 12(a), 12'(a) and 13(a), 13'(a), respectively.

4. *BAYES MODELS* FOR THE DATA

Using the PIC model selection criterion we set out to find the best *Bayes model* for the time series described in the last section. Two classes of models were considered. The first was the 'ARMA(p,q) + trend(r)' model given in (2) and the second was the simpler 'AR(p) + trend(r)' model.

The algorithm for determining the trend degree and lag orders of the ARMA model is the one given in Phillips and Ploberger (1992). This algorithm involves the following steps:

Step 1. Set maximum orders for the AR, MA and trend components.

Step 2. Run a long autoregression with maximum trend degree and use PIC or BIC to select the AR order ($\hat{p}$).

Step 3. Select the trend degree ($\hat{r}$) in the model chosen in Step 2 using PIC or BIC. Calculate the residuals $\hat{\varepsilon}_t$ from this regression.

Step 4. Run an array of ARMA(p,q) + trend($\hat{r}$) regressions using $\hat{\varepsilon}_{t-j}$ in place of ε_{t-j} for the MA variable. Choose the orders $(\hat{p},\hat{q})$ using either PIC or BIC.

Step 5. If $\hat{p} > 0$, compare the *Bayes model* selected in Steps 1–4, viz. 'ARMA$(\hat{p},\hat{q})$ + trend($\hat{r}$)' with a *Bayes model* of the same order having a unit autoregressive root. Choose the restricted 'ARMA$(\hat{p},\hat{q})$ with unit root + trend($\hat{r}$)' model if the posterior odds criterion PIC favours this model (i.e. is greater than unity) over the reference model 'ARMA$(\hat{p},\hat{q})$ + trend($\hat{r}$)'. If $\hat{p} = 0$, then there is no autoregressive component and hence no autoregressive unit root.

The algorithm for selecting the best *Bayes model* in the 'AR(p) + trend(r)' class is the same as the above, but simply omits the MA component and hence Step 4. One of our interests is to discover whether this simpler class of models is adequate for most economic time series.

These algorithms of model selection were applied to the thirteen Australian macroeconomic time series described earlier. The empirical results are shown in Table 2. All of these series are found to be stochastically nonstationary. Twelve of the series have a unit autoregressive root while one series, GDP, has a mildly explosive long-run autoregressive coefficient of 1.001 that is preferred to a competing model with a unit root.

Only one of the series (the GDP price deflator) is found to have trend degree $r = 1$. Since the best *Bayes model* for this series also has a unit root, the implied

Table 2: Best *Bayes models* for Australian Macro Time Series

	Block A				Block B			
	Model class = ARMA(p,q) + trend(r)				Model class = AR(p) + trend(r)			
	Model selected				Model selected			
Series	Dynamics	r	ρ[a]	Odds[b]	Dynamics	r	ρ[a]	Odds[b]
C	ARMA(2,1)	–1	1.000	3313.723	AR(3)	–1	1.000	102.212
RC	AR(1)	0	0.994	11.221				
GDP	AR(2)	–1	1.001	0.000				
RGDP	AR(1)	0	0.992	8.936				
PGDP	ARMA(2,1)	1	0.993	84.493	AR(4)	–1	1.001	255.238
CPI	AR(4)	–1	1.001	472.010				
U	AR(4)	–1	1.005	117.851				
WR	AR(3)	–1	1.002	3.998				
RWR	AR(1)	0	0.986	19.489				
SP500	ARMA(1,1)	–1	1.001	174.232	AR(2)	–1	1.001	745.438
Int1	ARMA(1,1)	–1	0.992	78.325	AR(2)	–1	0.999	52.635
Int2	AR(2)	–1	0.997	196.366				
Int3	AR(1)	–1	0.996	92.544				

[a] Long-run autoregressive coefficient; [b] Posterior odds in favour of a unit root
Note: If the model selected in Block A is an AR(p), the Block B result is identical and hence not repeated

model for the series is a stochastic trend around a quadratic. Three of the series (real consumption, real GDP and the real wage rate) are found to have trend degree $r = 0$, leading to a stochastic trend with drift as the best *Bayes model* for these series.

The dynamics are generally well modelled by autoregressions. But for four series (consumption, the GDP deflator, stock prices and the short term interest rate) low order ARMA models are chosen in place of autoregressions. The choice of dynamic model has no effect on the decision in favour of a unit root for these series. Block B of Table 2 shows the model choice outcomes in the 'AR(p) + trend(r)' class and these can be compared with the outcomes selected in the 'ARMA(p,q) + trend(r)' class given in Block A of the table. There is only one important change from restricting the model class to be autoregressive. For the

GDP deflator series an 'AR(4) + trend(–1)' process is selected as distinct from an 'ARMA(2,1) + trend(1)' process when the model class is wider. Note that the long run autoregressive coefficient is larger for the AR(4) model than the ARMA(2,1) (1.001 as distinct from 0.993) but that a unit root *Bayes model* is chosen in each case. Looking at the graph of this series in Figure 6(a), it is apparent that both models can be rationalised in terms of the historical trajectory. The 'AR(4) + trend(–1)' Bayes model with a unit root is, in fact, the more parsimonious of the two (3 parameters as distinct from 4).

5. *BAYES MODEL* FORECAST PERFORMANCE

The four final years of the sample data (1988–1991) were used for an *ex post* forecasting exercise. This involves sixteen observations for the quarterly series and forty-eight observations for the monthly series. The best *Bayes model* sequence $\{H(Q_t^B)\}_{t=1988(1)}^{1991(4 or 12)}$ was determined recursively using the PIC criterion. For the quarterly data an 'AR(p) + trend(r)' class was used with $p \leq 5$, $r \leq 1$. For the monthly series the parameters were prescribed as $p \leq 12$, $r \leq 1$. The autoregressive model class was chosen in place of the ARMA class because most of the series seemed to be well modelled within this class as discussed in the previous section. The best *Bayes model* sequence was compared with a fixed format *Bayes model* sequence in terms of their respective one-period ahead forecasting capabilities. For the quarterly data an 'AR(4) + linear trend' fixed format model was used. For the monthly data series, we used both 'AR(4) + linear trend' and 'AR(12) + linear trend' fixed format rules.

Figures 1–13 show the one-period ahead forecast performance of these *Bayes model* sequences over the period 1988–1991 inclusive. For each series Figure (a) displays the data and the relevant forecast period, and Figure (b) shows the period by period forecast errors from the two rival models (the solid line is the Bayes model error and the dashed line is the fixed model error while the origin is given by a dotted line). Figure (c) gives details of the evolving form of the best *Bayes model*: the solid line on the graph shows the autoregressive lag order selected (0–6 lags), the dashed line shows the trend degree (–1 = no intercept; 0 = fitted intercept; 1 = fitted linear trend), and the dotted line shows whether or not a unit autoregressive root is selected (–1 = yes, 0 = no). Figure (d) gives a recursive plot of the forecast encompassing test statistic dQ^B/dQ^F over the forecast period. Table 3 tabulates these details, gives the root mean squared error (RMSE) of forecasts for the two models over the forecast period, and records the evolving format of the best *Bayes model*.

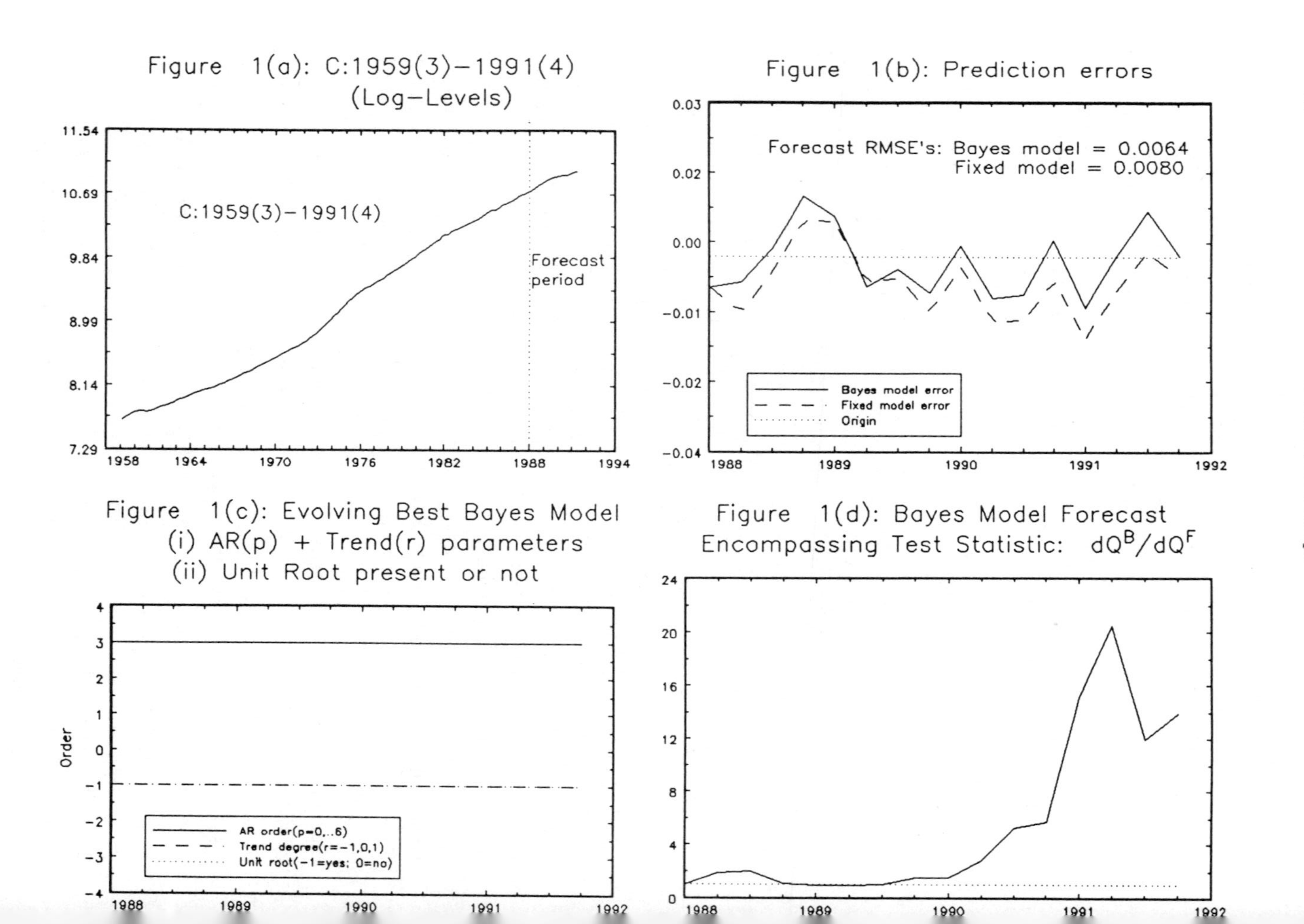

Figure 1(a): C:1959(3)−1991(4)
(Log−Levels)
C:1959(3)−1991(4)
Forecast period
Figure 1(b): Prediction errors
Forecast RMSE's: Bayes model = 0.0064
Fixed model = 0.0080
Bayes model error
Fixed model error
Origin
Figure 1(c): Evolving Best Bayes Model
(i) AR(p) + Trend(r) parameters
(ii) Unit Root present or not
Order
AR order(p=0,..6)
Trend degree(r=−1,0,1)
Unit root(−1=yes; 0=no)
Figure 1(d): Bayes Model Forecast
Encompassing Test Statistic: dQ^B/dQ^F

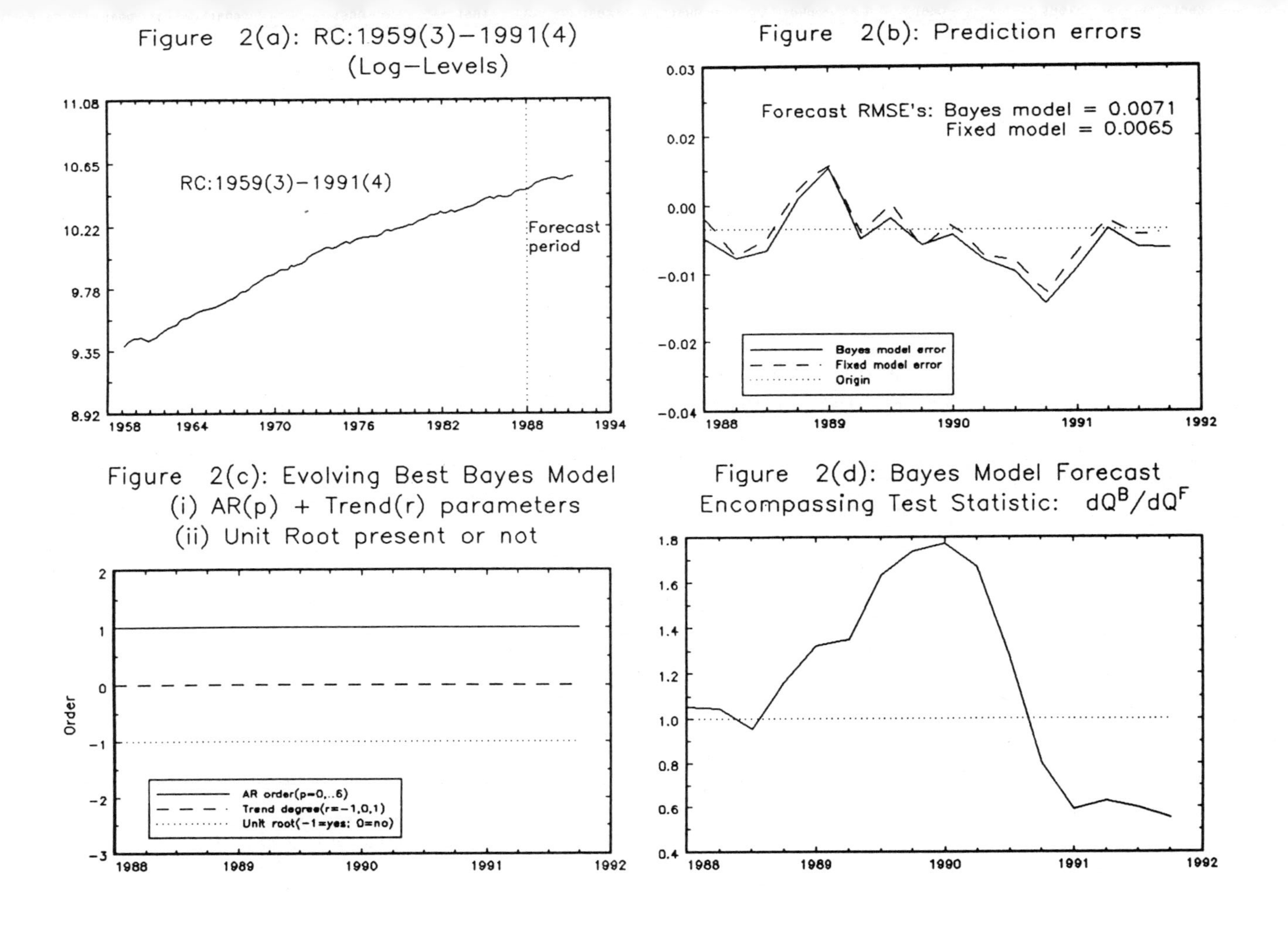

Figure 2(a): RC:1959(3)−1991(4)
(Log−Levels)
RC:1959(3)−1991(4)
Forecast period
11.08
10.65
10.22
9.78
9.35
8.92
1958
1964
1970
1976
1982
1988
1994
Figure 2(b): Prediction errors
Forecast RMSE's: Bayes model = 0.0071
Fixed model = 0.0065
Bayes model error
Fixed model error
Origin
0.03
0.02
0.00
−0.01
−0.02
−0.04
1988
1989
1990
1991
1992
Figure 2(c): Evolving Best Bayes Model
(i) AR(p) + Trend(r) parameters
(ii) Unit Root present or not
AR order(p=0,..6)
Trend degree(r=−1,0,1)
Unit root(−1=yes; 0=no)
Order
2
1
0
−1
−2
−3
1988
1989
1990
1991
1992
Figure 2(d): Bayes Model Forecast
Encompassing Test Statistic: dQ^B/dQ^F
1.8
1.6
1.4
1.2
1.0
0.8
0.6
0.4
1988
1989
1990
1991
1992

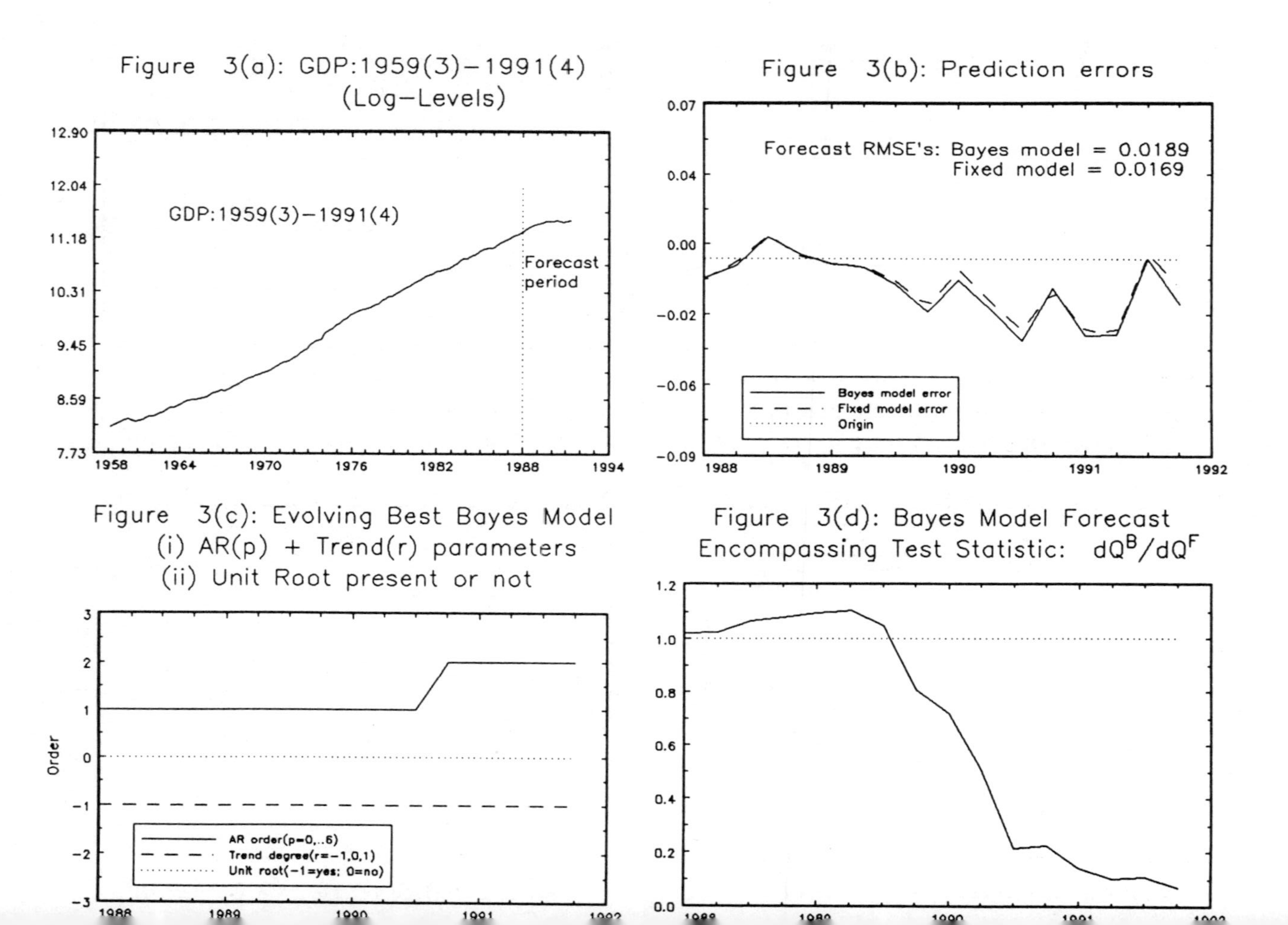
Figure 3(a): GDP:1959(3)–1991(4)
(Log–Levels)
GDP:1959(3)–1991(4)
Forecast
period
12.90
12.04
11.18
10.31
9.45
8.59
7.73
1958
1964
1970
1976
1982
1988
1994
Figure 3(b): Prediction errors
Forecast RMSE's: Bayes model = 0.0189
Fixed model = 0.0169
Bayes model error
Fixed model error
Origin
0.07
0.04
0.00
–0.02
–0.06
–0.09
1988
1989
1990
1991
1992
Figure 3(c): Evolving Best Bayes Model
(i) AR(p) + Trend(r) parameters
(ii) Unit Root present or not
Order
AR order(p=0,..6)
Trend degree(r=–1,0,1)
Unit root(–1=yes; 0=no)
3
2
1
0
–1
–2
–3
Figure 3(d): Bayes Model Forecast
Encompassing Test Statistic: dQ^B/dQ^F
1.2
1.0
0.8
0.6
0.4
0.2
0.0

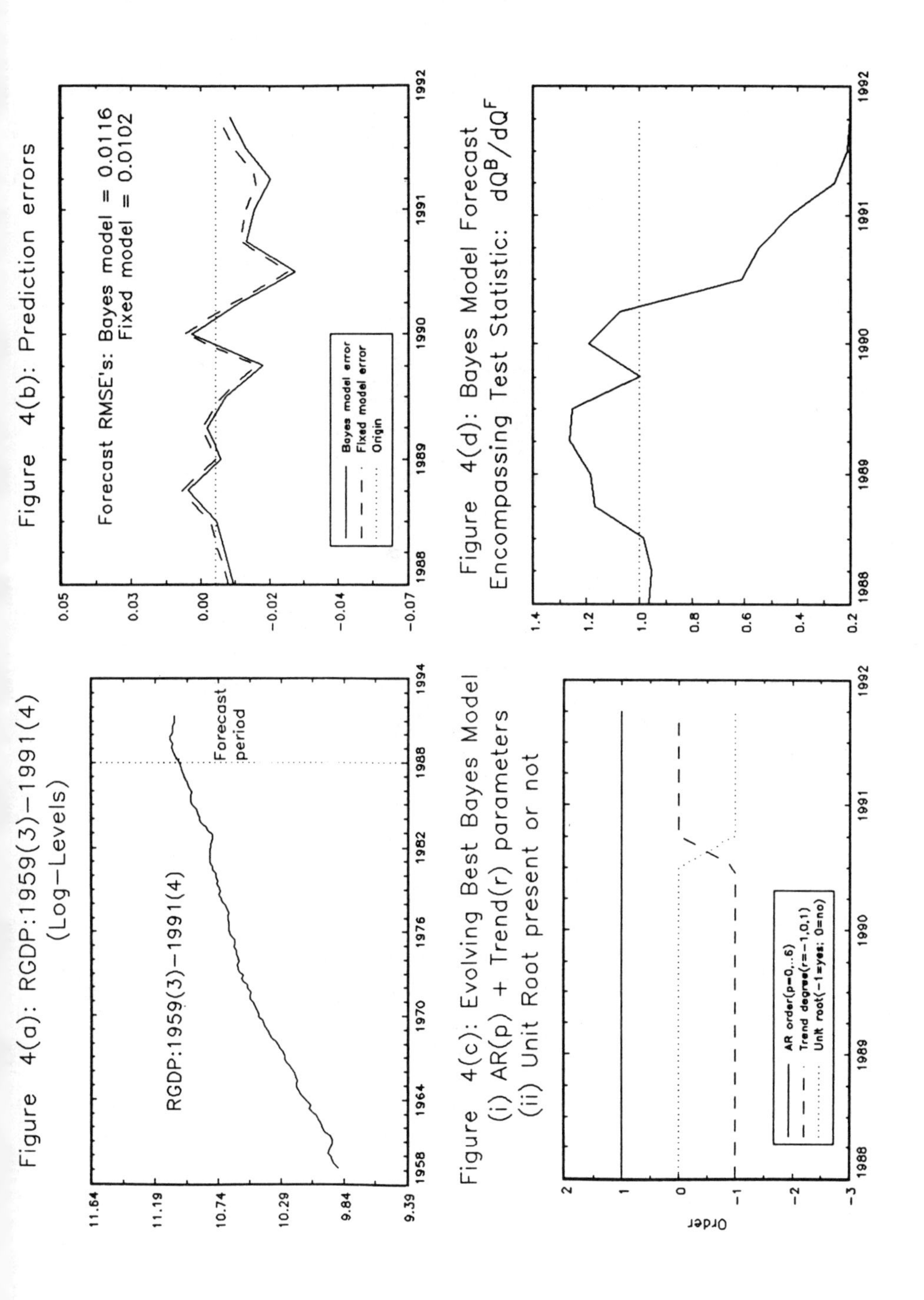
Figure 4(a): RGDP:1959(3)–1991(4)
(Log–Levels)
RGDP:1959(3)–1991(4)
Forecast
period
Figure 4(b): Prediction errors
Forecast RMSE's: Bayes model = 0.0116
Fixed model = 0.0102
Bayes model error
Fixed model error
Origin
Figure 4(c): Evolving Best Bayes Model
(i) AR(p) + Trend(r) parameters
(ii) Unit Root present or not
AR order(p=0,..6)
Trend degree(r=-1,0,1)
Unit root(-1=yes; 0=no)
Order
Figure 4(d): Bayes Model Forecast
Encompassing Test Statistic: dQ^B/dQ^F

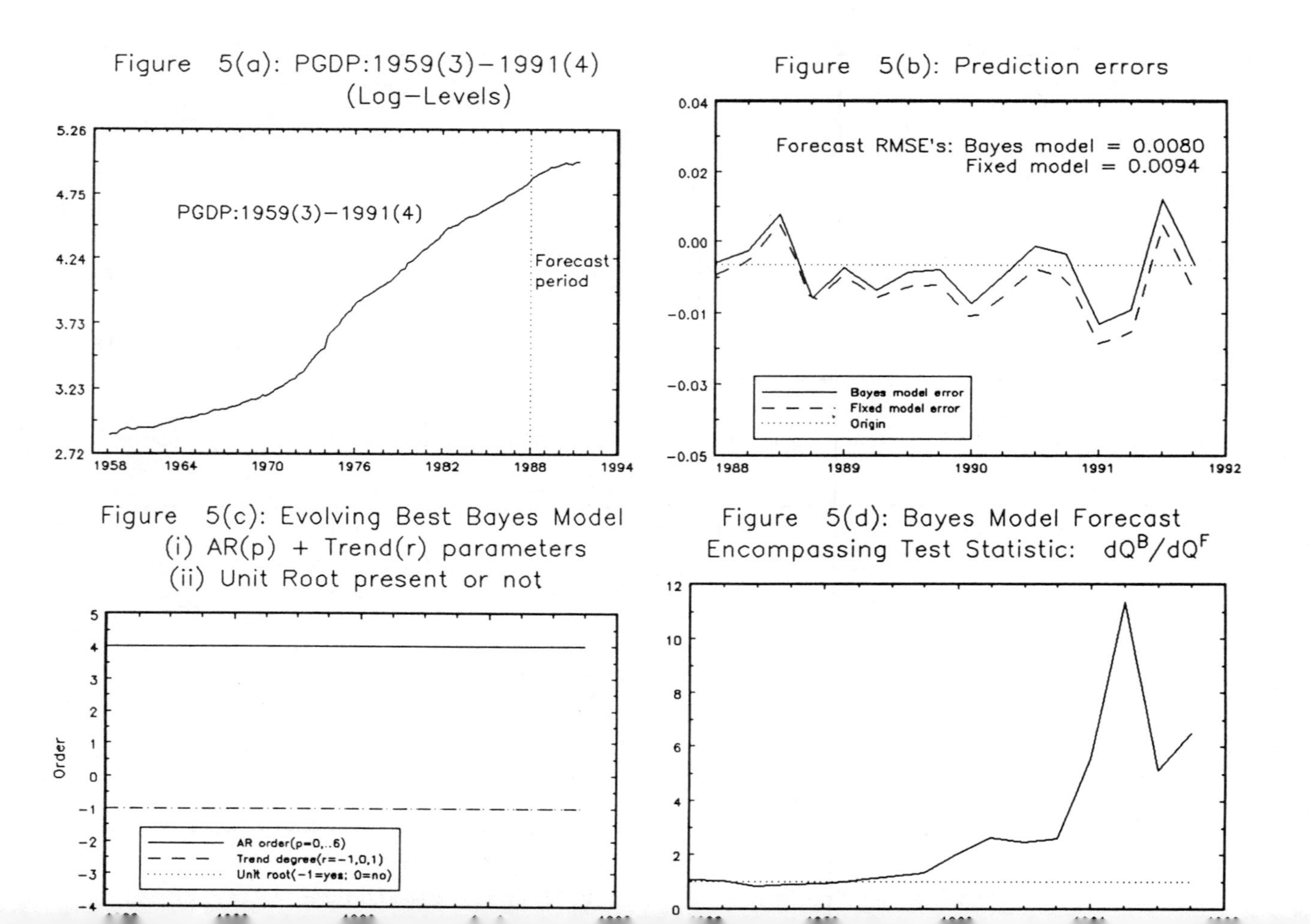
Figure 5(a): PGDP:1959(3)–1991(4)
(Log–Levels)
PGDP:1959(3)–1991(4)
Forecast
period
5.26
4.75
4.24
3.73
3.23
2.72
1958
1964
1970
1976
1982
1988
1994
Figure 5(b): Prediction errors
Forecast RMSE's: Bayes model = 0.0080
Fixed model = 0.0094
Bayes model error
Fixed model error
Origin
0.04
0.02
0.00
−0.01
−0.03
−0.05
1988
1989
1990
1991
1992
Figure 5(c): Evolving Best Bayes Model
(i) AR(p) + Trend(r) parameters
(ii) Unit Root present or not
AR order(p=0,..6)
Trend degree(r=−1,0,1)
Unit root(−1=yes; 0=no)
Order
5
4
3
2
1
0
−1
−2
−3
−4
Figure 5(d): Bayes Model Forecast
Encompassing Test Statistic: dQ^B/dQ^F
12
10
8
6
4
2
0

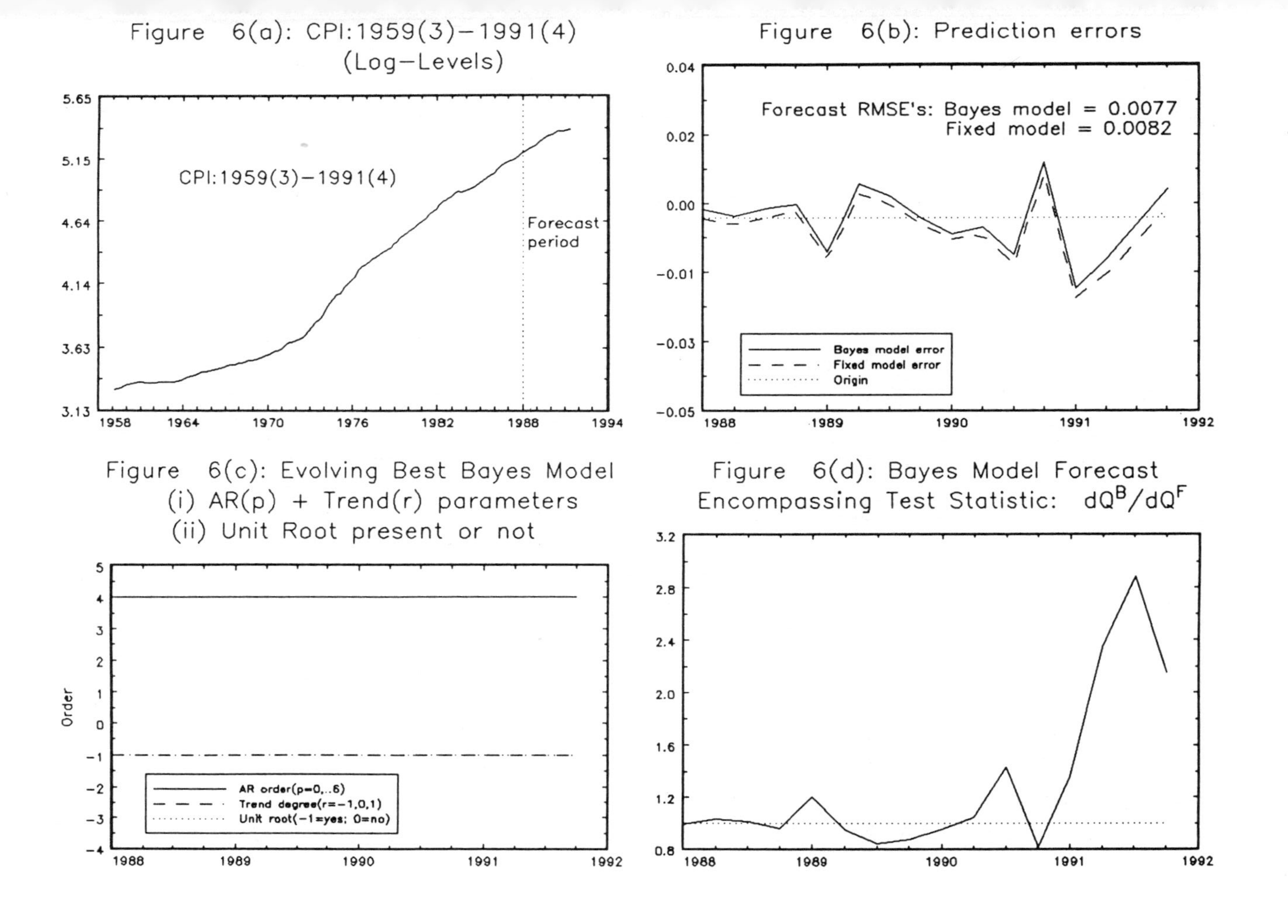

Figure 6(a): CPI:1959(3)–1991(4) (Log–Levels)

Figure 6(b): Prediction errors

Figure 6(c): Evolving Best Bayes Model (i) AR(p) + Trend(r) parameters (ii) Unit Root present or not

Figure 6(d): Bayes Model Forecast Encompassing Test Statistic: dQ^B/dQ^F

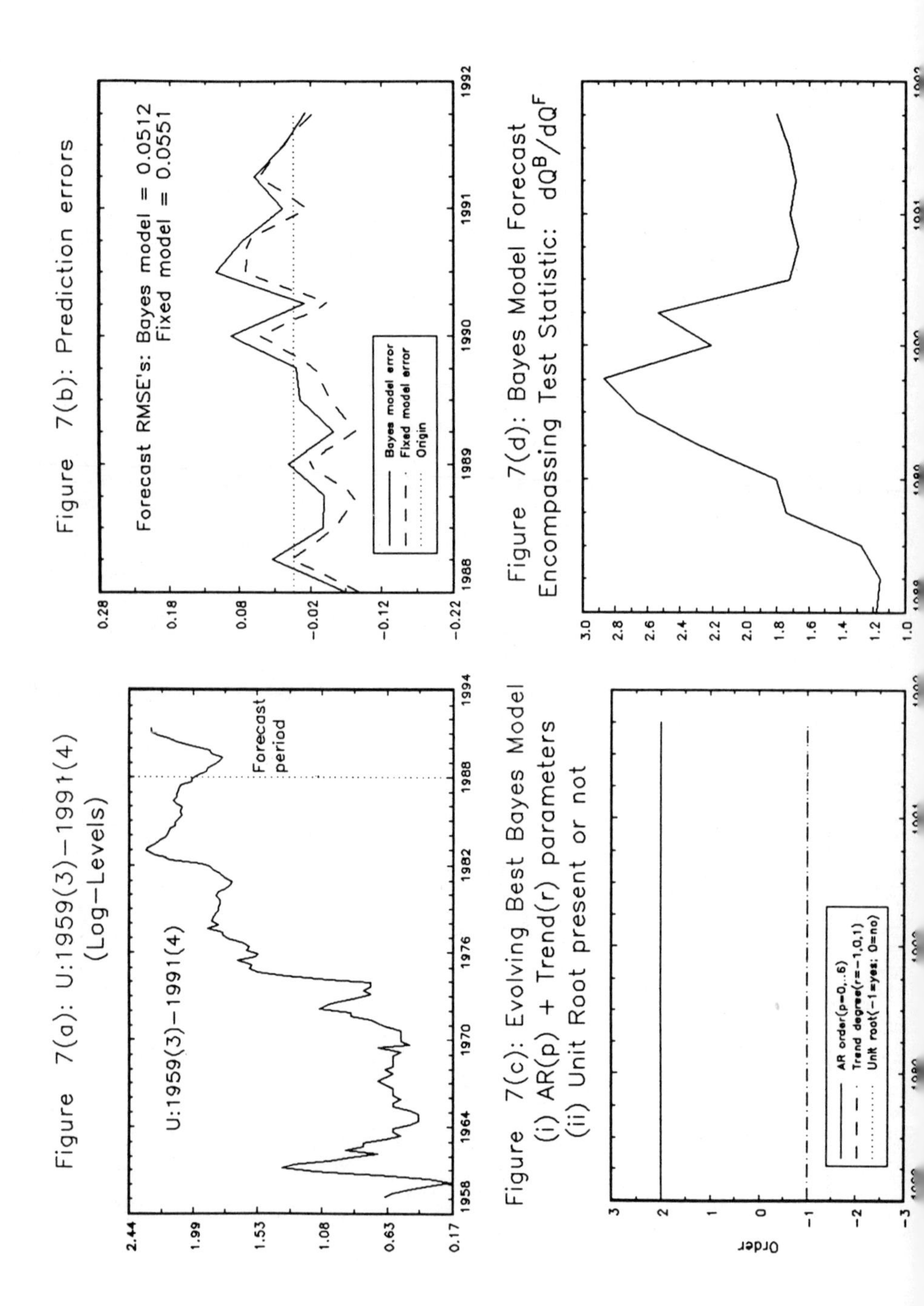

Figure 7(a): U:1959(3)–1991(4) (Log–Levels)

Figure 7(b): Prediction errors

Figure 7(c): Evolving Best Bayes Model (i) AR(p) + Trend(r) parameters (ii) Unit Root present or not

Figure 7(d): Bayes Model Forecast Encompassing Test Statistic: dQ^B/dQ^F

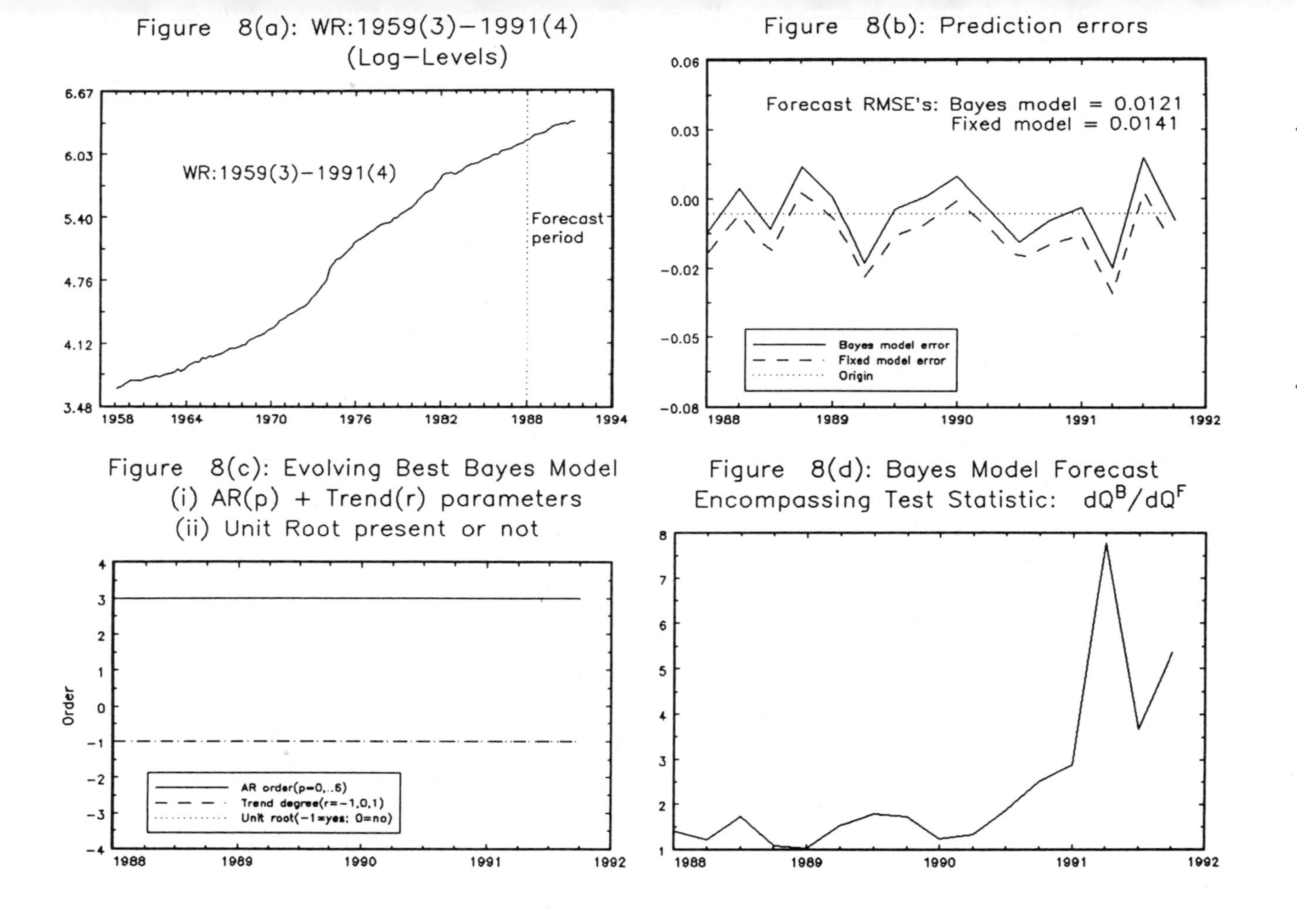

Figure 8(a): WR:1959(3)−1991(4) (Log−Levels)

Figure 8(b): Prediction errors

Figure 8(c): Evolving Best Bayes Model (i) AR(p) + Trend(r) parameters (ii) Unit Root present or not

Figure 8(d): Bayes Model Forecast Encompassing Test Statistic: dQ^B/dQ^F

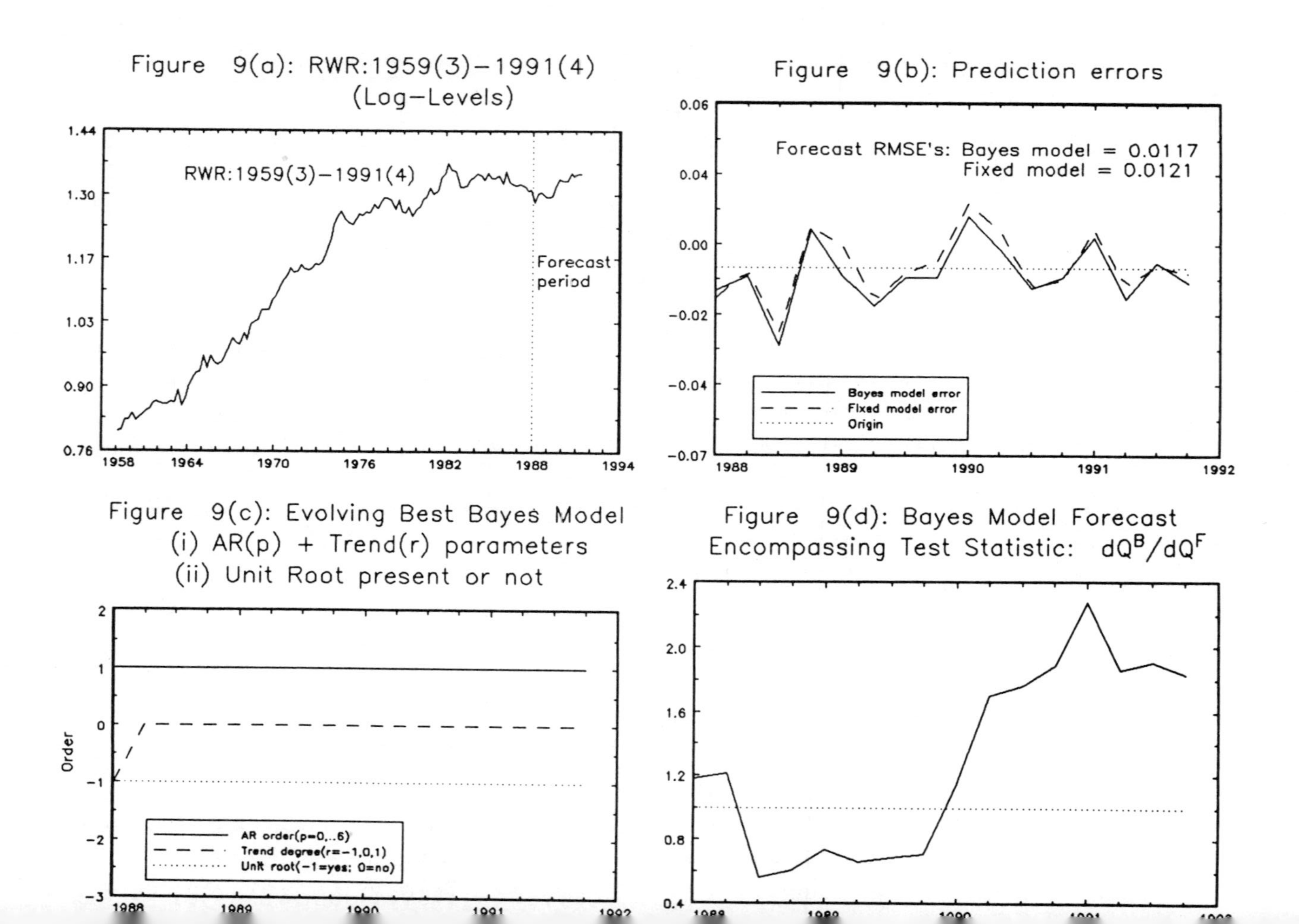
Figure 9(a): RWR:1959(3)–1991(4)
(Log–Levels)
RWR:1959(3)–1991(4)
Forecast period
1.44
1.30
1.17
1.03
0.90
0.76
1958
1964
1970
1976
1982
1988
1994
Figure 9(b): Prediction errors
Forecast RMSE's: Bayes model = 0.0117
Fixed model = 0.0121
Bayes model error
Fixed model error
Origin
0.06
0.04
0.00
−0.02
−0.04
−0.07
1988
1989
1990
1991
1992
Figure 9(c): Evolving Best Bayes Model
(i) AR(p) + Trend(r) parameters
(ii) Unit Root present or not
Order
2
1
0
−1
−2
−3
AR order(p=0,..6)
Trend degree(r=−1,0,1)
Unit root(−1=yes; 0=no)
1988
1989
1990
1991
1992
Figure 9(d): Bayes Model Forecast
Encompassing Test Statistic: dQ^B/dQ^F
2.4
2.0
1.6
1.2
0.8
0.4

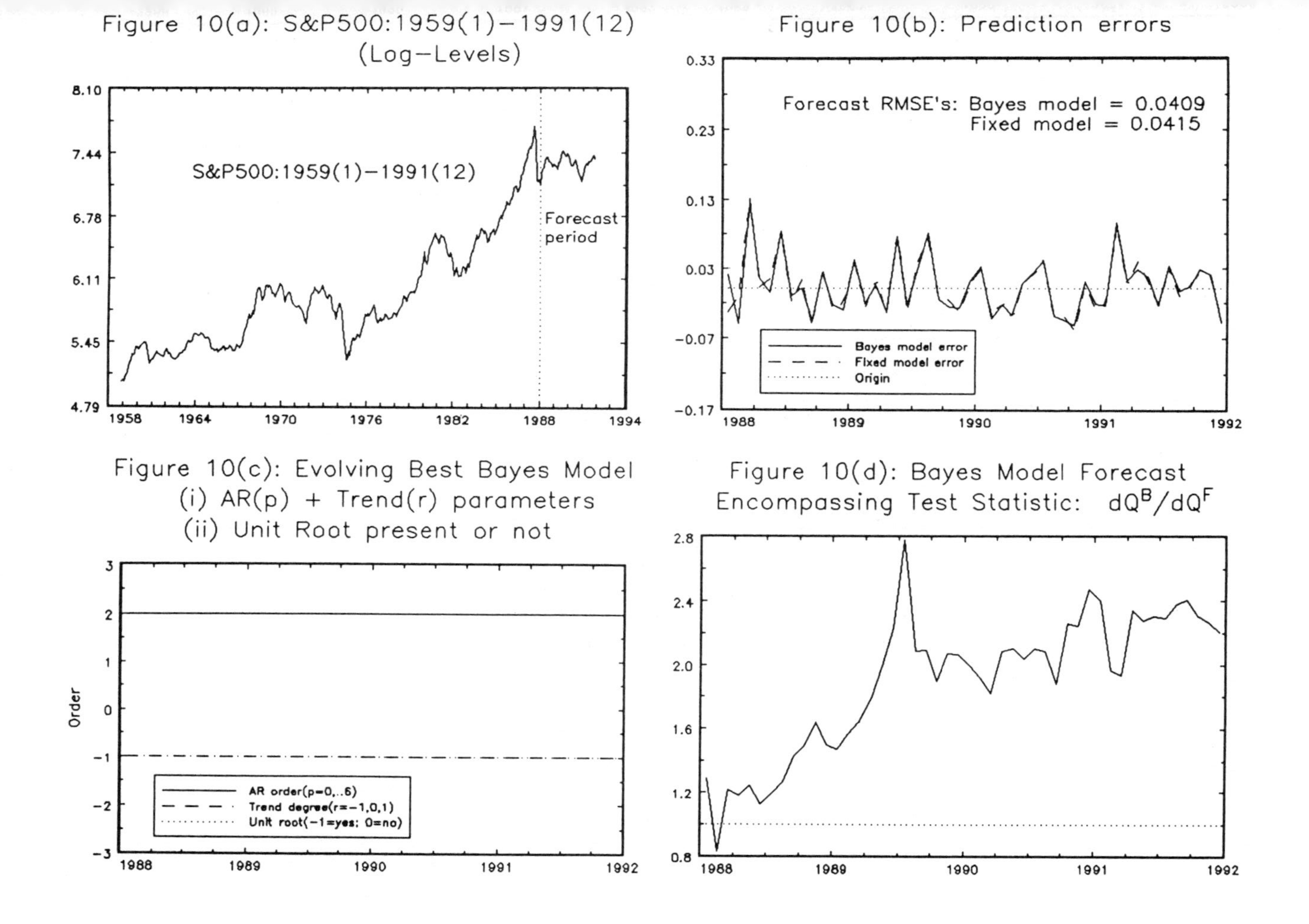

Figure 10(a): S&P500:1959(1)−1991(12)
(Log−Levels)
S&P500:1959(1)−1991(12)
Forecast
period
8.10
7.44
6.78
6.11
5.45
4.79
1958
1964
1970
1976
1982
1988
1994
Figure 10(b): Prediction errors
Forecast RMSE's: Bayes model = 0.0409
Fixed model = 0.0415
Bayes model error
Fixed model error
Origin
0.33
0.23
0.13
0.03
−0.07
−0.17
1988
1989
1990
1991
1992
Figure 10(c): Evolving Best Bayes Model
(i) AR(p) + Trend(r) parameters
(ii) Unit Root present or not
AR order(p=0,..6)
Trend degree(r=−1,0,1)
Unit root(−1=yes; 0=no)
Order
3
2
1
0
−1
−2
−3
1988
1989
1990
1991
1992
Figure 10(d): Bayes Model Forecast
Encompassing Test Statistic: dQ^B/dQ^F
2.8
2.4
2.0
1.6
1.2
0.8
1988
1989
1990
1991
1992

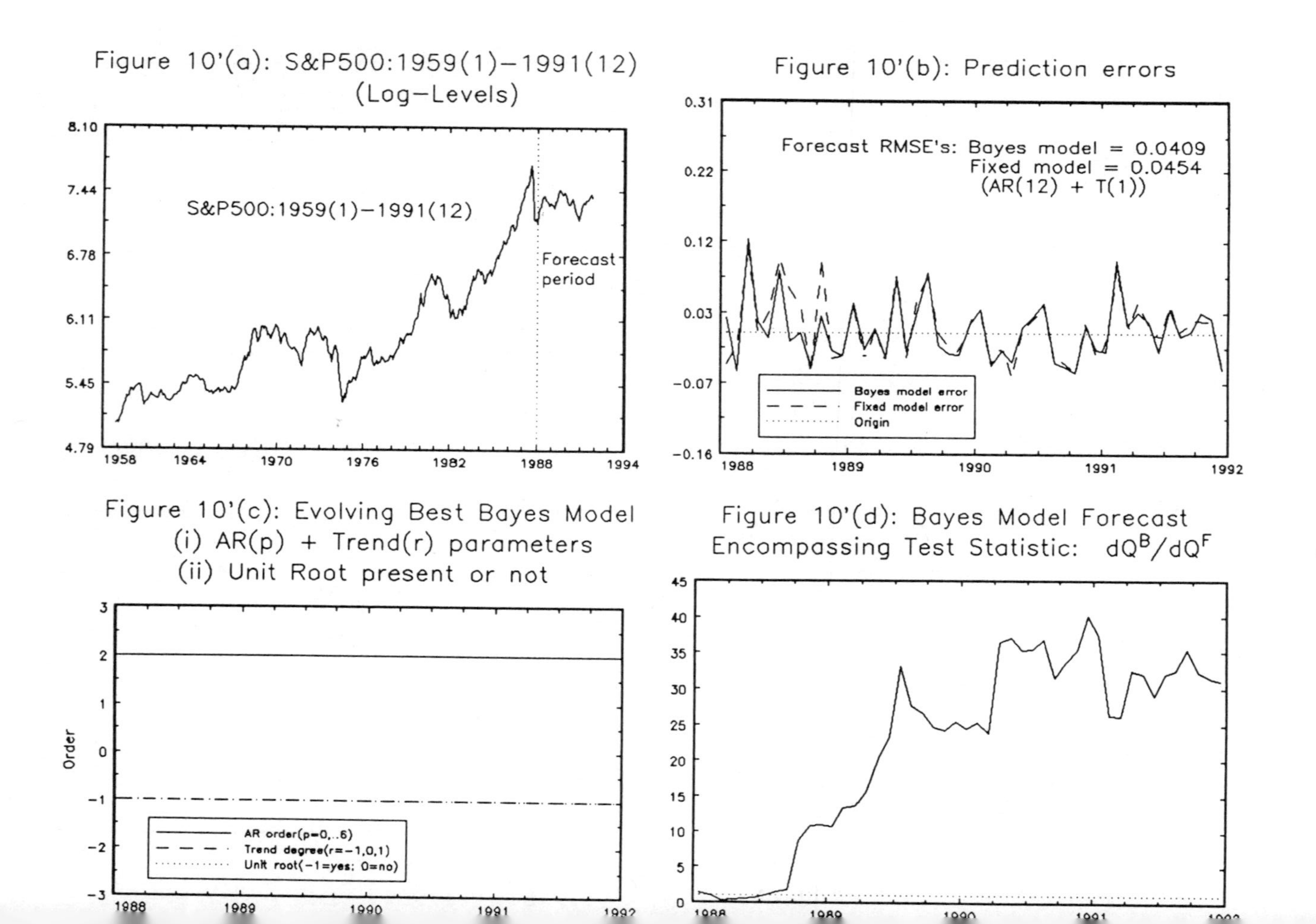

Figure 10'(a): S&P500:1959(1)–1991(12) (Log–Levels)
S&P500:1959(1)–1991(12)
Forecast period
Figure 10'(b): Prediction errors
Forecast RMSE's: Bayes model = 0.0409
Fixed model = 0.0454
(AR(12) + T(1))
Bayes model error
Fixed model error
Origin
Figure 10'(c): Evolving Best Bayes Model
(i) AR(p) + Trend(r) parameters
(ii) Unit Root present or not
AR order(p=0,..6)
Trend degree(r=-1,0,1)
Unit root(-1=yes; 0=no)
Order
Figure 10'(d): Bayes Model Forecast
Encompassing Test Statistic: dQ^B/dQ^F

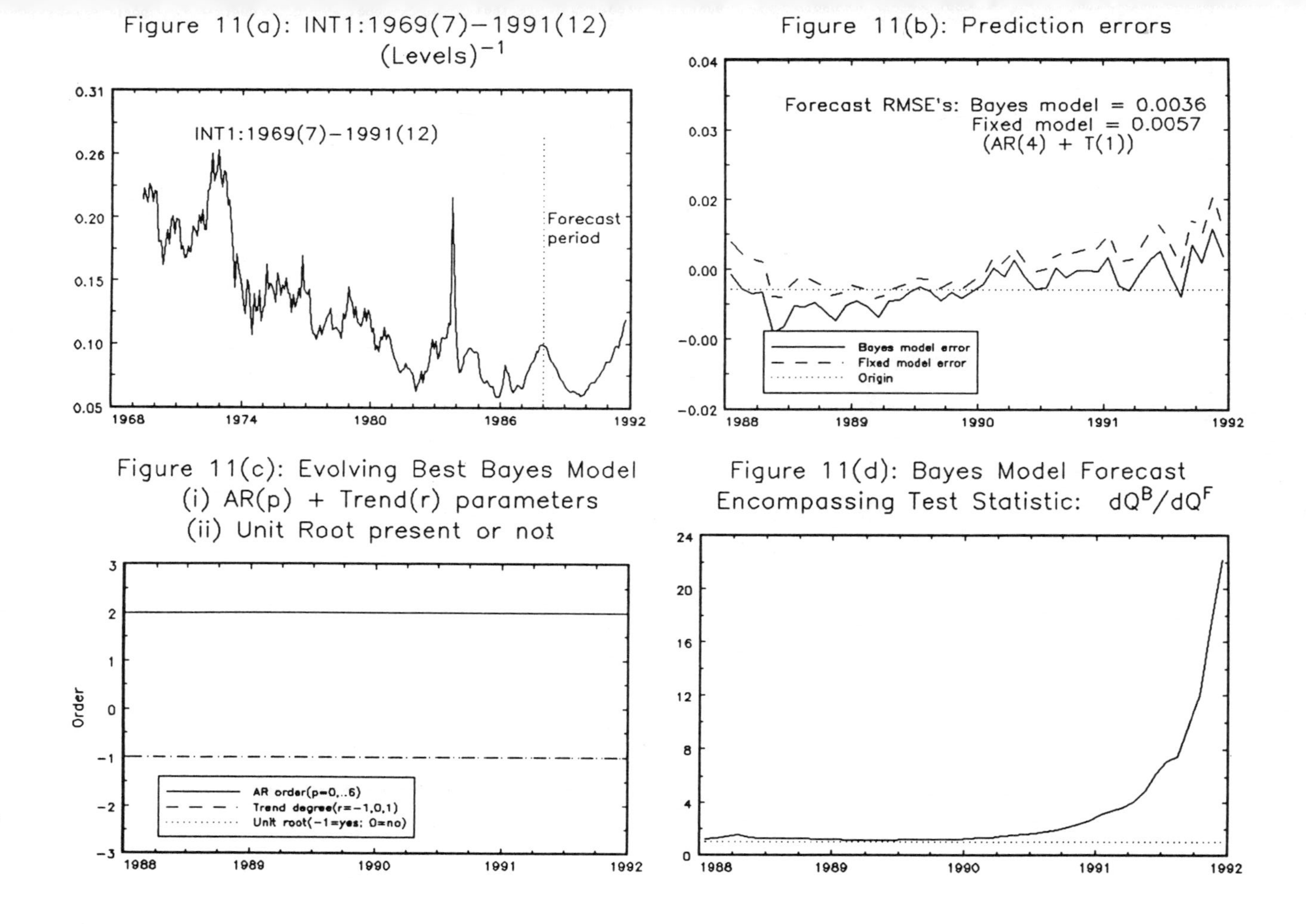

Figure 11(a): INT1:1969(7)–1991(12) $(\text{Levels})^{-1}$

Figure 11(b): Prediction errors

Figure 11(c): Evolving Best Bayes Model (i) AR(p) + Trend(r) parameters (ii) Unit Root present or not

Figure 11(d): Bayes Model Forecast Encompassing Test Statistic: dQ^B/dQ^F

Figure 11'(a): INT1:1969(7)–1991(12)
Levels

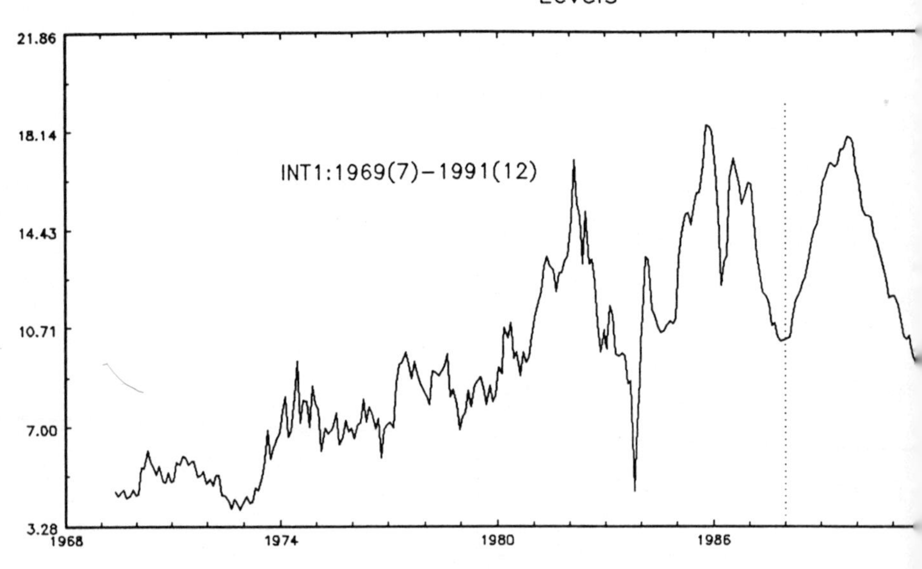

Figure 11'(b): Prediction errors

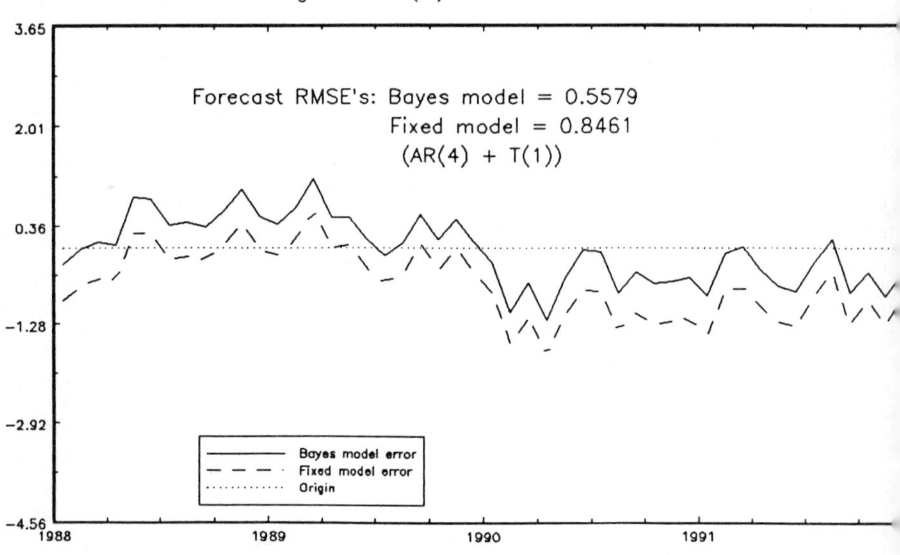

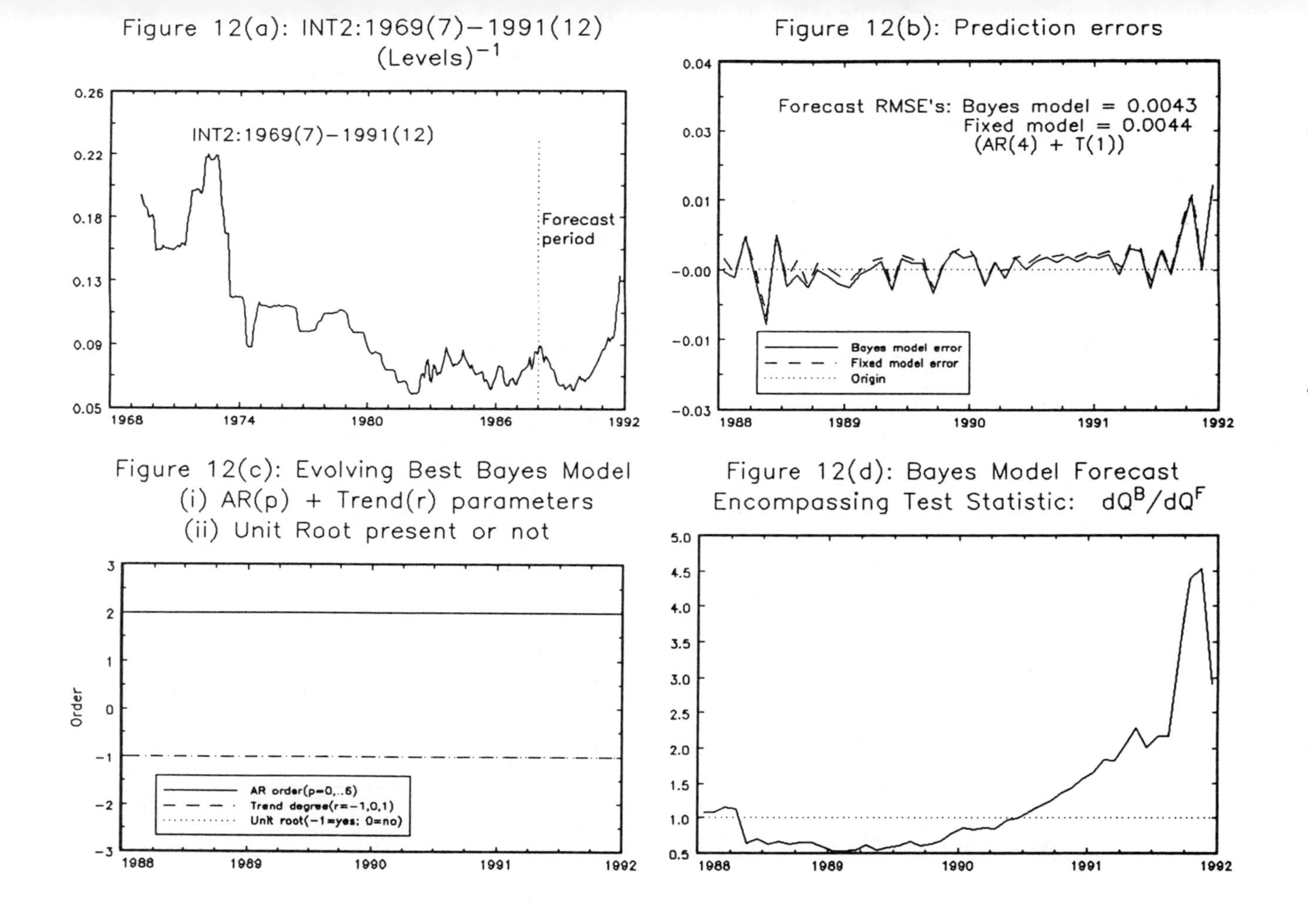
Figure 12(a): INT2:1969(7)−1991(12)
(Levels)$^{-1}$
INT2:1969(7)−1991(12)
Forecast
period
0.26
0.22
0.18
0.13
0.09
0.05
1968
1974
1980
1986
1992
Figure 12(b): Prediction errors
Forecast RMSE's: Bayes model = 0.0043
Fixed model = 0.0044
(AR(4) + T(1))
Bayes model error
Fixed model error
Origin
0.04
0.03
0.01
−0.00
−0.01
−0.03
1988
1989
1990
1991
1992
Figure 12(c): Evolving Best Bayes Model
(i) AR(p) + Trend(r) parameters
(ii) Unit Root present or not
Order
AR order(p=0,..6)
Trend degree(r=−1,0,1)
Unit root(−1=yes; 0=no)
3
2
1
0
−1
−2
−3
1988
1989
1990
1991
1992
Figure 12(d): Bayes Model Forecast
Encompassing Test Statistic: dQ^B/dQ^F
5.0
4.5
4.0
3.5
3.0
2.5
2.0
1.5
1.0
0.5
1988
1989
1990
1991
1992

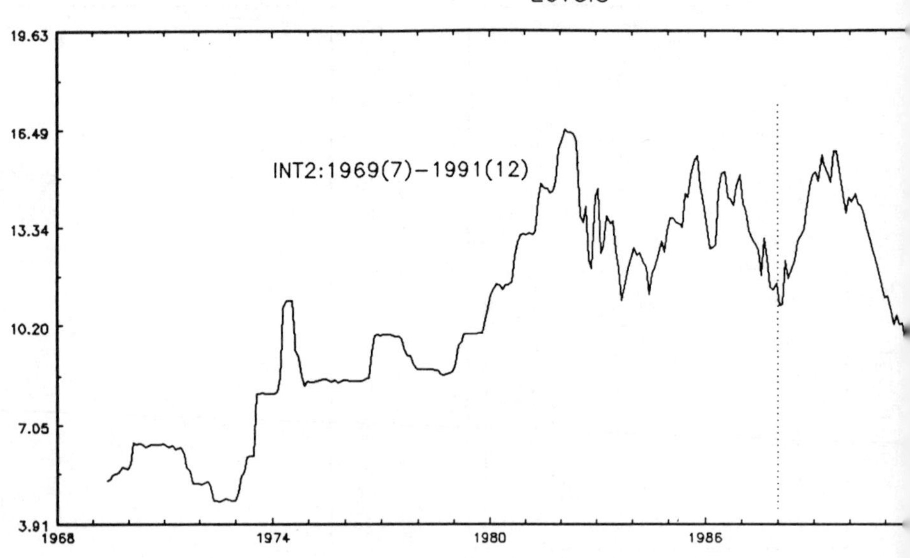
Figure 12'(a): INT2:1969(7)–1991(12)
Levels
INT2:1969(7)–1991(12)
19.63
16.49
13.34
10.20
7.05
3.91
1968
1974
1980
1986

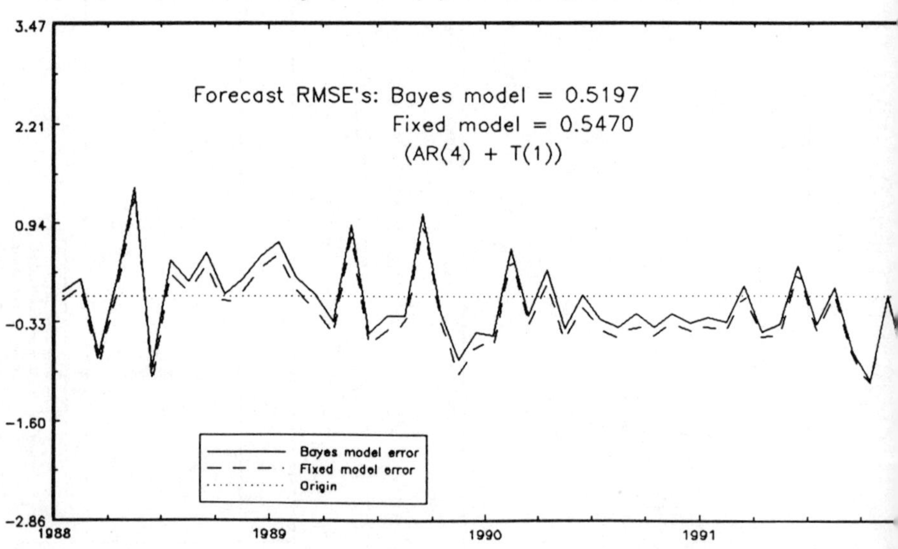
Figure 12'(b): Prediction errors
Forecast RMSE's: Bayes model = 0.5197
Fixed model = 0.5470
(AR(4) + T(1))
Bayes model error
Fixed model error
Origin
3.47
2.21
0.94
-0.33
-1.60
-2.86
1988
1989
1990
1991

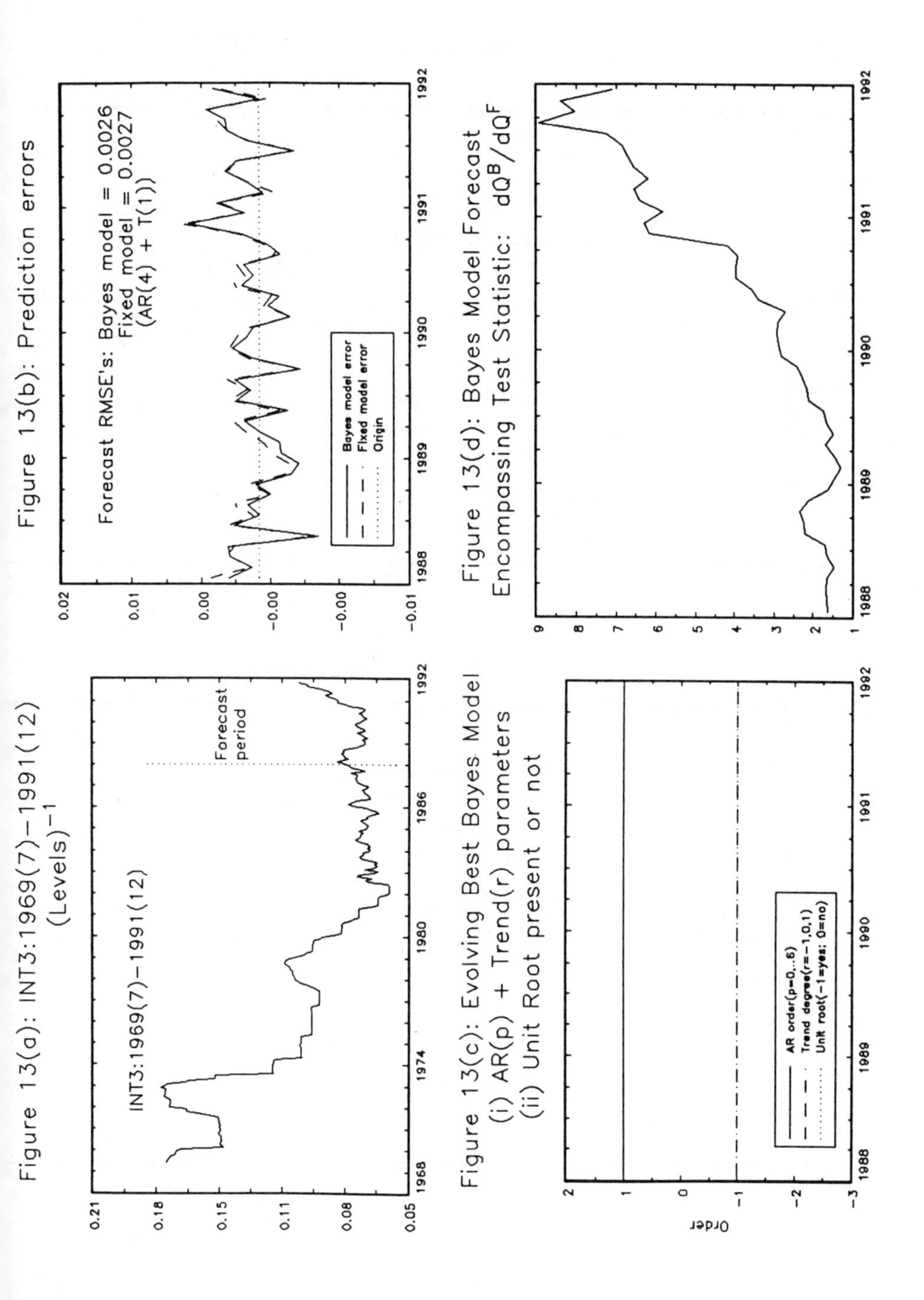

Figure 13(a): INT3:1969(7)–1991(12) (Levels)$^{-1}$

Figure 13(b): Prediction errors

Figure 13(c): Evolving Best Bayes Model (i) AR(p) + Trend(r) parameters (ii) Unit Root present or not

Figure 13(d): Bayes Model Forecast Encompassing Test Statistic: dQ^B/dQ^F

Figure 13'(a): INT3:1969(7)–1991(12) Levels

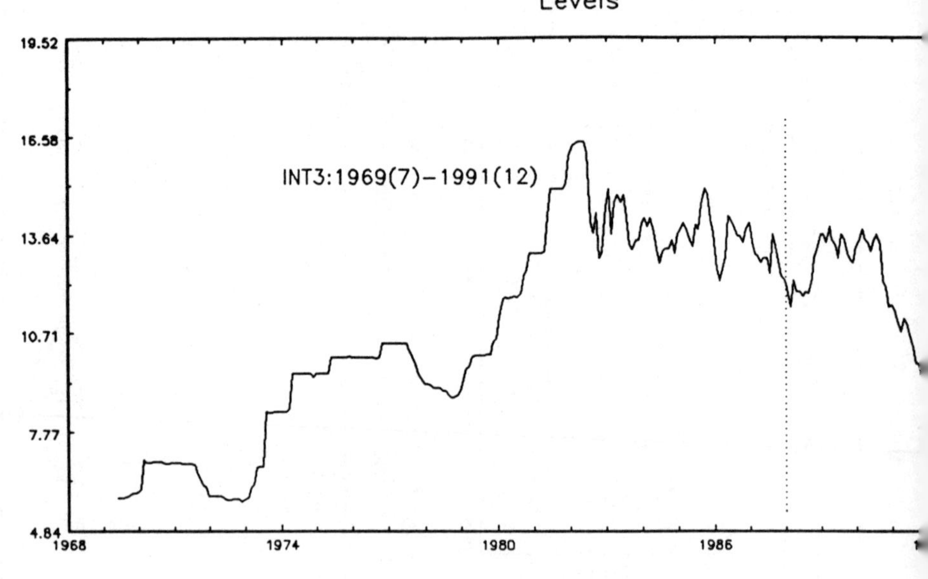

Figure 13'(b): Prediction errors

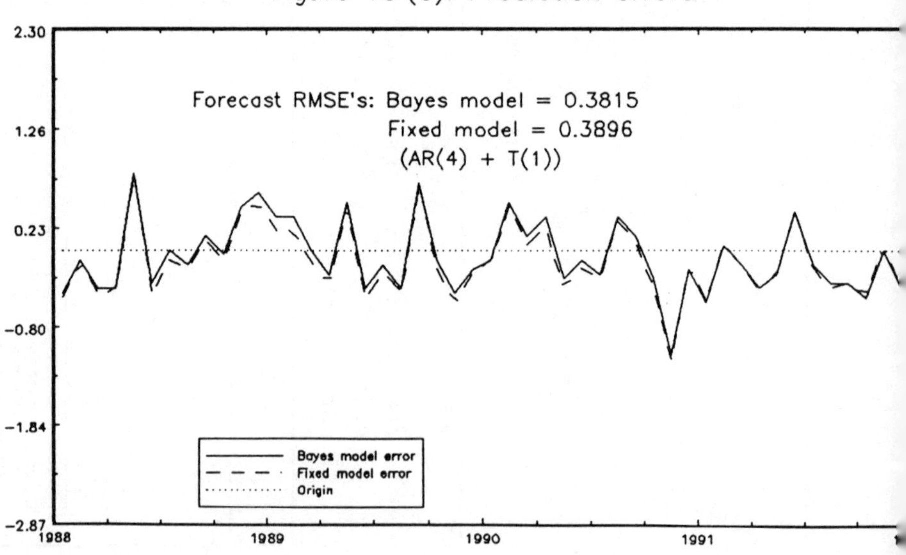

Table 3: Forecasting Exercises for Australian Macroeconomic Time Series, 1988–91

Series	Forecast RMSE *Bayes model*	Forecast RMSE Fixed model	Number of model changes (date)	Best *Bayes model*	Parameter count ratio *Bayes model* /fixed model	Forecast encompassing test dQ^B/dQ^F in 1991
C	0.0064	0.0080	0	$AR(3)^{-1}$	2/6	13.8880
RC	0.0071	0.0065	0	$AR(1)^{-1}+T(0)$	1/6	0.5491
GDP	0.0189	0.0169	1('90(4))	AR(1); AR(2)	1/6; 2/6	0.0679
RGDP	0.0116	0.0102	1('90(3))	AR(1); $AR(1)^{-1}+T(0)$	1/6; 1/6	0.2006
PGDP	0.0080	0.0094	0	$AR(4)^{-1}$	3/6	6.5094
CPI	0.0077	0.0082	0	$AR(4)^{-1}$	3/6	2.1490
U	0.0512	0.0551	0	$AR(2)^{-1}$	1/6	1.7985
WR	0.0121	0.0141	0	$AR(3)^{-1}$	2/6	5.3556
RWR	0.0117	0.0121	1('88(2))	$AR(1)^{-1}$; $AR(1)^{-1}+T(0)$	0/6; 1/6	1.8320
SP500	0.0409	0.0415	0	$AR(2)^{-1}$	1/6	2.2081
		0.0454[b]			1/14	31.1012
$(Int1)^{-1}$	0.0036	0.0057	0	$AR(2)^{-1}$	1/6	22.1763
Int1[a]	0.5579	0.8461				
$(Int2)^{-1}$	0.0043	0.0044	0	$AR(2)^{-1}$	1/6	2.9060
Int2[a]	0.5197	0.5470				
$(Int3)^{-1}$	0.0026	0.0027	0	$AR(1)^{-1}$	0/6	7.1099
Int3[a]	0.3815	0.3896				

Notes: [a] forecasts for the 'Inti' series were obtained from models for the series in reciprocals, i.e. '$(Inti)^{-1}$'
[b] forecast RMSE for fixed model of form 'AR(12) + T(1)'; $AR(p)^{-1}$ = AR(p) model with a unit root

autoregressive root is selected (–1 = yes, 0 = no). Figure (d) gives a recursive plot of the forecast encompassing test statistic dQ^B/dQ^F over the forecast period. Table 3 tabulates these details, gives the root mean squared error (RMSE) of forecasts for the two models over the forecast period, and records the evolving format of the best *Bayes model.*

The main outcomes from this empirical forecasting exercise are as follows:

(i) For none of the series and for no subperiod is the fixed format 'AR(4) + trend(1)' model a chosen *Bayes model.* Three series (real consumption, real wage rate, real GDP) are chosen to have a unit root with drift. All series except for GDP and real GDP show evidence of a unit root throughout the entire forecasting period. Moreover, the best *Bayes model* for real GDP has a unit root from the 1990(3) quarter and, as noted in the discussion of Table 1, the *Bayes model* for GDP has a mildly explosive long-run autoregressive coefficient. Thus, all series are found to be stochastically nonstationary.

(ii) The best *Bayes model* sequence encompasses the forecasts of the fixed model for all three of the series, these being real consumption, GDP and real GDP. Note from the recursive graphs shown in Figures 2(d), 3(d) and 4(d) that the best *Bayes model* forecasts encompass those of the fixed model for these series also in the first half of the forecast period. For some series the forecast dominance of the best *Bayes model* sequence is substantial and uniform over the forecast period. This is especially notable for consumption, where $dQ^B/dQ^F = 13.888$, the short run interest rate (Int1) where $dQ^B/dQ^F = 22.1763$, and the long run interest rate (Int3) where $dQ^B/dQ^F = 7.1099$.

(iii) From Table 3 it is clear that the best *Bayes models* have a substantial advantage in parsimony over the fixed models. For all series the *Bayes models* have at most 50 percent of the parameters of the fixed model and for ten of the thirteen series the parameter ratio is at most 1/6. Note that the presence of a unit root in the best *Bayes models* for the different series also plays a role in reducing the parameter count. For the long-run interest rate (Int3), the parameter ratio is 0/6 yet the best *Bayes model* – here a martingale – uniformly dominates the fixed model in terms of the forecast encompassing test.

(iv) Root mean squared errors (RMSEs) of forecasts over the period 1988–1991 are given in Table 3. In the graphs, Figure (b) for each series tracks the forecast error generated by each model over the forecast period. By the traditional RMSE criterion the best *Bayes model* is the superior model for ten of the series (consumption, GDP deflator, CPI, unemployment rate, wage rate, real wage rate, stock prices, and the three interest rates). For real consumption, GDP and real GDP the best *Bayes model* has a larger RMSE. For these same series, the *Bayes model* forecasts do not encompass those of the fixed model. So the two criteria reach the same conclusion on which model is superior for each of the thirteen series.

(v) It is worth noting that for some of the series the forecast performance of the best *Bayes model* is quite remarkable given its economical form. Thus, for the consumption series, the *Bayes model* reduces the RMSE of forecast by twenty percent. Looking at Figure 1(b) it is apparent that the *Bayes model* forecasts are substantially and almost uniformly better than those of the fixed model from 1989(3)–1991(4). The *Bayes models* also do very well for the GDP deflator and CPI series. The most dramatic improvement in forecasts comes for the short run interest rate series (Int1). For this series the *Bayes model* (an AR(2) with only one fitted parameter) reduces the RMSE of the fixed model by thirty-six percent from 0.0057 to 0.0036. Figure 11(b) shows that for the subperiod 1990(1)–1991(4) the *Bayes model* is uniformly superior to the fixed model, which consistently underpredicts through this subperiod (leading to a persistently positive forecast error). The reason for this underprediction by the fixed model is clear from the graph of the series in Figure 11(a): a model with a linear trend, like the fixed model, is misspecified. Even though the trend coefficient in the model is revised each period with the latest observation this is not enough to prevent a serious and persistent forecast error. The more parsimonious best *Bayes model* is more flexible, adapts more quickly and convincingly outperforms the fixed model in this case.

(vi) As discussed in section 3, models for the interest rate series are constructed in reciprocals of levels to make the volatility of the series more homogeneous over the sample. Forecasts for both reciprocals of levels and levels are then generated for these series. The results are tabulated in Table 3 and shown in Figure 11', 12' and 13'. In spite of their parsimony, the best *Bayes models* do exceedingly well and dominate the fixed model for all three series both in levels and in reciprocals of levels. In terms of forecasts the odds in favour of the best *Bayes model* are 22.18:1, 2.91:1 and 7.11:1 for Int1, Int2 and Int3 respectively.

(vii) For the monthly series we also considered a fixed model with the format 'AR(12) + trend(1)' to allow for calendar year effects. In each case, this fixed model performed worse than the 'AR(4) + trend(1)' model. Results for this choice of fixed model are shown only for stock prices – see Figures 10'(a)–(d) and Table 3. The best *Bayes model* remains the same in this case and now does even better than before in comparison with the fixed model.

6. CONCLUSION

This paper shows that best *Bayes models* of parsimonious form can be constructed for Australian macroeconomic time series that do very well in competition with fixed format models. For ten out of the thirteen series considered here the *Bayes models* not only improve on the forecasts of more richly parameterised models but also encompass those forecasts. In effect, the predictive distribution of the best *Bayes model* explains the forecasts delivered by the rival model. According to our Bayesian forecast-encompassing statistic and given the actual forecast history, the posterior odds favour the *Bayes models*, sometimes by a factor as high as 30:1, as in the case of the short-term interest rate series.

The models we have considered in this paper are scalar time series models. However, all of the ideas we have employed extend in a natural way to multivariate time series; the statistical theory for this extension will be provided in a subsequent paper. And we hope to conduct some empirical exercises with these multivariate methods on Australian macroeconomic data at a later date.

REFERENCES

HANNAN, E. J. and M. DEISTLER (1988), *The Statistical Theory of Linear Systems*, John Wiley and Sons, New York.

HANNAN, E. J. and J. RISSANEN (1982), 'Recursive Estimation of ARMA Order', *Biometrika*, 69, pp. 273–80 [Corrigenda, *Biometrika*, 1983, p. 70].

MILLS, J. A. and K. PRASAD (1992), 'A Comparison of Model Selection Criteria', *Econometric Reviews*, 11, pp. 201–33.

PAULSEN, J. (1984), 'Order Determination of Multivariate Autoregressive Time Series with Unit Roots', *Journal of Time Series Analysis*, 5, pp. 115–27.

PHILLIPS, P. C. B. (1991), 'Bayesian Routes and Unit Roots: de Rebus Prioribus Semper est Disputandum', *Journal of Applied Econometrics*, 6(4), pp. 435–74.

— (1992), 'Bayesian Model Selection and Prediction with Empirical Applications', *mimeo*, Yale University.

PHILLIPS, P. C. B. and W. PLOBERGER (1991), 'Time Series Modeling with a Bayesian Frame of Reference: I. Concepts and Illustrations', Discussion Paper No. 980, Cowles Foundation.

— (1992), 'Posterior Odds Testing for a Unit Root with Data-based Model Selection', Discussion Paper No. 1017, Cowles Foundation, to appear in *Econometric Theory*, 1994.

PÖTSCHER, B. M. (1989), 'Model Selection Under Nonstationarity: Autoregressive Models and Stochastic Linear Regression Models', *Annals of Statistics*, 17, pp. 1257–74.

RISSANEN, J. (1986), 'Stochastic Complexity and Modeling', *Annals of Statistics*, 14, pp. 1080–100.

— (1987*a*), 'Stochastic Complexity', *Journal of the Royal Statistical Society, B*, pp. 223–39 and 252–65.

— (1987*b*), 'Stochastic Complexity and the MDL Principle', *Econometric Reviews*, pp. 85–102.

SCHWARZ, G. (1978), 'Estimating the Dimension of a Model', *Annals of Statistics*, 6, pp. 461–64.

TSAY, R. S. (1984), 'Order Selection in Nonstationary Autoregressive Models', *Annals of Statistics*, 12, pp. 1425–33.

4

A review of methods of estimating cointegrating relationships

Colin Hargreaves[*]

Six estimators of cointegrating relations are compared — OLS, Augmented OLS, Fully-Modified, Three-Step, Johansen MLE and Box-Tiao. Monte Carlo simulations were run using a four-dimensional system to assess the effects of a very wide range of problems. The results found that the Johansen estimator was best as long as the sample was reasonable large (about 100) and the model was accurately specified but not otherwise. Given we are rarely sure of our specifications, the results implied using the Johansen method to try and determine the number of cointegrating vectors and if there is one, then to start with OLS, assess the model, move up to FM and assess any changes in the results. If the researcher erroneously underestimated the cointegrating rank, the Johansen estimator faired worse than the single-equation estimators. Accuracy of estimating three-dimensional cointegrating spaces within four space is described separately in Hargreaves (1992) where the multidimensional OLS described here and augmented OLS fair very well.

1. INTRODUCTION

It is now over 12 years since the path-breaking paper by Clive Granger(1981) on cointegration. However, it was not really until 1987 that Engleand Granger's paper brought the concept to the attention of most economists let alone specialised econometricians. It seemed then that to estimate a cointegrating relationship all one needed to know was how to carry out a simple regression using ordinary least

[*] I would like to thank David Hendry, Adrian Pagan and Peter Phillips for their advice and comments; responsibility for the views expressed here and any errors is however mine alone.

squares; one could always appeal to the super-consistency results to support one's technique. However, even before the 1987 paper appeared in print, many econometricians were finding problems with simple OLS with integrated processes (Phillips and Durlauf, 1986) and were beginning to put forward alternative estimators.

A number of problems exist with OLS. Firstly, there clearly are difficulties if one tries to estimate accurately a 'long-run' relationship with only a small sample over a short time period. Furthermore, while the OLS estimator may be super-consistent, i.e. of order($1/T$) instead of the usual $1/\sqrt{T}$, it can be biased in finite samples especially of the size typical in economics. The distribution is non-symmetric, because of a unit root term within it, plus there is additional bias created not necessarily by a contemporaneous covariation but rather a 'long-run' covariation between the regressors, x, and the error term, u. This 'long-run' covariation is defined as the sum of all the covariances of x_{t-k} with u_t for $k = 0$ to infinity. Because of the analogy with the standard assumption that $\mathrm{Cov}(x_t, u_t) \neq 0$ in normal OLS, this bias is sometimes referred to as a simultaneity or endogeneity bias.

Given these problems with OLS many alternative estimators have emerged. Amongst the many estimators of long-run coefficients, including transforms of OLS, there are:

OLS as proposed in Engle and Granger (1987);

Augmented Least Squares (with leads and lags, sometimes nonlinear) – Bewley (1979), Hendry and Richard (1982, 1983), Stock (1987), Bardsen (1989), Phillips and Loretan (1991), Saikonnen (1991);

Instrumental Variables - Phillips and Hansen (1990);

'Fully-Modified' Estimator – Park and Phillips (1988, 1989), Phillips and Hansen (1990);

Non-parametric Canonical Cointegrating – Park (1989);

Three-Step Estimator – Engle and Yoo (1991);

Spectral Regression – Phillips (1991*b*);

Principal Components – Stock and Watson (1989), Phillips and Ouliaris (1988);

Canonical Cointegration – Bossaerts (1988);

Differenced VAR Maximum Likelihood – Johansen (1988, 1991), Ahn and Reinsel (1988, 1990);

Modified Box-Tiao – Bewley, Orden and Fisher (1991), based on Box and Tiao (1977).

The equivalence of many of the OLS transforms is shown in Wickens and Breusch (1988) and a good comparative discussion of these estimators is given in Chapter 2 of Banerjee *et al.* (1993). Then there are a set of regression routines with modifications to make them as close as possible to full maximum likelihood estimation. The spectral regression method is so different to the others that it is not discussed further here. The final four methods are 'system' methods based on finding the relevant eigenvalues and vectors of whatever each author defines as the most 'appropriate' matrix. However these methods require additional information compared to simple OLS and hence, as we shall see, the improvement comes at a cost; if the estimator's 'model' is misspecified, the estimator may perform very poorly.

While this paper specifically concentrates on methods of estimating cointegrating relationships, a brief mention of methods of testing for cointegration is appropriate. Firstly there are the battery of tests based on the residuals of the OLS regression. These include the Dickey-Fuller and Augmented Dickey-Fuller (Said and Dickey, 1984), the Cointegrating Regression Durbin Watson and a range of other tests, all described in Engle and Granger (1987). Related to these are the non-parametrically adjusted versions of the ADF tests, which are variously called Phillips or Phillips-Perron tests or Z_a and Z_t tests (Phillips, 1987; Phillips and Perron, 1988). As with the estimation methods there are then a series of tests based on the eigen values of the relationship between the vector z_t or Δz_t to z_{t-1}. The Stock and Watson (1988) test adjusts for an estimate of the long-run variance-covariance matrix while the Johansen parametrically adjusts by the addition of lagged differences, exogenous variables and trends. There is no test associated with the modified Box-Tiao other than ADF style tests on the cointegrating vectors defined by this estimator.

There are also three approaches to testing for cointegration that are not directly related to an estimation procedure. Park, Ouliaris and Choi's (1988) spurious regressors test checks that additional powers of time trends do not have any explanatory power when added to the cointegrating regression. Hansen (1990) applies the ADF and Z tests to the errors of the cointegrating regression on the first order quasi differences after performing a Cochrane-Orcutt style iteration to estimate the autocorrelation coefficient. Finally the Phillips and Ouliaris's (1988) trace and variance-ratio tests look for singularities in the long-run variance-covariance matrix. All of these (except the method which analyses the residuals from a modified Box-Tiao cointegrating vector) are reviewed in Gregory (1991). If anything, Gregory comes out in favour of the original and relatively simple ADF and Z_α tests but he states that overall 'no one test dominates in terms of size and power' and hence 'instances of test conflict are likely to be more numerous than is usually encountered' (p.20). It is possible that conflicting results arise from poor estimation methods especially with the methods that test the residuals of potentially cointegrating regressions. This is one reason why here we concentrate purely on estimation without entering the further questions of test procedures.

Another reason for concentrating on estimation is the relative lack of analysis of two particular estimation methods, one fairly new and one very old. The first is the Engle and Yoo (1991) Three Step procedure and the second is the old Box - Tiao (1977) procedure, championed in modified form by Bewley, Orden and Fisher (1991). Great claims are made for each and certainly Bewley, Orden and Fisher have produced some Monte Carlo evidence to suggest that their estimator does not have the same outlier behaviour of the Johansen MLE reduced rank regression estimator.

Phillips (1992) shows that, in the 'leading' case of the data being generated by n independent random walks, the distribution of the Johansen estimator is 'matrix Cauchy and has no finite first moment. Cauchy-like tail behaviour is shown to persist when the leading case hypothesis is relaxed, thereby providing an analytical explanation for the outlier behaviour observed' (p. 2). Phillips compares the Johansen model to the triangular system model (analysed in Phillips, 1991*a*) wherein the restriction is imposed that given r cointegrating relationships between p variables there must be a subset of r variables that are not cointegrated. The statistical identification gained clearly also parallels the difference between 2SLS and LIML estimation where the former defines the estimation space with respect to a particular subset of variates. Phillips comments how, just as LIML appears preferable to 2SLS (see Anderson, 1982; Phillips, 1983; Phillips, 1990; Hillier, 1988), the invariance of the Johansen method to equation normalisation may prove to be an advantage overall.

Now a simultaneous model can be expressed in various ways, as in for instance the structural form, the reduced form and the final form; here the whole model is expressed in one way or another. The cointegration literature generally assumes that the model can be written as a finite unrestricted VAR which in turn can be expressed as a restricted VAR where the error correction mechanisms are used like exogenous variables on the RHS. One can also express a model in a form which separates out the long run from the short run; the definition of one 'run' entails a definition of the other. Whatever the form one needs a set of identifying restrictions which can vary according to the formulation used. Where one distinguishes the short run from the long run, one could impose them wholly on the short run, wholly on the long or a mixture of the two. Implicitly what the standard Johansen estimator is doing is applying all restrictions to the short run; the long run is totally unrestricted and hence may well appear to be unidentified. For estimation, the key question is whether you require that your error-correction mechanisms are equations of the model or not, as, if they are, then identifying restrictions must apply directly apply to them. Then testing restrictions on the long run has to be understood in the context of the implied restrictions on the short run.

To try and express this more definitively, take a simple model where, given a p-dimensional vector z_t,

$$Az_t = Bz_{t-1} + u_t, \qquad \mathrm{Var}(u_t) = \Sigma \qquad \text{and} \qquad A_{ii} = 1 \text{ for all } i.$$

This could be identified by making either A or Σ the unit matrix, the former being a VAR. One could equally impose some set of cross-equation restrictions implied by a certain theory. One can rewrite the system as

$$A\Delta z_t = B^* z_{t-1} + u_t$$

where $B^* = B - A$. Under the assumption of a unit root, B^* does not have full rank when A is the identity matrix. This leads to reduced rank regression and the estimation of the cointegrating vector space which is only statistically identified by the imposition of certain mathematical constraints that are nothing to do with any economic model. One aspect of this is that the normalisation is arbitrary and hence there is no implicit causal direction even though there is in most long-run economic theory. For instance, in a long-run money demand function,

$$m = f(m^*) + e = f(y, p, r) + e$$

where m^* is desired long-run money stocks, an OLS estimate of the error-correction, e, is clearly an amount of money above or below equilibrium but one could not definitely say this with a Johansen estimate. With a one-dimensional cointegrating space, one can impose the normalisation as one wants and hopefully obtain an interpretable function. When there are two cointegrating vectors, a single OLS regression is liable to obtain a linear combination of the two and a Johansen estimate of the two vectors will only define the two-dimensional plane in which they lie (any rotation of the vectors to define the space is possible). This problem of statistical identification is taken up further in Phillips and Park (1992); they argue that one must be very careful when analysing hypotheses about particular coefficients in multidimensional cointegrating spaces estimated by reduced rank regression.

We take up these issues of statistical identification of the cointegrating space, using Monte Carlo simulations within four dimensional spaces. It would clearly be a gigantic task to use each and every estimator that has been proposed and so only six estimators are used. Given Hansen (1990)'s claim that Park's canonical cointegrating estimator 'duplicates the existing FM procedure', the former estimator is omitted in preference to the fully-modified Phillips-Hansen procedure. Within the single equation framework, we also use what Phillips and Loretan (1991) describe as the linear parametric 'Hendry-style' methodology (OLS plus varying orders of leads and lags). We use the Engle and Yoo 3-Step estimator (Engle and Yoo, 1991) since it is relatively new and of a different nature to the 'Hendry' and FM estimators. On the multivariate side we use the Johansen estimator (Johansen, 1988) as it has proven very popular and the modified Box-Tiao (Bewley and Orden, 1991) as the quality of its results do not seem to have been fully recognised.

We initially consider fairly simple experiments with just one cointegrating vector in four dimensional space and then experiments where a second cointegrating vector exists but is not estimated. Hargreaves (1993) goes on to discuss the accuracy of estimating multidimensional cointegrating spaces, reporting results for three cointegrating dimensions in four-dimensional space. The latter leads to the problem of identifying an OLS style estimator of a multi-dimensional space; we propose an easy-to-use method that can find 'orthogonal regression' estimates of multivariate cointegration spaces using most simple regression packages. Hargreaves (1993) also proposes a method of measuring the accuracy of estimation of three-dimensional cointegrating spaces.

There are clearly different specifications of the system implicit within the multivariate versus univariate approaches. It is hard to make a 'fair' comparison when one estimator was just not created to model the complicated system set up by our data generation process. However equally if one uses the simpler data generation process implied by the 'simpler' estimator with exogenous independent regressors, this will not reveal the capacity of the more complex approaches. Since in practice one may not know the actual complexity of the true data generation process, simulations are run attempting to show the properties of the estimators within each other's 'province'.

Section 2 briefly defines the estimators and tries to give a non-mathematical interpretation of the substantive differences between the approaches. Section 3 defines the data-generation processes used for the Monte Carlo simulations and explains the measures used to analyse the results. In Section 4.1, we compare the results for the different estimators given a uni-dimensional cointegration space within a four dimensional system and Section 4.2 describes results for two dimensional cointegrating spaces. Section 5 concludes.

2. DEFINITIONS OF THE SIX ESTIMATORS CONSIDERED

2.1 Ordinary least squares for multidimensional cointegrating spaces

Let z_t be a p-dimensional process which we partition as follows

$$z_t = \begin{bmatrix} y_t \\ x_t \end{bmatrix}\begin{matrix} 1 \\ p-1 \end{matrix}.$$

Then letting the T observations be stacked into matrices Y and X where each observation is one row, the standard OLS estimator of β in the model

$$y_t = x_t'\beta + u_t$$

is

$$\hat{\beta} = (X'X)^{-1}X'Y.$$

A second dimension of the cointegrating space can be estimated by a second OLS regression which is orthogonal to the estimates of the first cointegrating regression. A parallel can be made to principal components where the first component is that linear function with the maximum variance and the second component is that with maximum variance subject to being orthogonal to the first. However here we are normalising on a specific variable and so instead of minimising the distances from the observation points to the line as in principal components, we are minimising the distances to the line in the direction of the dependent variable. Thus the ith cointegrating vector consists of a 1 as coefficient of y and say β_i as the vector of coefficients on x. Thus $[1, -\hat{\beta}_1']$ would be the estimate of the first cointegrating vector. The estimate of the second dimension, $[1, -\beta_2']$, is a restricted OLS where the restriction is that

$$H\beta_2 = 1$$

where

$$H = \hat{\beta}_1{}'.$$

Thus the restriction is

$$\begin{bmatrix} 1 & -\hat{\beta}_1' \end{bmatrix}\begin{bmatrix} 1 \\ -\hat{\beta}_2 \end{bmatrix} = 0,$$

i.e. the two cointegrating dimensions are orthogonal. Since we do not wish to involve the constant term in this orthogonality, the corresponding element of$H$ is zero. The estimator of the second cointegrating relation is then just

$$\hat{\beta}_2 = \hat{\beta}_1 - (X'X)^{-1}H'\left(H(X'X)^{-1}H'\right)^{-1}\left(H\hat{\beta}_1 - 1\right).$$

This can be extended to further cointegrating relations by simply stacking the previous $\hat{\beta}$'s in H.

2.2 The 'Hendry-style' dynamic regression method

Here the OLS model is augmented by possible lags and leads of differenced integrated variables which essentially 'mop-up' the dynamics. Thus we have

$$y_t = \beta x_t + \sum_{k=-l_1}^{l_2} \left(\gamma_k \Delta y_{t+k} + \lambda_k \Delta x_{t+k}\right) + u_t,$$

where l_1 is the number of lags and l_2 is the number of leads. A second cointegrating relation is then defined as above for OLS but where H is non-zero only for the corresponding elements in β and then equal to the values found for the first cointegrating regression.

2.3 The Phillips-Hansen fully modified estimator

This method specifically assumes that the regressor variables are not cointegated and there differences are stationary. The model is

$$y_t = \beta x_t + u_{1t},$$

with

$$x_t = x_{t-1} + u_{2t}.$$

Here u_{1t} and u_{2t} are stationary processes; if u_{2t} is white noise then x_t follows a random walk. Although this method is usually considered just for the case where y_t is a single variable as in section 2.1, it may well be r dimensional such that we are looking at a system of relationships between an r dimensional y and a $(p-r)$ dimensional x. This system can be related to a general simultaneous model in the following way,

$$\begin{bmatrix} A_1 & A_2 \\ B_1 & B_2 \end{bmatrix} \begin{bmatrix} y_t \\ x_t \end{bmatrix}_{p-r}^{r} \Rightarrow \begin{bmatrix} I & A_1^{-1}A_2 \\ 0 & I \end{bmatrix} \begin{bmatrix} y_t \\ x_t \end{bmatrix}_{p-r}^{r} = \begin{bmatrix} u_{1t} \\ \Delta^{-1}u_{2t} \end{bmatrix}_{p-r}^{r},$$

which is the same as the first two equations in this section with

$$\beta = -A_1^{-1}A_2.$$

Δ is the difference operator. With B_1 set to zero here one is assuming that there is no feedback from y to x and hence one is really estimating the reduced form if $r > 1$. If there is only one cointegrating relation then $r = 1$ and $A_1 = 1$.

Now let u_{1t} follow a stationary AR(1) process such that

$$u_{1t} = \rho u_{1t-1} + e_t, \qquad |\rho| < 1,$$

and let

$$\begin{bmatrix} e_t \\ u_{2t} \end{bmatrix} \sim \text{NID}\left[\begin{bmatrix} 0 \\ 0 \end{bmatrix}, \begin{bmatrix} \sigma_1^2 & \varphi\sigma_1\sigma_2 \\ \varphi\sigma_1\sigma_2 & \sigma_2^2 \end{bmatrix}\right].$$

e_t and u_{2t} are both stationary and so u_{1t} is stationary as well. Since the first difference of x_t is stationary, x_t itself is I(1). As y_t is a linear function of an I(1) variable and an I(0) variable, it also is I(1). Since the linear combination $(1,-\beta')$ is I(0), y_t and x_t are cointegrated by definition.

Now economically this is a very strange model. It is unlikely that there are many systems where all the variables are I(1) and just functions of other I(1) variables. Other stationary variables may well also determine both the y_t and x_t. One must regard this as an approximation which catches the characteristics of the variables in a similar way to a Box-Jenkins time series ARIMA model paralleling a causal structural model. One assumes that the effects of the other stationary variables are modelled equivalently by the ARMA process on the errors. One would seldom believe the data generation process assumed here to be the 'true' data generation process. Rather it is a device used to efficiently estimate the cointegrating relationship.

Stacking the errors into one vector u_t, i.e.

$$u_t = \begin{bmatrix} u_{1t} \\ u_{2t} \end{bmatrix}_{p-r}^{r},$$

let us form the cumulative sum of u_t which is a multivariate random walk. One can now divide both the variable and the time interval by T so that one has a process over a time interval from 0 to 1 which changes at each Tth step. If one lets T tend to infinity one has a continuous variable in the $[0,1]$ interval observed at each position τ, where $\tau = t/T$. This is known as a Brownian motion (or Wiener process) which can be written as

$$W_T(\tau) = \frac{1}{T}\sum_{t=1}^{[T\tau]} u_t \Rightarrow B(\tau) = \begin{bmatrix} B_1 \\ B_2 \end{bmatrix}_{p-r}^{r} = BM(\Omega) \quad \text{for} \quad \tau \in [0,1] \text{ and } T \to \infty.$$

This is a continuous random walk on the interval $[0,1]$. The covariance matrix, Ω, of the Brownian motion is often referred to as the 'long-run' covariance matrix and is defined as

$$\Omega = \lim_{T\to\infty} \frac{1}{T} \sum_{t=1}^{T} \sum_{j=1}^{T} E(u_j u_t').$$

This 'long-run' variance is the sum of all the covariances backwards and forwards of u_j with u_t. This can be decomposed into a contemporaneous variance and sums of autocovariances, i.e.

$$\begin{aligned}\Omega &= \lim_{T\to\infty} \frac{1}{T} \left\{ \sum_{t=1}^{T} E(u_t u_t') + \sum_{t=2}^{T} \sum_{j=1}^{t-1} E(u_j u_t') + \sum_{t=2}^{T} \sum_{j=1}^{t-1} E(u_t u_j') \right\} \\ &= E(u_0 u_0') + \sum_{t=1}^{\infty} E(u_0 u_t') + \sum_{t=1}^{\infty} E(u_t u_0') \\ &= \Sigma + \Lambda + \Lambda'\end{aligned}$$

where Σ is the contemporaneous covariance matrix. Defining Δ as

$$\Delta = \Sigma + \Lambda ,$$

and partitioning Ω and Δ conformably with u_t,

$$\Omega = \begin{bmatrix} \Omega_{11} & \Omega_{12} \\ \Omega_{21} & \Omega_{22} \end{bmatrix}_{p-r}^{r} \quad and \quad \Delta = \begin{bmatrix} \Delta_{11} & \Delta_{12} \\ \Delta_{21} & \Delta_{22} \end{bmatrix}_{p-r}^{r},$$

the data generating process leads to

$$\Sigma = \begin{bmatrix} \dfrac{\sigma_1^2}{1-\rho^2} & \varphi\sigma_1\sigma_2 \\ \varphi\sigma_1\sigma_2 & \sigma_2^2 \end{bmatrix},$$

$$\Lambda = \begin{bmatrix} \dfrac{\rho\sigma_1^2}{(1-\rho)(1-\rho^2)} & 0 \\ \dfrac{\rho\varphi\sigma_1\sigma_2}{1-\rho} & 0 \end{bmatrix},$$

and

$$\Omega = \begin{bmatrix} \sigma_1^2/(1-\rho)^2 & \varphi\sigma_1\sigma_2/(1-\rho) \\ \varphi\sigma_1\sigma_2/(1-\rho) & \sigma_2^2 \end{bmatrix}.$$

The asymptotic distribution of the OLS estimator of β is then

$$T(\hat{\beta}-\beta) \Rightarrow \left(\int B_2^2\right)^{-1} \left\{ \frac{\sigma_1\sqrt{1-\varphi^2}}{1-\rho} \int B_2 dW + \frac{\varphi\sigma_1}{(1-\rho)\sigma_2} \int B_2 dB_2 + \frac{\varphi\sigma_1\sigma_2}{1-\rho} \right\},$$

where $W = BM(I)$ is independent of B_2. This expression is the sum of three terms of which only the first remains in full maximum likelihood estimation. The first term is independent of the latter two as B_2 and W are independent and so the variance of the OLS estimator must be larger. If x_t is strictly exogenous, then φ is zero and the latter two parts vanish. These latter two terms are a serial correlation unit root term and an 'endogeneity' term. Fully modified OLS tries to estimate and asymptotically remove these terms.

One can define the 'long-run' variance of u_{1t}, conditional upon u_{2t} as

$$\Omega_{1.2} = \Omega_{11} - \Omega_{12}\Omega_{22}^{-1}\Omega_{21}$$

and the bias due to the endogeneity of the regressors as

$$\Delta_{21}^{+} = \Delta_{11} - \Delta_{22}\Omega_{22}^{-1}\Omega_{21}.$$

To estimate these covariance parameters, one obtains estimates of the errors by ordinary least squares with deterministic time trends in the x_t processes if thought appropriate. The method used for the simulations in this paper is that described in Hansen (1991) where he uses a pre-whitened kernel estimator with the automatic bandwidth recommended by Andrews and Monahan (1990). A Gauss program written by Bruce Hansen to do this was used for the Monte Carlo simulations. This also required the choice of a kernel and the Parzen kernel was used.

Having found $\hat{\Omega}$ and $\hat{\Delta}$ and partitioned them as before, one then sets

$$\hat{\Omega}_{1.2} = \hat{\Omega}_{11} - \hat{\Omega}_{12}\hat{\Omega}_{22}^{-1}\hat{\Omega}_{21}$$

and

$$\hat{\Delta}_{21}^{+} = \hat{\Delta}_{11} - \hat{\Delta}_{22}\hat{\Omega}_{22}^{-1}\hat{\Omega}_{21}.$$

One defines the transformed dependent variable

$$y_t^{+} = y_t - \hat{\Omega}_{12}\hat{\Omega}_{22}^{-1}\hat{u}_{2t}$$

and then the fully modified (FM) estimator is

$$\hat{\beta}^{fm} = \left(\sum_{t=1}^{n}\left(y_t^{+}x_t' - \begin{pmatrix}0 & \hat{\Delta}_{21}'^{+}\end{pmatrix}\right)\right)\left(\sum_{t=1}^{n}x_t x_t'\right)^{-1}.$$

While conditional on $\sigma(B_2(\tau), 0 < \tau \le 1)$, the estimator is asymptotically Normally distributed but, unconditionally, it is a mixture of normals like all 'unit root' estimators. However the approach does lead to asymptotically standard normal 't' statistics which can be used for standard hypothesis tests although this does not translate simply into confidence intervals since the distribution of the estimator is non-Normal and generally quite skewed.

2.4 The Engle-Yoo 3-Step estimator

Engle and Yoo (1991, p. 257) consider the general Gaussian error-correction model,

$$\Delta z_t = \Gamma_1 \Delta z_{t-1} + \ldots + \Gamma_{k-1} \Delta z_{t-k+1} - \alpha\beta' z_{t-1} + \varepsilon_t,$$

where $\varepsilon_t \sim N(0, \Psi)$, z is p-dimensional and α and β are both $p \times r$ where $r < p$. The log-likelihood is proportional to

$$L = \sum_{t=1}^{T} L_t(\varphi_1, \varphi_2; \Delta z_t) = -\frac{T}{2}\log(|\Psi|) - \frac{1}{2}\sum_{t=1}^{T} \varepsilon_t' \Psi^{-1} \varepsilon_t.$$

The parameters of the model are here split between those in φ_1 (namely the Γs, α and Ψ) and those in φ_2 (only β). Let us define the information matrix

$$Q = \sum_{t=1}^{T} E_{t-1}\left[\frac{\partial L_t}{\partial \varphi}\frac{\partial L_t}{\partial \varphi'}\right],$$

where the expectation is taken with respect to all past and current$z$'s which may be treated as predetermined. If one partitions Q conformably with φ_1 and φ_2 as

$$Q = \begin{bmatrix} Q_{11} & Q_{12} \\ Q_{12}' & Q_{22} \end{bmatrix},$$

then the diagonal elements of Q_{11} are of order T, and the diagonal elements of Q_{22} are of order T^2. The elements in the off-diagonal blocks, Q_{12}, are of at most order T, but are normalised by $T^{-3/2}$ and hence tend to zero after normalisation, making Q asymptotically block diagonal. Hence 'estimates of φ_1 which are asymptotically equivalent to FIML can be obtained simply by updating estimates of φ_1 taking consistent estimates of φ_2 as given' (p.258[1]). Just as the Two-Step estimator of Engle and Granger (1987) used consistent estimates of φ_2 to produce fully efficient estimates of φ_1 which are consistent, then these estimates of φ_1 can be used to re-estimate φ_2 in a third step. Given the block diagonality, one need only use a simple regression on the equations for φ_2 alone and hence

$$\hat{\varphi}_2 = \tilde{\varphi}_2 + Q_{22}^{-1}\frac{\partial L}{\partial \varphi_2},$$

where

$$Q_{22} = \sum_{t=1}^{T} E_{t-1}\left[\frac{\partial L_t}{\partial \varphi_2}\frac{\partial L_t}{\partial \varphi_2'}\right].$$

---

[1] The letter φ is used instead of θ here as θ is used elsewhere.

Engle and Yoo then proceed to give the algebra for the simple case of one cointegrating vector where the regressors are weakly exogenous in the sense of Engle, Hendry and Richard (1983) for the first equation's parameters. For the more common case where the regressors are not weakly exogenous, they recommend the following procedure given one cointegrating relation:

1) After finding the first-step standard OLS estimates, $\tilde{\beta}$, of the cointegrating relationships, find the two-step estimates, $\tilde{\alpha}$, of α and the residual vectors, $\tilde{e}_i$, of the i'th second stage regression ($i = 1,...,p$).

2) Use the residual vectors to find estimates of the error variance of each second-step regression, i.e. $\tilde{\sigma}_i^2 = \tilde{e}_i'\tilde{e}_i / T$, and use this to create standardised residuals of each regression, i.e. $\tilde{e}_i / \tilde{\sigma}_i$.

3) Create the vector $x' = (x_0, ... ,x_{T-1})$ where these are the regressors of the one cointegrating relationship; y is the 'dependent variable' and the vector $z' = [y \;\; x']$. These are then weighted by $\tilde{\alpha}_i' / \tilde{\sigma}_i$ and stacked into one long matrix, X. Also stack all the standardised residual vectors into one large vector, E. Thus we have

$$\underset{pT\times 1}{\tilde{E}} = \begin{bmatrix} \tilde{e}_1 / \tilde{\sigma}_1 \\ \tilde{e}_2 / \tilde{\sigma}_2 \\ \vdots \\ \tilde{e}_p / \tilde{\sigma}_p \end{bmatrix} \quad \text{and} \quad \underset{pT\times(p-1)}{X} = \begin{bmatrix} x\tilde{\alpha}_1' / \tilde{\sigma}_1 \\ x\tilde{\alpha}_2' / \tilde{\sigma}_2 \\ \vdots \\ x\tilde{\alpha}_p' / \tilde{\sigma}_p \end{bmatrix}$$

4) Now

$$\frac{\partial L}{\partial \theta_2} = X'\tilde{E} \quad \text{and} \quad Q_{22} = X'X$$

and the three-step estimator is

$$\hat{\beta} = \tilde{\beta} + (X'X)^{-1} X'\tilde{E}$$

with

$$(X'X)^{-1/2}\left(\hat{\beta} - \beta\right) \rightarrow N(0, I) \qquad \text{and} \qquad \tilde{E}'\tilde{E} / pT \equiv 1$$

Thus like the FM estimator, while the coefficients are not asymptotically Normally distributed, one can create 't' statistics which are. Further stacking of vectors is required when there are r cointegrating relations. Clearly this three-step method could be iterated, revising the θ_1 estimate in a 4th step and then the θ_2 estimate again in a 5th step and so on, but Engle and Yoo advise that this is probably not worthwhile since the third step involves the regression of the I(0) $\tilde{e}_i$ on the I(1) x_{2t}. Hence the adjustment will be quite small making iteration not worthwhile.

2.5 The Johansen VAR maximum likelihood estimator

Given our p-dimensional z vector, a vector autoregression of order k could be written as

$$z_t = \Pi_1 z_{t-1} + \ldots + \Pi_k z_{t-k} + \mu + \varepsilon_t,$$

where the ε_i are iid $N_p(0,\Psi)$ and μ is a constant term; centred seasonal dummies can easily be added to the above but are not here as we do not want to raise problems of seasonal cointegration. Without loss of generality one can add other exogenous variables to the system and also deterministic time trends.

This model can be written in difference terms, similar to that used in the Engle-Yoo 3-Step estimator above, except for a change of the levels variable from z_{t-1} to z_{t-k}, i.e.

$$\Delta z_t = \Gamma_1 \Delta z_{t-1} + \ldots + \Gamma_{k-1} \Delta z_{t-k+1} + \Pi z_{t-k} + \mu + \varepsilon_t,$$

where

$$\Gamma_i = -\mathrm{I} + \Pi_1 + \ldots + \Pi_i, \qquad i = 1, \ldots, (k-1),$$

and

$$\Pi = -(\mathrm{I} - \Pi_1 - \ldots - \Pi_k).$$

Hence Π is the multivariate distributed lag function of the first equation in this section with the lag operator, L, set equal to 1, which is the usual way of determining long-term multipliers. Three situations now arise related to the rank of Π. Firstly if Π has full rank (equal to p), then it can be shown that z_t must be stationary. If the rank of Π is zero, then Π is a null matrix and there is no cointegration. If the rank of Π is equal to $r < p$ then Π can be written as the product of two matrices, α and β, i.e.

$$\Pi = \alpha\beta'.$$

The cointegrating space is defined by β and the 'adjustment' factors are defined by α.

The differenced model can be written as

$$Z_{ot} = \Gamma Z_{1t} + \Pi Z_{kt} + \varepsilon_t,$$

where

$$Z_{ot} = \Delta z_t, \qquad Z_{1t} = [\Delta z_{t-1}, \ldots, \Delta z_{t-k+1}, 1], \quad \text{and} \qquad Z_{kt} = z_{t-k}.$$

If one partials out the effect of Z_{1t} on both Z_{ot} and Z_{kt}, one could then estimate Π by regressing the adjusted Z_{ot} on Z_{kt}. These adjusted values are simply the residuals of two multivariate regressions on Z_{1t}, *i.e.*

$$R_o = Z_o - Z_1(Z_1'Z_1)^{-1}Z_1'Z_o$$

and

$$R_k = Z_k - Z_1(Z_1'Z_1)^{-1}Z_1'Z_k.$$

where Z_o, Z_k, R_o and R_k are $T \times p$ and Z_1 is $T \times [p(k-1)+1]$. Now let

$$S_{ok} = T^{-1}R_o'R_k, \qquad S_{kk} = T^{-1}R_k'R_k, \qquad \text{and} \qquad S_{ko} = S_{ok}'.$$

Then, for the regression

$$R_o = \Pi R_k + \varepsilon,$$

we have

$$\hat{\Pi} = S_{ok}S_{kk}^{-1}.$$

The concentrated likelihood function is

$$|\Psi|^{-T/2}\exp\left\{-\frac{1}{2}(R_o - R_k\Pi')'\Psi^{-1}(R_o - R_k\Pi')\right\}.$$

Now

$$\hat{\Psi} = S_{oo} - S_{ok}S_{kk}^{-1}S_{ko}$$

and the maximum of the likelihood function becomes

$$L_{\max}^{-2/T} = |\hat{\Psi}| = |S_{oo}|\prod_{i=1}^{r}(1-\hat{\lambda}_i),$$

where the $\hat{\lambda}_i$ are the eigenvalues found from the characteristic equation

$$\left|\hat{\lambda}S_{kk} - S_{ko}S_{oo}^{-1}S_{ok}\right| = 0.$$

However instead of Π, we really want α and β but these are not identified, i.e we can premultiply β by an $r \times r$ square full rank matrix, R, and postmultiply α by its inverse and still have the same Π. That is

$$\Pi = \alpha\beta = \alpha R^{-1}R\beta' = \breve{\alpha}\breve{\beta}'.$$

Given that Π does not have full rank, this is a reduced rank regression or equally a multivariate canonical correlation analysis. To obtain estimates of α and β, the Johansen method uses the eigenvalues and vectors obtained from the characteristic equation above. The p eigenvectors

$$\hat{V} = (\hat{v}_1, \ldots, \hat{v}_p)$$

are normalised such that

$$\hat{V}'S_{kk}\hat{V} = I.$$

The eigenvalues are ordered from the largest down to the smallest with the eigenvectors ordered in parallel and then the cointegrating relations are estimated as the first r eigenvectors. Another way of writing this is to say that two sets of constraints have been imposed on the cointegrating vectors in β. The first is that in the upper left r dimensional block in the equation above which can be written as

$$\hat{\beta}'S_{kk}\hat{\beta} = I_r,$$

which entails $r(r+1)/2$ restrictions. The second set of constraints comes from determining the cointegrating relations as the first r eigenvectors (first in the sense of those having the largest eigenvalues). Here we have

$$\hat{\beta}'S_{ko}S_{oo}^{-1}S_{ok}\hat{\beta} = diag\left(\hat{\lambda}_1,\ldots,\hat{\lambda}_r\right) = \hat{\Lambda},$$

which entails $r(r-1)/2$ restrictions. Thus in total we have r^2 restrictions as for the FM estimator but it should be stressed here that while these are on the 'true model' β with the FM estimator, they are only on the estimated $\hat{\beta}$ with the Johansen method.

A major advantage of the Johansen procedure is that it leads to a whole battery of hypothesis tests. A test of whether there are r cointegrating vectors in p dimensional space is given by

$$Q_r = -T\sum_{i=r+1}^{p}\ln\left(1-\hat{\lambda}_i\right).$$

This likelihood ratio statistic, Q_r, has a non-standard distribution which is tabulated by simulation. Osterwald and Lenum (1992) gives tables for $p-r$ = 1,...,10. Various test of restrictions on β can be performed by comparing the eigenvalues with and without the restrictions imposed. Given c restrictions the eigenvalues are reestimated under the constaint that

$$\beta = H\varphi,$$

where H is $p \times (p-c)$ and φ is $(p-c) \times r$. If the restriction fits the data, the new estimated eigenvalues, $\hat{\lambda}_i^c$, should not be significantly different from the old, $\hat{\lambda}_i$. Given the likelihood is a simple function of the eigenvalues, this leads directly to a likelihood ratio test distributed asymptotically as χ^2 with $r \times c$ degrees of freedom, i.e.

$$Q\left(H_c|H_o\right) = -T\sum_{i=1}^{r}\ln\left(\left(1-\hat{\lambda}_i^c\right)/\left(1-\hat{\lambda}_i\right)\right)\xrightarrow[a]{}\chi^2_{r\times c},$$

where H_c is the hypothesis tested and H_0 is the condition that there are r cointegrating vectors. The results can differ quite dramatically if one changes the dimensionality of the cointegrating space. Using this approach one can easily test for a known β, for the exclusion of a variable from the cointegrating space and for relational tests such as long-run price and income homogeneity in a money demand function. For an applied example showing a variety of these tests, see Hargreaves (1991) and Juselius and Hargreaves (1992).

While the availability of these tests appears to make this approach very useful, Phillips (1992) raises a number of queries. Clearly if the eigenvalues were all the same, there would be no criteria by which one could choose the first r; equally if the kth and $(k+1)$th eigenvalues are equal in size and the space seemed to be of dimension k, one would not be able to tell whether the kth or $(k+1)$th vector was in the space. This may seem relatively trivial as in the probability limit, all the eigenvalues are distinguishable and hence

$$\left(\hat{\lambda}_1,\ldots,\hat{\lambda}_r\right) \xrightarrow[p]{} (\lambda_1,\ldots,\lambda_r),$$

but this does not necessarily imply that

$$\hat{\beta} \xrightarrow[p]{} \beta.$$

Phillips (1992) goes on to say that 'Reduced rank regression methods produce consistent estimates of the dimension of the cointegration space and the space itself. However, without further identifying information ... this class of methods cannot be used to validly estimate individual structural relations or to test hypotheses about the coefficients in such relations when the dimension of the cointegration space $r \geq 2$.' Since this is precisely the situation for which many have been using the Johansen method, this raises serious questions as to how we should proceed.

Finally one practical problem with the Johansen estimator is the difficulty of expressing what is going on inside the 'black box' which it probably is to most users. While most economic research assistants understand what OLS is trying to do and have little problem discussing regression results with their superiors, the same does not apply to the Johansen method with its requirement for matrix algebra and a knowledge of what are eigenvalues and vectors and arbitrary rotations of these. Partly as a result of this, there may be a tendency for researchers to table the estimated cointegrating relations and some test results without fully detailing all the other coefficients estimated on the lagged Δz_t's etc; see for example Johansen (1992) and Juselius and Hargreaves (1992). However one must remember that these estimates are conditional on the rest of the model used, the number of lags of Δz_t, whether deterministic trends were included inside and/or outside the cointegrating space, whether other *stationary* variables were added on the right hand side and so on, i.e. the whole specification of the model.

This is saying the same as the estimate of one coefficient in a regression will generally depend on what other variables are included in the regression. However you rarely see only the first three out of, say, five slope coefficients quoted from a regression; the model is small enough so that one can quote the whole model without obfuscation. In this way, OLS cointegrating regression estimates are more like unconditional estimates; there is only one set. With the Johansen method two different researchers may well lead to different estimates depending on the number of lags they have used. In this sense one can refer to them as conditional cointegrating relation estimates. Compared to the way the Johansen estimates change with different lag lengths, the FM estimates do not seem to change as dramatically when one changes the type of filter etc used to estimate the 'long-run' variance. This is further compounded as at least the FM estimates are definitely consistent while Phillips and Park (1992) states that hypothesis testing about individual coefficients in Johansen's $\hat{\beta}$ can lead to inconsistent tests. Before commenting further on this approach, let us define another very similar approach.

2.6 The modified Box-Tiao estimator

Box and Tiao (1977) noted that if one finds the canonical correlations between z_t and z_{t-1}, the most predictable linear combinations would be highly persistent functions that may be nearly or actually non-stationary while the least predictable combinations would reflect stable stationary processes. To find those linear combinations of z_t most highly correlated with z_{t-1} one solves an eigenvalue routine exactly like the Johansen method; in fact the Johansen method can be described similarly as finding canonical correlations but in that case between Δz_t and z_{t-1} instead of z_t and z_{t-1}.

Suitably normalised, the eigenvalues in these canonical analyses are akin to R^2 statistics. Now one has to separate the eigenvalues into those defining stationary and those defining non-stationary sub-spaces. In the Box-Tiao method the highest eigenvalues denote the most persistent highly autocorrelated functions and hence probably nonstationary processes. Thus in the Box-Tiao method one uses the *lowest* eigenvalues to determine the likely stationary linear functions. Vice versa, since the Johansen method uses Δz_t which is by creation I(0), the highest correlations to Δz_t must define stationary processes and so there the highest rather than the lowest eigenvalues determine the cointegrating relations.

A major difference however is that, since Box-Tiao's dependent variable is non-stationary, the null hypothesis of no cointegration here involves an estimate of a unit root and hence very complex distribution theory, not as yet leading to any useful testing procedures. To test whether the cointegrating relation defined by the lowest eigenvalue is really cointegrating, one has to turn to ADF and other unit root tests. Thus the whole battery of tests available within the Johansen approach based on the Normality of the error is not available.

Bewley, Orden and Fisher (1991) modify the original Box-Tiao method to allow for deterministic trends and other variables explaining the short-term dynamics. As in Johansen, we can reduce a large model down to

$$z_t = Az_{t-1} + \Gamma X_t + \varepsilon_t,$$

where X_t is a vector including possibly leads and/or lags of differences of z_t, other stationary explanatory regressors and deterministic trends, all of which are to be partialled out just as Z_1 was in the Johansen method and, equally, as lags and leads are in the Hendry approach. In fact just as in the uni-dimensional case, there are many transformations of the same model that essentially lead to the same results; this model is a transformation of the Johansen model.

After partialling out the X effects from z_t and z_{t-1} similar to the Johansen method to obtain the residuals in $T \times p$ matrices, R_0 and R_1, the above equation can be written

$$R_0 = R_1 A + \varepsilon = \hat{R}_0 + \varepsilon,$$

and then, as in Johansen, one finds the canonical variates that order linear combinations of R according to how predictable/persistent they are. This involves finding the eigenvectors and values of the characteristic equation

$$\left|\lambda(R_0'R_0) - \hat{R}_0'\hat{R}_0\right| = \left|\lambda(R_0'R_0) - R_0'R_1(R_1'R_1)^{-1}R_1'R_0\right| = 0.$$

Note that parallel to the Johansen formulation this could be written

$$\left|\lambda S_{00} - S_{01}S_{11}^{-1}S_{10}\right| = 0,$$

where 1 replaces k.

Bewley, Orden and Fisher (1991) argue that the Box-Tiao model can be simply reparametrized by post-multiplying by A^{-1} and then rearranging to form

$$R_1 = R_0 A^{-1} - \varepsilon A^{-1}.$$

This provides no problems since A^{-1} always exists. However if one tried to do this on the Johansen model, one would have to invert $(A - \mathrm{I})$ which does not exist when there are unit roots as assumed in the Johansen approach. Bewley, Orden and

Fisher argue that the nonexistence of this implicit transformation may be one reason for the frequent extreme outliers especially in small samples when using the Johansen method.

Another interesting comparison is afforded by remembering that canonical covariate analysis looks for the strongest relation between two linear combinations of the different vectors. The Box-Tiao method directly uses this idea to define the non-stationary combinations and then the stationary combinations are defined as what is left over. While the Box-Tiao method finds the canonical correlations between z_t and z_{t-1}, the Johansen method does the same between Δz_t and z_{t-1}, i.e. it finds those linear functions (of the variables) with the highest correlations to the differenced variables. While such a relationship should only exist if there is cointegration, none of the characteristics of correlations between integrated processes are being used to help identify the estimates. Instead we are only relying on correlations between stationary variables which may well be contaminated in small samples by other stationary short term dynamics.

However a major problem with the Box-Tiao method is the lack of a relevant distribution theory and hence the lack of testing procedures. The distribution theory is complex enough for the Johansen case with a stationary dependent variable. We have already seen how hard it is to derive theoretical results as shown in Phillips (1992) where the problem of no finite moments of integer order for some components is raised. He finds that the Johansen reduced rank regression estimator has a Cauchy-like distribution while the FM estimator has matrix t-distribution tails with at least finite integer moments to order $T-p+r$. For the Box-Tiao method we have a nonstationary dependent variable and so clearly a unit root is estimated within the model which may bias results. The distribution theory would be much more complicated and does not exist at present. As a result there is no standard way for testing for cointegration within the Box-Tiao approach and thus Bewley and Orden (1991) have to use unit root tests on the cointegrating error (ECM) to see whether a canonical variate is stationary. Given that standard chi-squared tests already exist for testing many hypotheses within the Johansen framework, it may well be argued that the Johansen method is much easier to use at present especially with all the implementations in computer packages now.

3. THE DESIGN OF THE MONTE CARLO EXPERIMENT

Having defined the various estimators considered let us now define the Monte Carlo experiment used to compare these estimators and the measures used to analyse the results.

3.1 The Data Generation Process (DGP)

Initially a four dimensional model was used, i.e.

$$Bz = B\begin{bmatrix} y \\ x \end{bmatrix}_3^1 = u \quad \text{with} \quad u_t = A_1 u_{t-1} + A_2 u_{t-2} + e_t + \theta e_{t-1},$$

where B, A_1, A_2 and θ are square 4×4 matrices (B of full rank) and z, u and ε are 4-dimensional vectors; x is three dimensional. Both the A_1 and A_2 matrices were always diagonal. The error process initially involved one stationary process and three stochastic trends, i.e.

$$\begin{aligned}
u_{1t} &= \rho u_{1t-1} && + \varepsilon_{1t} + \textstyle\sum_i \theta_{1i}\varepsilon_{it-1} \\
u_{2t} &= 1.2u_{2t-1} - 0.2u_{2t-2} && + \varepsilon_{2t} + \textstyle\sum_i \theta_{2i}\varepsilon_{it-1} \\
u_{3t} &= 1.5u_{3t-1} - 0.5u_{3t-2} && + \varepsilon_{3t} + \textstyle\sum_i \theta_{3i}\varepsilon_{it-1} \\
u_{4t} &= 1.0u_{4t-1} && + \varepsilon_{4t} + \textstyle\sum_i \theta_{4i}\varepsilon_{it-1}
\end{aligned}$$

where

$$|\rho| < 1, \text{ and } \varepsilon_t = iid\ \mathrm{N}(0, \Sigma).$$

Compared to the data generation process used by Banerjee *et al.* (1986); the dimension of the system is four instead of their two, the AR order of the u_{it}'s is two instead of one, some moving average errors have been added and the variance-covariance matrix of the ε_{it}'s, Σ, is not necessarily diagonal. In many of the simulations some of the algebraic parameters were set equal to zero. After normalisation on the first variable, the B matrix was initially set at

$$B = \begin{bmatrix} 1 & -2/7 & 5/7 & -1/7 \\ 1 & 3/4 & 7/4 & 1/4 \\ 1 & 4/3 & -2/3 & 1/3 \\ 1 & 5 & 3 & 1 \end{bmatrix},$$

which has full rank.

The first u_{1t} is a stationary autoregressive moving average process, ARMA(1,1), and hence the first row of the B matrix defines a linear combination of x_t that is stationary and hence this is a cointegrating vector. The second and third rows of B relate to u_{2t} and u_{3t} which are autoregressive integrated moving average processes, ARIMA(1,1,1)'s, and the fourth row of B relates to u_{4t} which is an integrated moving average process, IMA(1,1). To simplify the system at least a bit, we decided not to include different orders of deterministic time trends; this can be left for further research.

When Engle and Granger (1987, p. 263) extend the Banerjee *et al.* DGP by allowing a non-diagonal Σ, they point out that the system is not identified in the usual sense. The DGP used in Hansen and Phillips (1990), based on that used by Banerjee and Engle & Granger, is a two dimensional system with one y and one x variable. In our algebra, one has

$$Bz = \begin{bmatrix} 1 & -2 \\ -1 & 3 \end{bmatrix} \begin{bmatrix} y \\ x \end{bmatrix} = \begin{bmatrix} u_{1t} \\ u_{2t} \end{bmatrix} = \begin{bmatrix} \rho & 0 \\ 0 & 1 \end{bmatrix} \begin{bmatrix} u_{1t} \\ u_{2t} \end{bmatrix} + \begin{bmatrix} \varepsilon_{1t} \\ \varepsilon_{2t} \end{bmatrix},$$

with

$$\begin{bmatrix} \varepsilon_{1t} \\ \varepsilon_{2t} \end{bmatrix} \equiv iid\ N[0, \Sigma],\ |\rho| < 1,\ |r| < 1,\ \Sigma = \begin{bmatrix} 1 & \varphi\sigma \\ \varphi\sigma & \sigma^2 \end{bmatrix}.$$

If one writes this out as a two equation system substituting out the u_{it}'s, one has

$$\begin{aligned} y_t - 2x_t - \rho y_{t-1} + 2\rho x_{t-1} &= \varepsilon_{1t} \\ -y_t + 3x_t - \ y_{t-1} - 3x_{t-1} &= \varepsilon_{2t} \end{aligned}$$

One has three identifying restrictions on the first equation, viz a normalisation on y_t, var(ε_1)=1 and a between parameter constraint that

$$\text{coef}(x_{t-1})/\text{coef}(x_t) = \text{coef}(y_{t-1})/\text{coef}(y_t).$$

On the second equation, there are three restrictions,

$$\text{coef}(y_t) = -\text{coef}(y_{t-1}) = -1$$

and

$$\text{coef}(x_t) = \text{coef}(x_{t-1}).$$

On estimation, one would not restrict var(ε_t) but the system would still be identified as long as one assumes as above that the first equation is cointegrating and the second is an ARMA in the first differences. These assumptions create a series of between-parameter within-equation restrictions whose identifying ability depends on the fact that $\rho \neq 1$.

In Phillips and Hansen (1990) and Phillips and Loretan (1991), they make the B matrix upper triangular and the time-series dynamics are driven by an MA(1) for the cointegrating vector and an IMA(1) for the regressor and not an ARIMA(1,1,1) as above. The resulting system is identified even with a covarying error. Their system is a marvellous one for showing the effect of endogeneity as it is controlled by just two parameters, the covariance of the errors and the relationship of ε_{1t} to ε_{2t-1}. However their DGP is slightly unrealistic as few economic systems would have a regressor just explained by a random walk whereas in our DGP the variables in the system are related in both cointegrating vectors and in driving trends. We would expect the driving trends to be functions of various variables such as an interest differential or yield curve and not just simple random walks. Also Phillips and Hansen's cointegrated vector followed a moving average process while ours is an ARMA process. This implies that the 'memory' of their equilibrium vector is extremely short while the equilibrium error may well have a fairly long 'memory' and may even be nearly nonstationary.

If one writes our DGP in the Johansen VAR style, one has

$$Bz_t = A_1 Bz_{t-1} + A_2 Bz_{t-2} + \varepsilon_t + \theta\varepsilon_{t-1},$$

which can be written in ECM form as

$$\Delta z_t = -B^{-1}(I - A_1)B\Delta z_{t-1} - B^{-1}(I - A_1 - A_2)Bz_{t-2} + \varepsilon_t + \theta\varepsilon_{t-1},$$

i.e.

$$\Delta z_t = \Gamma_1 \Delta z_{t-1} - \Pi z_{t-2} + \varepsilon_t + \theta\varepsilon_{t-1}.$$

with Γ_1 and Π defined appropriately. For the base case with one cointegrating vector

$$\mathrm{I} - A_1 - A_2 = \begin{bmatrix} (1-\rho) & 0 & 0 & 0 \\ 0 & 0 & 0 & 0 \\ 0 & 0 & 0 & 0 \\ 0 & 0 & 0 & 0 \end{bmatrix},$$

and so

$$B^{-1}(\mathrm{I} - A_1 - A_2)B = (1-\rho)B^{\cdot 1}B_{1.}$$

where $B_{1.}$ is the first row of the B matrix (i.e. the cointegrating vector) and $B^{\cdot 1}$ is the first column of the inverse of B. To study the effect of endogeneity later, we make the first element in each row of B equal to zero except in the first row which is the cointegrating vector. This makes

$$(1-\rho)B^{\cdot 1} = \begin{bmatrix} \alpha \\ 0 \\ 0 \\ 0 \end{bmatrix}$$

and hence the latter three variables are exogenous with respect to the long-run cointegrating vector and hence, in this case, nothing is gained by estimating the whole system of equations as opposed to just the first equation alone. A number of other changes to the DGP are described below with their ensuing results.

Given the number of possible combinations of different parameter settings here, it was decided to settle initially on a reference base case wherein $\rho = 0$,

$\theta = 0$, and a sample size of $T = 100$. After running this base case, parameters were then changed and the results could be compared with the base case. This system allows one to have up to three cointegrating vectors in a four dimensional space and many forms of misspecification can also be analysed. The programming of the simulations was carried out in Gauss 2.2, for which Aptech Systems Inc have the copyright.

3.2 Measures used to analyse the simulation results

Various standard measures are used such as mean and median bias and the standard errors of the results, although it should be realised that some of these estimators do not have finite sample moments. As a measure of skewness we used

$$\frac{q_{97.5} - q_{50.0}}{q_{50.0} - q_{2.5}} - 1,$$

where q_i is the ith quartile. As a measure of kurtosis we used

$$\frac{q_{97.5} - q_{2.5}}{q_{99.5} - q_{0.5}} - \frac{1.96}{2.575}.$$

Both measures were drawn from Bewley, Orden and Fisher (1991) and both should equal zero for the standard Normal distribution.

4. MONTE CARLO SIMULATION RESULTS FOR A FOUR DIMENSIONAL SYSTEM

4.1 With one cointegrating vector

For the base case with 100 observations and the full B matrix given above (full rank with no zeros anywhere), u_1 was just white noise, u_2 and u_3 were ARIMA(1,1,0) processes and u_4 was a simple random walk; since u_1 is the 'equilibrium error' for the one cointegrating vector, making u_1 white noise should make the cointegrating vector stand out very clearly as there is absolutely no persistence at all in its error. It should be noted that in all the simulations in this section it was assumed that each estimator 'knew' the correct model in terms of the number of lags except when MA errors were included ($\theta \neq 0$) and they 'knew' that there was just one cointegrating vector; knowing the true model is particularly favourable to the multivariate estimators. The single equation estimators always use the first Z variable, y, as the dependent variable.

Given that some of these estimators do not have finite moments, one should possibly only consider the median bias and not the mean bias. For this base case,

Table 1. Monte Carlo Results for Base Simulation

(a) Mean Bias

	OLS	Hendry	EY3	FM	Box Tiao	Johansen
β_1	–0.069	–0.049	–0.042	–0.0136	–0.0013	–0.003
β_2	–0.033	–0.018	–0.029	–0.0107	–0.0003	–0.0007
β_3	–0.022	–0.016	–0.015	–0.0043	–0.0004	–0.001

(b) Median Bias

	OLS	Hendry	EY3	FM	Box Tiao	Johansen
β_1	–0.058	–0.043	–0.037	–0.0122	0.0006	–0.0014
β_2	–0.028	–0.015	–0.024	–0.0094	0.0007	–0.0001
β_3	–0.019	–0.014	–0.013	–0.0036	0.0005	–0.0006

(c) Standard Deviation

	OLS	Hendry	EY3	FM	Box Tiao	Johansen
β_1	0.060(0.037)	0.063	0.057	0.048	0.051	0.055
β_2	0.039(0.025)	0.043	0.043	0.033	0.035	0.036
β_3	0.017(0.010)	0.017	0.016	0.013	0.014	0.015

(d) Skewness

	OLS	Hendry	EY3	FM	Box Tiao	Johansen
β_1	–0.456	–0.380	–0.211	–0.213	–0.143	–0.157
β_2	–0.375	–0.018	–0.357	–0.012	–0.022	–0.147
β_3	–0.489	–0.418	–0.288	–0.231	–0.162	–0.174

(e) Kurtosis

	OLS	Hendry	EY3	FM	Box Tiao	Johansen
β_1	–0.038	–0.061	–0.132	–0.012	–0.091	–0.248
β_2	–0.012	–0.116	–0.043	–0.043	–0.172	–0.119
β_3	–0.068	–0.043	–0.098	–0.065	–0.159	–0.154

the results in Table 1 show that the median bias, of the Hendry-style dynamic OLS and the Engle-Yoo 3-step estimators, is distinctly less than for simple OLS but the FM estimator is distinctly better still. However when one moves to the multivariate Box-Tiao and Johansen estimators, there is yet further dramatic improvement but little to choose between the two estimators in terms of median bias as different coefficients are less biased for each estimator. Many Monte Carlo studies only consider two dimensional systems with only one coefficient to estimate after normalisation; this possibly creates a spuriously clear result as only one coefficient is considered. We see here that with more than one coefficient to estimate a comparison of the estimators can produce different results across the different coefficients.

When we look at the standard deviations of the Monte Carlo estimates we find that the FM estimator dominates and strangely the Hendry-style estimator is even worse than simple OLS. In brackets after the OLS standard deviations, we show the average standard errors produced by the standard OLS formula. As expected we see that they are badly under-estimated.

The coefficients of the 'regressors' in the cointegrating relation were $\beta_1 = 2/7$ (0.2857), $\beta_2 = -5/7$ (–0.7143) and $\beta_3 = 1/7$ (0.1429) and hence the cointegrating vector is [1, –0.2857, 0.7143, –0.1429]. We refer to the cointegrating vector as β and the values within as $[1, -\beta_1, -\beta_2, -\beta_3]$ after normalisation. Relative to its size, β_2 has the least bias and standard deviation but admittedly β_3 has lower bias than β_1 even though it is half the size. When we turn to skewness, the relatively high signal β_2 is generally less skewed, sometimes dramatically so as for the Hendry, FM and Box-Tiao estimators. Here we start to see that the basic difference between the two multivariate estimators is in their tail behaviour. Although the Johansen estimator is dominated by the Box-Tiao estimator, one cannot say this with reference to the FM estimator; however it is very noticeable how much higher is the skewness of the distribution of the Johansen estimator of β_2. Not having moments of finite integer order, the Johansen estimator tends to have much fatter tails and we find that it has the fattest tails of *all* the estimators except possibly the Box-Tiao. Again there is a difference in comparison across different coefficients.

Now consider a move to a much smaller sample size of only 50 instead of 100 observations. Here many similar patterns emerge (see Table 2). The first three estimators show about twice as much median bias and the FM estimator shows about 3 to 4 times as much bias as before though still less than the other single equation estimators. The increase in bias for the Johansen estimator varies dramatically across the three coefficients from 2 to 20 times the bias with 100 observations but strikingly the Box-Tiao estimator's bias increases least of all with the bias actually decreasing for β_3. Relative to the other estimators we expected both multivariate estimators to be more sensitive to the smaller sample size because of the large number of autoregressive coefficients estimated in their models; given this, the results for the Box-Tiao method are strikingly good.

The standard deviations of the estimators generally increase fairly similarly across the estimators except for the Johansen estimator whose standard deviations for β_1 and β_2 have blown right out; those for the 3-step estimator also increased markedly. One cannot make a general rule out of this as the standard deviation for β_3 did not increase by anywhere near as much. On skewness and kurtosis, there is

Table 2. Monte Carlo Results with a Sample Size of 50

(a) Mean Bias

	OLS	Hendry	EY3	FM	Box Tiao	Johansen
β_1	−0.13	−0.1	−0.044	−0.0541	−0.0147	0.0851
β_2	−0.059	−0.032	−0.027	−0.0311	−0.0079	−0.0542
β_3	−0.042	−0.032	−0.018	−0.0174	−0.0049	0.0011

(b) Median Bias

	OLS	Hendry	EY3	FM	Box Tiao	Johansen
β_1	−0.112	−0.076	−0.058	−0.0423	−0.0012	−0.0038
β_2	−0.053	−0.027	−0.037	−0.027	−0.0011	0.002
β_3	−0.037	−0.025	−0.022	−0.0136	−0.0003	−0.0016

(c) Standard Deviation

	OLS	Hendry	EY3	FM	Box Tiao	Johansen
β_1	0.119(0.069)	0.212	1.158	0.111(0.064)	0.186	2.69
β_2	0.075(0.048)	0.153	0.796	0.070(0.045)	0.115	2.16
β_3	0.034(0.019)	0.058	0.374	0.031(0.018)	0.051	0.171

(d) Skewness

	OLS	Hendry	EY3	FM	Box Tiao	Johansen
β_1	−0.440	−0.335	−0.187	−0.342	−0.258	−0.100
β_2	−0.225	−0.045	−0.119	−0.160	−0.301	−0.214
β_3	−0.502	−0.392	−0.209	−0.349	−0.275	−0.151

(e) Kurtosis

	OLS	Hendry	EY3	FM	Box Tiao	Johansen
β_1	−0.110	−0.123	−0.592	−0.084	−0.265	−0.483
β_2	−0.053	−0.122	−0.611	−0.009	−0.199	−0.534
β_3	−0.104	−0.129	−0.607	−0.08	−0.323	−0.466

a trade-off between them such that while the skewness has hardly increased for the 3-step and Johansen methods, the kurtosis has; for the FM and Box-Tiao estimators it is the other way around. For the single equation estimators the skewness and kurtosis appear least for the high signal-to-noise coefficient, β_2, and the bias is least relative to the value of β_2.

For the next simulation, we returned the sample size to 100 but increased the variance of the 'equilibrium error', u_1, fivefold (see Table 3). For the OLS estimator we find a dramatic increase in the bias such that the estimates are really very poor but the kurtosis is the same and the skewness has in fact decreased with

Table 3. Monte Carlo Results with a High ECM Error Variance compared to the Base Case

(a) Mean Bias

	OLS	Hendry	EY3	FM	Box-Tiao	Johansen
β_1	−0.746	−0.714	−9.122	−0.367	−0.033	−0.082
β_2	−0.359	−0.24	−3.393	−0.213	−0.012	−0.027
β_3	−0.238	−0.236	−2.107	−0.116	−0.011	−0.026

(b) Median Bias

	OLS	Hendry	EY3	FM	Box-Tiao	Johansen
β_1	−0.742	−0.618	−1.225	−0.318	0.003	−0.007
β_2	−0.355	−0.229	−0.478	−0.189	0.003	0
β_3	−0.237	−0.199	−0.378	−0.102	0.002	−0.003

(c) Standard Deviation

	OLS	Hendry	EY3	FM	Box-Tiao	Johansen
β_1	0.359(0.102)	3.414	133.0	0.349(0.184)	0.296	1.206
β_2	0.286(0.078)	2.475	56.5	0.251(0.138)	0.190	0.362
β_3	0.087(0.026)	0.797	28.7	0.091(0.046)	0.083	0.381

(d) Skewness

	OLS	Hendry	EY3	FM	Box-Tiao	Johansen
β_1	−0.109	−0.505	−0.315	−0.419	−0.413	−0.468
β_2	−0.064	−0.135	−0.252	−0.304	−0.279	−0.328
β_3	−0.005	−0.570	−0.093	−0.401	−0.412	−0.499

(e) Kurtosis

	OLS	Hendry	EY3	FM	Box-Tiao	Johansen
β_1	0.023	−0.486	−0.684	0.046	−0.154	−0.299
β_2	0.033	−0.575	−0.702	−0.069	−0.180	−0.234
β_3	0.078	−0.488	−0.688	0.035	−0.203	−0.310

the extra variation. For the single equation estimators, the bias of the FM estimator is still the least but it has still increased much more than for the multivariate estimators. Between the multivariate estimators, the Box-Tiao has a smaller bias for two out of three coefficients and it dominates in terms of variation, skewness and kurtosis. It is possibly a worry that the actual standard deviation of the Monte Carlo estimates was much more than the average estimate thereof but this may be a function of the non-Normality of the coefficient distribution.

Table 4. Monte Carlo Results with a Low ECM Error Variance compared to the Base Case

(a) Mean Bias

	OLS	Hendry	EY3	FM	Box-Tiao	Johansen
β_1	–0.0028	–0.0020	–0.0023	–0.0002	–0.0001	–0.0004
β_2	–0.0014	–0.0008	–0.0016	–0.0004	0.0001	0
β_3	–0.0009	–0.0006	–0.0008	–0.0001	0	–0.0001

(b) Median Bias

	OLS	Hendry	EY3	FM	Box-Tiao	Johansen
β_1	–0.0022	–0.0014	–0.002	–0.00021	0.00013	–0.00029
β_2	–0.0013	–0.0006	–0.0013	–0.00025	0.00013	–0.00002
β_3	–0.0008	–0.0006	–0.0007	–0.00008	0.00009	–0.00011

(c) Standard Deviation

	OLS	Hendry	EY3	FM	Box-Tiao	Johansen
β_1	0.008(0.008)	0.009	0.009	0.009(0.008)	0.010	0.011
β_2	0.006(0.005)	0.006	0.006	0.006(0.005)	0.007	0.007
β_3	0.002(0.002)	0.003	0.002	0.003(0.002)	0.003	0.003

(d) Skewness

	OLS	Hendry	EY3	FM	Box-Tiao	Johansen
β_1	–0.249	–0.128	–0.154	–0.025	–0.072	–0.100
β_2	–0.168	–0.062	–0.194	–0.061	0.035	–0.067
β_3	–0.167	–0.104	–0.153	0.022	–0.082	–0.086

(e) Kurtosis

	OLS	Hendry	EY3	FM	Box-Tiao	Johansen
β_1	–0.113	–0.109	–0.083	–0.152	–0.107	–0.231
β_2	–0.060	–0.070	–0.037	–0.124	–0.148	–0.122
β_3	–0.066	–0.116	–0.086	–0.108	–0.169	–0.133

Table 4 show the results when instead of increasing the 'equilibrium error' fivefold, we decreased it fivefold. Before we found the FM bias increase relatively more than the other OLS estimators; in a similar manner it has now decreased more, so much so that the bias is less than for the Johansen estimates of β_1 and β_3, and it completely dominates the Johansen method in terms of variation, skewness and kurtosis. It is hard to distinguish the FM and Box-Tiao estimators when one considers all coefficients and criteria.

Table 5. Monte Carlo Results with y not in the Common Trends

(a) Mean Bias

	OLS	Hendry	EY3	FM	Box-Tiao	Johansen
β_1	−0.051	−0.037	−0.041	−0.0078	−0.0013	−0.0026
β_2	−0.026	−0.013	−0.029	−0.0086	0.0002	−0.0005
β_3	−0.018	−0.013	−0.011	−0.003	−0.0004	−0.0009

(b) Median Bias

	OLS	Hendry	EY3	FM	Box-Tiao	Johansen
β_1	−0.043	−0.031	−0.038	−0.0050	0.0005	−0.0007
β_2	−0.022	−0.010	−0.024	−0.0062	0.0006	−0.0002
β_3	−0.016	−0.011	−0.010	−0.0027	0.0005	−0.0005

(c) Standard Deviation

	OLS	Hendry	EY3	FM	Box-Tiao	Johansen
β_1	0.038(0.023)	0.039	0.038	0.030(0.022)	0.032	0.035
β_2	0.034(0.022)	0.038	0.039	0.029(0.021)	0.031	0.034
β_3	0.016(0.010)	0.016	0.015	0.013(0.010)	0.014	0.015

(d) Skewness

	OLS	Hendry	EY3	FM	Box-Tiao	Johansen
β_1	−0.534	−0.487	−0.343	−0.210	−0.156	−0.166
β_2	−0.340	−0.138	−0.324	−0.226	−0.057	−0.083
β_3	−0.458	0.406	−0.262	−0.196	−0.148	−0.143

(e) Kurtosis

	OLS	Hendry	EY3	FM	Box-Tiao	Johansen
β_1	−0.002	−0.076	−0.072	−0.114	−0.065	−0.143
β_2	−0.030	−0.101	−0.107	−0.106	−0.079	−0.205
β_3	−0.069	−0.090	−0.085	−0.104	−0.017	−0.139

We then altered the B matrix so that there was no feedback from the dependent variable used in the single equation estimators to the other variables; more specifically the B matrix was made block triangular, i.e.

$$B = \begin{bmatrix} 1 & -2/7 & 5/7 & -1/7 \\ 0 & 3/4 & 7/4 & 1/4 \\ 0 & 4/3 & -2/3 & 1/3 \\ 0 & 5 & 3 & 1 \end{bmatrix}.$$

Since this is the structure behind the FM estimator, we expected it to improve more than the other estimators but while this is so in terms of median bias, the effect was not all that great. All the single equation estimators improved a little (see Table 5). While the kurtosis was in fact worse now for the FM estimator the estimated standard deviation was much closer to the actual. For the multivariate estimators the normalisation on the first variable is quite arbitrary and hence triangularisation relative to that variable should be irrelevent and so it turned out. We also carried out a simulation in which the β matrix was an identity matrix outside the cointegrating vector, i.e. independent x. Given that here there was nothing to be gained from multivariate estimation, these two estimators both had median biases no better than the single equation estimators. However this is a very unlikely economic model and so it was not pursued.

The other experiments performed with one cointegrating vector changed the time series properties of the 'equilibrium error', u_1. The first made the 'equilibrium error' highly autocorrelated with ρ equal to 0.9 (see Table 6). Given the problem of identifying the cointegrating vector now as ρ is so high, it is possibly not surprising that the results are generally worse but that the OLS estimator now fairs very well in comparison to the other estimators and in fact has the lowest variation such that you may well be more likely to be closer to the true value with OLS. The FM estimator fairs well. While the multivariate estimators and the 3-step have the least bias, they have very high variation relative to the other estimators such that you are possibly less likely to be closer to the true value with the simpler estimators. In comparing the Box-Tiao with the Johansen estimator, the Box-Tiao dominates almost entirely on all criteria with all coefficients.

As noted before the above results all assumed that the estimator was used knowing the correct order of the autoregressive process. Clearly OLS and FM are distinct in that they do not use this information. The Johansen estimator in particular can suffer when the wrong order is used as one is then estimating p^2 more or less coefficients than appropriate for each order over or under-estimated. As it should only lead to inefficiency rather than bias, the general advice seems to be to use a higher order VAR than required rather than a VAR of too low an order.

Here we changed the error process to include independent moving averages so that the autoregressive representation is (infinitely) underspecified; θ was diagonal (0.5). However the results (see Table 7) are not very different. The 3-step estimator is now much better than the OLS and Hendry estimators but still no better than the FM estimator. We were expecting the multivariate estimators to fair much worse but this did not prove to be the case. An MA coefficient of 0.5 may not be high enough to see the effect but it is still high in economic model terms.

Table 6. Monte Carlo Results with Highly Autocorrelated Equilibrium Error ($\rho = 0.9$)

(a) Mean Bias

	OLS	Hendry	EY3	FM	Box-Tiao	Johansen
β_1	−0.187	−0.251	2.742	−0.126	−0.156	−1.60
β_2	−0.087	−0.088	2.632	−0.077	−0.035	−2.22
β_3	−0.059	−0.080	0.905	−0.040	−0.030	−0.38

(b) Median Bias

	OLS	Hendry	EY3	FM	Box-Tiao	Johansen
β_1	−0.180	−0.215	0.119	−0.118	−0.032	−0.042
β_2	−0.091	−0.084	0.006	−0.081	−0.031	−0.053
β_3	−0.057	−0.067	0.028	−0.037	−0.009	−0.025

(c) Standard Deviation

	OLS	Hendry	EY3	FM	Box-Tiao	Johansen
β_1	0.226(0.060)	0.472	62.694	0.337(0.164)	13.296	32.864
β_2	0.155(0.042)	0.379	65.713	0.234(0.114)	4.707	53.030
β_3	0.061(0.016)	0.134	21.851	0.091(0.045)	3.195	7.935

(d) Skewness

	OLS	Hendry	EY3	FM	Box-Tiao	Johansen
β_1	−0.132	−0.258	−0.058	−0.197	−0.234	−0.119
β_2	0.108	−0.008	0.042	−0.018	0.093	0.442
β_3	−0.098	−0.272	−0.012	−0.144	0.016	−0.105

(e) Kurtosis

	OLS	Hendry	EY3	FM	Box-Tiao	Johansen
β_1	−0.037	−0.275	−0.652	−0.098	−0.512	−0.641
β_2	−0.069	−0.312	−0.623	−0.140	−0.578	−0.575
β_3	−0.028	−0.280	−0.637	−0.109	−0.572	−0.650

The lack of change for the multivariate estimators may be a function of the diagonality of the error processes above, a simulation was performed where the θ matrix was a full square matrix with 0.5 on the diagonal as above but now with 0.4 in all off-diagonal positions. The results shown in Table 8 show little change for the single equation estimators. While the Johansen faired slightly worse than before and is little better than the FM estimator now, the Box-Tiao faired much worse and is highly skewed.

Table 7. Monte Carlo Results with Independent Moving Average Errors ($\theta = 0.5$)

(a) Mean Bias

	OLS	Hendry	EY3	FM	Box-Tiao	Johansen
β_1	–0.040	–0.052	0.016	0.005	–0.0020	–0.004
β_2	–0.020	–0.020	–0.005	–0.003	–0.0002	–0.002
β_3	–0.013	–0.016	0.003	0.001	–0.0006	–0.001

(b) Median Bias

	OLS	Hendry	EY3	FM	Box-Tiao	Johansen
β_1	–0.033	–0.043	0.011	0.003	–0.0007	–0.0008
β_2	–0.016	–0.016	–0.003	–0.002	0.0004	0.0001
β_3	–0.011	–0.014	0.002	0.001	–0.0002	–0.0009

(c) Standard Deviation

	OLS	Hendry	EY3	FM	Box-Tiao	Johansen
β_1	0.048(0.028)	0.063	0.065	0.048(0.036)	0.048	0.071
β_2	0.032(0.019)	0.043	0.042	0.034(0.024)	0.033	0.046
β_3	0.013(0.008)	0.018	0.018	0.013(0.010)	0.013	0.022

(d) Skewness

	OLS	Hendry	EY3	FM	Box-Tiao	Johansen
β_1	–0.417	–0.407	0.264	0.081	–0.171	–0.206
β_2	–0.306	–0.104	–0.059	–0.029	–0.087	–0.099
β_3	–0.446	–0.468	0.213	0.059	–0.205	–0.247

(e) Kurtosis

	OLS	Hendry	EY3	FM	Box-Tiao	Johansen
β_1	–0.068	–0.105	–0.207	–0.116	–0.116	0.289
β_2	–0.021	–0.136	–0.028	–0.091	–0.125	–0.194
β_3	–0.080	–0.013	–0.169	–0.131	–0.134	–0.231

When we move to negative moving averages we now find that both the FM and Box-Tiao fair much worse still. Table 9 reports the results for a simulation where all the moving average coefficients were the negative of the previous simulation (i.e. –0.5 on the diagonal and –0.4 off the diagonal). We now find that the Hendry-style estimator is the best single equation estimator with the Johansen clearly the best overall.

Finally in this section we look at two simulations where, instead of the errors being interrelated over time by a multivariate moving average, the errors are

Table 8. Monte Carlo Results with Positively Interrelated Moving-Average Errors

(a) Mean Bias

	OLS	Hendry	EY3	FM	Box-Tiao	Johansen
β_1	–0.0336	–0.0065	–0.0318	–0.0066	–0.0313	–0.0071
β_2	–0.0468	–0.0045	–0.0580	–0.0184	–0.0108	–0.0034
β_3	–0.0120	–0.0037	–0.0123	–0.0021	–0.0089	–0.0021

(b) Median Bias

	OLS	Hendry	EY3	FM	Box-Tiao	Johansen
β_1	–0.0288	–0.0052	–0.0290	–0.0044	–0.0250	–0.0049
β_2	–0.0405	–0.0040	–0.0525	–0.0142	–0.0066	–0.0008
β_3	–0.0102	–0.0030	–0.0106	–0.0014	–0.0071	–0.0016

(c) Standard Deviation

	OLS	Hendry	EY3	FM	Box-Tiao	Johansen
β_1	0.039(0.024)	0.032	0.045	0.033(0.024)	0.041	0.035
β_2	0.049(0.028)	0.039	0.062	0.040(0.028)	0.048	0.041
β_3	0.011(0.006)	0.009	0.012	0.009(0.006)	0.011	0.009

(d) Skewness

	OLS	Hendry	EY3	FM	Box-Tiao	Johansen
β_1	–0.385	–0.194	–0.165	–0.187	–0.448	–0.239
β_2	–0.373	–0.033	–0.309	–0.298	–0.256	–0.225
β_3	–0.499	–0.279	–0.357	–0.279	–0.506	–0.192

(e) Kurtosis

	OLS	Hendry	EY3	FM	Box-Tiao	Johansen
β_1	–0.075	–0.124	–0.071	–0.113	–0.159	–0.161
β_2	–0.090	–0.101	–0.054	–0.117	–0.097	–0.094
β_3	–0.042	–0.124	–0.063	–0.062	–0.091	–0.171

simply correlated with each other contemporaneously. In the first simulation the correlation matrix of the errors was set so that

$$r_{ij} = (+0.8)^{|i-j|}$$

and in the second simulation –0.8 was used. The major difference between the two simulations was in the performance of the Hendry-style estimator which performed very well when the errors were positively correlated but poorly when some negative correlations were introduced. Overall the Johansen estimator

Table 9. Monte Carlo Results with Negatively Interrelated Moving Average Errors

(a) Mean Bias

	OLS	Hendry	EY3	FM	Box-Tiao	Johansen
β_1	−0.2184	−0.0580	−0.2114	−0.1138	−0.0985	−0.0079
β_2	−0.0948	−0.0199	−0.0964	−0.0506	−0.0404	−0.0056
β_3	−0.0654	−0.0182	−0.0642	−0.0330	−0.0276	−0.0022

(b) Median Bias

	OLS	Hendry	EY3	FM	Box-Tiao	Johansen
β_1	−0.2100	−0.0515	−0.2117	−0.0945	−0.0584	−0.0056
β_2	−0.0890	−0.0178	−0.0911	−0.0410	−0.0233	−0.0031
β_3	−0.0629	−0.0161	−0.0648	−0.0274	−0.0161	−0.0013

(c) Standard Deviation

	OLS	Hendry	EY3	FM	Box-Tiao	Johansen
β_1	0.109(0.048)	0.066	0.209	0.096(0.049)	0.179	0.057
β_2	0.072(0.032)	0.042	0.094	0.057(0.033)	0.085	0.036
β_3	0.031(0.013)	0.018	0.057	0.027(0.014)	0.050	0.016

(d) Skewness

	OLS	Hendry	EY3	FM	Box-Tiao	Johansen
β_1	−0.277	−0.355	0.110	−0.546	−0.723	−0.107
β_2	−0.244	−0.161	−0.198	−0.442	−0.645	−0.203
β_3	−0.297	−0.381	0.115	−0.560	−0.740	−0.138

(e) Kurtosis

	OLS	Hendry	EY3	FM	Box-Tiao	Johansen
β_1	−0.044	−0.050	−0.249	−0.004	−0.238	−0.112
β_2	−0.004	−0.043	−0.093	−0.045	−0.263	−0.116
β_3	0.034	−0.044	−0.276	−0.007	−0.236	−0.078

performed clearly best with little difference between the two simulations. Now the Box-Tiao did not fair anywhere near as well as the Johansen and in fact is possibly worse than the FM estimator. The FM estimator showed significant skewness but very little kurtosis.

Throughout these simulations the estimated standard errors for the OLS and FM estimators were compared with those shown by the Monte Carlo results. While we expected the OLS estimates to be serious underestimates we expected the FM estimator would produce much better estimates. However it was

Table 10. Monte Carlo Results with Contemporaneously Positively Correlated Errors

(a) Mean Bias

	OLS	Hendry	EY3	FM	Box-Tiao	Johansen
β_1	–0.0861	–0.0004	–0.1093	–0.0461	–0.0456	–0.0002
β_2	–0.1827	–0.0005	–0.2444	–0.0959	–0.0509	–0.0008
β_3	–0.0295	–0.0014	–0.0379	–0.0142	–0.0175	–0.0002

(b) Median Bias

	OLS	Hendry	EY3	FM	Box-Tiao	Johansen
β_1	–0.0790	–0.0006	–0.1058	–0.0356	–0.0345	0.0008
β_2	–0.1662	–0.0006	–0.2265	–0.0722	–0.0329	0.0001
β_3	–0.0267	–0.0012	–0.0361	–0.0108	–0.0134	0.0000

(c) Standard Deviation

	OLS	Hendry	EY3	FM	Box-Tiao	Johansen
β_1	0.063(0.048)	0.029	0.076	0.052(0.031)	0.080	0.030
β_2	0.117(0.032)	0.051	0.142	0.098(0.053)	0.131	0.053
β_3	0.017(0.013)	0.008	0.020	0.014(0.008)	0.053	0.008

(d) Skewness

	OLS	Hendry	EY3	FM	Box-Tiao	Johansen
β_1	–0.377	–0.005	–0.230	–0.561	–0.451	–0.105
β_2	–0.415	–0.040	–0.333	–0.621	–0.463	–0.082
β_3	–0.437	–0.151	–0.313	–0.618	–0.604	–0.053

(e) Kurtosis

	OLS	Hendry	EY3	FM	Box-Tiao	Johansen
β_1	0.011	–0.093	0.012	–0.053	–0.144	–0.110
β_2	–0.000	–0.095	0.005	–0.062	–0.124	–0.118
β_3	0.021	–0.102	0.040	–0.025	–0.142	–0.102

noticeable that the FM estimator also led to a downwards biased standard error.

Many more different simulations could have been performed to show the small sample effects of different misspecifications but clearly, when one starts considering misspecifications, the number of possible experiments is enormous. However this raises possibly the major difference between the multivariate and the FM estimator. The Johansen and modified Box-Tiao methods remove nuisance parameters parametrically by estimating an appropriate order difference VAR. If the order is wrongly estimated then this can lead to serious error. The FM estimator does an implicitly similar thing if one describes the estimation of the long-run error variance as a VAR on the error processes. However no order

Table 11. Monte Carlo Results with Contemporaneously 'Negatively' Correlated Errors

(a) Mean Bias

	OLS	Hendry	EY3	FM	Box-Tiao	Johansen
β_1	–0.1761	–0.1929	–1.2065	–0.0267	0.0344	–0.0044
β_2	–0.0315	–0.0338	0.0014	–0.0049	0.0384	–0.0007
β_3	–0.0576	–0.0584	–0.3356	–0.0099	0.0134	–0.0013

(b) Median Bias

	OLS	Hendry	EY3	FM	Box-Tiao	Johansen
β_1	–0.1537	–0.1684	0.0423	–0.0179	0.0338	–0.0011
β_2	–0.0279	–0.0272	–0.0007	–0.0031	0.0324	–0.0004
β_3	–0.0505	–0.0511	0.0052	–0.0066	0.0130	–0.0001

(c) Standard Deviation

	OLS	Hendry	EY3	FM	Box-Tiao	Johansen
β_1	0.152(0.078)	0.653	72.35	0.082(0.055)	0.151	0.073
β_2	0.034(0.019)	0.070	2.11	0.020(0.013)	0.035	0.017
β_3	0.044(0.022)	0.173	19.17	0.024(0.016)	0.042	0.020

(d) Skewness

	OLS	Hendry	EY3	FM	Box-Tiao	Johansen
β_1	–0.455	–0.560	–0.203	–0.358	0.008	–0.153
β_2	–0.334	–0.448	–0.003	–0.292	0.913	–0.078
β_3	–0.481	–0.583	–0.206	–0.420	0.093	–0.186

(e) Kurtosis

	OLS	Hendry	EY3	FM	Box-Tiao	Johansen
β_1	0.012	–0.173	–0.644	–0.044	–0.081	–0.099
β_2	–0.050	–0.122	–0.613	–0.107	–0.061	–0.116
β_3	–0.002	–0.181	–0.638	–0.051	–0.088	–0.104

selection is involved here as implicitly all possible autocovariances are estimated and then weighted and combined appropriately. In this study we use prewhitening with an automatic bandwith on a Parzen filter. The key point is that there is no possible mis-specification problem for the order of the VAR.

To summarise the results so far, it would appear that the FM estimator is the best single equation estimator overall. In terms of median bias, the multivariate estimators are both better still but the Johansen estimator can vary dramatically with very fat tails. Between the two multivariate estimators, the Box-Tiao clearly seemed to estimate the coefficients more accurately overall than the Johansen estimator when the error structure was simple. However immediately one added a

Table 12. Estimation of CV with low variance in a 2-dimensional cointegrating space

(a) Median Bias

	OLS	Hendry	EY3	FM	Box-Tiao	Johansen
β_1	0.0027	0.0031	0.00054	0.00419	–0.0114	0.00072
β_2	0.0070	0.0057	0.00644	0.00480	–0.0107	0.00154
β_3	0.0017	0.0016	0.00099	0.00178	–0.0033	0.00048

(b) Standard Deviation

	OLS	Hendry	EY3	FM	Box-Tiao	Johansen
β_1	0.0113(0.0106)	0.0167	0.0122	0.0121(0.0098)	1.045	4.385
β_2	0.0107(0.0101)	0.0163	0.0114	0.0114(0.0094)	0.991	4.107
β_3	0.0032(0.0030)	0.0055	0.0034	0.0035(0.0028)	0.401	1.605

multivariate moving average error structure or correlated errors, the Box-Tiao estimator performed far worse than the Johansen estimator and no better, if not worse in fact, than the FM estimator. Since such error structures would be common in economic systems, the Johansen estimator seems best overall.

4.2 With two cointegrating vectors

Simulations were also run comparing the estimation of a two-dimensional cointegrating space in four dimensions using different variances for the cointegrating errors. Since a cointegrating regression looks for that vector with the lowest variance one would expect it to home in on the cointegrating vector with the lowest cointegrating error variance but one may well obtain a linear function of the two vectors weighted according to some function of the variance-covariance matrix of the cointegrating errors.

Firstly we look at the results when the first two rows of the B matrix above are cointegrating vectors where the equilibrium errors (u_1 and u_2) are independent white noise in each case but the variation in the second error is 25 times that of the first error. Here one would expect the single equation estimators to home in on the vector with the lowest variance and for the multivariate estimators one would expect the 'first' estimated cointegrating vector also to be dominantly that vector with the lower variance. The results in Table 12 assume that the vector with lower variance is the 'true' vector being sought; they show that there was little difference amongst the single equation estimators but that the Johansen dominated them all in terms of median bias but faired very badly in terms of variation. Curiously the modified Box-Tiao did not do at all well being dominated by the single-equation estimators in terms of both bias and variation. But since actually both vectors were cointegrating, this means that if there are two cointegrating relationships between the same set of variables, all the estimators will tend to that vector with the lowest variance although this may not actually be the vector one is trying to estimate.

We then ran an experiment where the second cointegrating vector did not include the dependent variable used when estimating by single equation methods the first cointegrating vector, i.e. the B matrix was

$$B = \begin{bmatrix} 1 & -2/7 & 5/7 & -1/7 \\ 0 & 3/4 & 7/4 & 1/4 \\ 1 & 4/3 & -2/3 & 1/3 \\ 1 & 5 & 3 & 1 \end{bmatrix},$$

where only the first two rows are cointegrating vectors. We also ran this experiment with the regressors completely exogenous to the dependent variable, i.e. with

$$B = \begin{bmatrix} 1 & -2/7 & 5/7 & -1/7 \\ 0 & 0 & 7/4 & 1/4 \\ 1 & 4/3 & -2/3 & 1/3 \\ 1 & 5 & 3 & 1 \end{bmatrix}.$$

We present the results for median bias in Table 13(a) and (b). The difficulty with these situations is that all linear combinations of the two cointegrating vectors are stationary. However the interesting thing is that the second cointegrating vector has no explanatory power of the 'dependent variable' as we made the equilibrium errors uncorrelated and so the second cointegrating vector should have a zero coefficient in an equation explaining the dependent variable used for the single equation estimators. There is however some feedback in the first of these two simulations through the inclusion of the dependent variable in the two nonstationary trends (the last two rows of B).

The results show how the single equation methods using the normalisation on the first variable appear to home in on the true relationship much better than the multivariate methods. When one moves from Table 13(a) to 13(b), one notes that the bias for the FM estimator only halves for β_1 and β_3 but for OLS and Hendry-style there is a much greater improvement. In Table 13(b) the multivariate estimators are completely dominated by the single equation estimators (except EY3). However given different results over the three coefficients estimated one cannot say that one of the single equation estimators dominates overall except that the 3-step estimator does not fair well.

Table 13. Estimation of a cointegrating vector where the 'regressors' are cointegrated amongst themselves.

(a) Median biases with the regressors not fully exogenous

	OLS	Hendry	EY3	FM	Box-Tiao	Johansen
β_1	–0.0260	–0.0213	0.0513	–0.0052	0.0282	–0.0414
β_2	–0.0057	–0.0008	0.1559	–0.0048	0.0730	–0.0787
β_3	–0.0072	–0.0059	0.0181	–0.0010	0.0079	–0.0147

(b) Median biases with the regressors fully exogenous

	OLS	Hendry	EY3	FM	Box-Tiao	Johansen
β_1	–0.0028	–0.0006	0.0780	–0.0026	0.0302	–0.0409
β_2	–0.0083	–0.0065	0.1516	–0.0031	0.0716	–0.0760
β_3	–0.0011	–0.0011	0.0241	–0.0005	0.0090	–0.0138

With these simulations, the variation of the multivariate estimators was enormous compared to the single equation estimators which were very similar to each other. The problem is that the multivariate estimators are not estimating a relationship with x_1 as a dependent variable but rather they are estimating the cointegrating space. The researcher who does not realise that there is a second cointegrating vector will be completely misled by the multivariate estimators.

To test this further we ran a simulation where the B matrix was

$$B = \begin{bmatrix} 1 & -2/7 & 5/7 & -1/7 \\ 0 & 0 & 7/4 & 1/4 \\ 1 & 4/3 & -2/3 & 1/3 \\ 1 & 5 & 3 & 1 \end{bmatrix}.$$

Again only the first two row vectors were cointegrating vectors. The point of this simulation is that after normalisation on y, linear combinations of the second cointegrating vector can only change the coefficients of $\beta_2(= -5/7)$ and $\beta_3(= 1/7)$ but not $\beta_1(= 2/7)$ since the coefficient opposite β_1 is zero in the second cointegrating vector. Here one would expect all the estimators to be much better estimators of β_1 than the latter two coefficients.

The most striking thing here is that if one compares the results in Table 14 to those in Table 13(a), which is the more appropriate than 13(b), one finds that the OLS and Hendry-style estimators have hardly changed at all and yet we expected their estimates of β_1 to be markedly better while those of β_2 and β_3 to be unchanged. The bias for β_1 is less for all estimators but, while there may be a

Table 14. Median Biases with a subset of the regressors cointegrated amongst themselves

	OLS	Hendry	EY3	FM	Box-Tiao	Johansen
β_1	−0.0205	−0.0159	−0.0115	−0.0026	−0.0001	0.0033
β_2	−0.0057	−0.0008	0.1559	0.0044	0.0730	−0.0787
β_3	−0.0072	−0.0059	0.0181	−0.0011	0.0079	−0.0147

slight effect for OLS, it is hardly noticeable. For the FM estimator the bias is halved and the Box-Tiao has by far the best improvement; the biases on the other two coefficients are unchanged. This shows as expected how β_1 is now estimated much more accurately than both β_2 and β_3.

What conclusions can be drawn from these results? Firstly we need a way of looking at the whole cointegrating space to assess the accuracy of the multivariate estimators when the researcher has the correct cointegrating rank; this is analysed in Hargreaves (1993). However if the researcher only expects one cointegrating vector and there are more, he is liable to obtain completely misleading results. Equally if on the basis of statistical tests, one mistakenly infers that there is only one cointegrating vector when there are actually two, further analysis may be completely misled. Secondly we find that the single equation estimators are surprisingly robust to there being more than one cointegrating vector and if that vector has low variance relative to the other they will do quite well. Overall the FM estimator faired very well in the simulations in this section and responded appropriately to restrictions on the cointegrating space sometimes doing better than expected.

5. CONCLUSION

When cointegration was first introduced, there was a great hope that here was a new method that, by separating the non-stationary from the stationary space, would help to identify economic relationships which are so often more about the long run than the short. The idea has in many ways led to much confusion but if nothing else we have become aware of the distributional affects of nonstationarity. Since Gregory (1991) was fairly equivocal about different tests for cointegration, we decided to analyse the estimators' estimates which have to be created before going on to hypothesis testing.

For the Monte Carlo experiments, we analysed a system with much higher dimension than conventionally analysed on Monte Carlo experiments in order to look at a system much more like those often analysed economically and to look at the identification of the cointegrating space in higher dimensional systems. We suspect that many Monte Carlo experiments on small systems give a false apparent precision when they only look at how a single coefficient, say, is estimated. When one looks across a set of coefficients, one often finds the ranking of the estimators changes from one coefficient to another. Thus a larger system gives a broader picture.

When the model specified matched exactly the model used by the Johansen estimator, this estimator performed well with a sample size of 100. However it often had much higher variation due to its lack of finite order sample moments. When we reduced the sample size to only 50 observations, a much more common size for an economic data set, the Johansen estimator was dominated by the Box-Tiao estimator; the same happened when the signal-to-noise ratio was low. With a high signal-to-noise ratio these two estimators were roughly equally good but then the FM estimator also worked very well. As the 'equilibrium error' became more autocorrelated, the Box-Tiao dominated the Johansen estimator but both suffered a much greater increase in variation than the single equation estimators. The Box-Tiao suffered greatly however when the errors were correlated or followed a multivariate moving average process. The high variation of the Johansen estimator is often ascribed to its normalisation after estimation; however a problem with the single equation estimators is that the normalisation used may be incorrect. In all the DGP's analysed here, no one normalisation could be said to be correct.

The situation became quite different when one tried to estimate a single cointegrating vector when in fact there were two cointegrating vectors. If the vector one is estimating has a much lower variance than the other cointegrating vector then one will do fairly well; vice versa one will do very badly. Here it became a question of bias versus variation. The Johansen estimator had the best median bias but the worst variation; the single equation estimators performed very well with much lower variation; the Box-Tiao stood in between. However when the dependent variable did not feature in the second cointegrating relationship, the single equation estimators suddenly performed much better than the multivariate estimators with possibly the FM estimator doing best. There is clearly a problem with the Johansen estimator if an unknown second cointegrating vector is not known a priori or detected by the tests. Of course one should add however that, if the researcher does detect two cointegrating vectors by using one of the multivariate methods, then after suitable rotation (if known) he may estimate the vectors very accurately.

One procedure might be to use the Johansen estimator to determine the cointegrating rank and then estimate the cointegrating vectors with the OLS method proposed pulling out the number of vectors determined by the Johansen test. Given the OLS estimates are fairly robust, one could then use the system errors to see if one has highly autocorrelated errors, correlated errors and or moving average errors. One can then move on to the Fully-Modified estimates, see if there is much change and try to explain the change, ensuring that it is not due to some spurious correlations at long lags caused probably be errors in the data.

The Johansen estimator is best only if the model is well specified *without* highly autocorrelated cointegrating errors. If one is unsure of the cointegrating dimensionality and is only trying to estimate one cointegrating vector, then the FM estimator is best. When one moves up to a multidimensional cointegrating space, the 'benchmark' multidimensional OLS proposed faired surprisingly well comparatively, so much so as to be clearly preferred immediately there were any problems (see Hargreaves, 1993).

REFERENCES

AHN, S. K. and G. C. REINSEL (1988), 'Nested Reduced Rank Autoregressive Models for Multiple Time Series', *Journal of the American Statistical Association,* Vol. 83, pp. 849–56.

—— (1990), 'Estimation for Partially Nonstationary Multivariate Autoregressive Models', *Journal of the American Statistical Association,* Vol. 85, pp. 813–23.

ANDERSON, T. W. (1982), 'Some Recent Developments on the Distributions of Single Equation Estimators', in W. Hildenbrand (ed.), *Advances in Econometrics,* Cambridge, Cambridge University Press.

ANDREWS, D. W. K. and J. C. MONAHAN (1990), 'An Improved Heteroskedasticity and Autocorrelation Consistent Covariance Matrix Estimator', Cowles Discussion Paper No. 942.

BANERJEE, A., J. J. DOLADO, J. W. GALBRAITH and D. F. HENDRY (1993), *Co-integration, Error Correction, and the Econometric Analysis of Non-stationary Data,* Oxford, Oxford University Press.

BANERJEE, A., J. J. DOLADO, D. F. HENDRY and G. W. SMITH (1986), 'Exploring Equilibrium Relationships in Econometrics through Static models: Some Monte Carlo evidence', *Oxford Bulletin of Economics and Statistics,* 48, pp. 253–77.

BARDSEN, G. (1989), 'The Estimation of Long-Run Coefficients from Error-Correction Models', *Oxford Bulletin of Economics and Statistics.*

BEWLEY, R. A. (1979), 'The Direct Estimation of the Equilibrium Response in a Linear Model', *Economic Letters.*

BEWLEY, R. and D. ORDEN (1991), 'Alternative Methods for Estimating Long-Run Responses with Application to Australian Import Demand', UNSW Econ. Disc. Paper, July, No 91/6.

BEWLEY, R., D. ORDEN and L. FISHER (1991), 'Box-Tiao and Johansen Canonical Estimators of Cointegrating Vectors', UNSW School of Economics Dioscussion Paper, July, No 91/5.

BOSSAERTS, P. (1988), 'Common nonstationary components of asset prices', *Journal of Economic Dynamics and Control,* Vol. 12, pp. 347–64.

BOX, G. E. P. and G. C. TIAO (1977), 'A canonical analysis of multiple time series', *Biometrika,* Vol. 64, pp. 355–65.

ENGLE, R. F. and C. W. J. GRANGER (1987), 'Co-integration and Error Correction: Representation, Estimation, and Testing', *Econometrica*, Vol. 55, No. 2, pp. 251–76.

ENGLE, R. F., D. F. HENDRY and J.-F. RICHARD (1983), 'Exogeneity', *Econometrica,* Vol. 55, pp. 251–76.

ENGLE, R. F. and S. YOO (1991), 'Cointegrated Economic Time Series: an Overview with New Results', Ch. 12, pp. 237–66, in R. F. Engle and C. W. J. Granger (eds.), *Long-Run Economic Relationships; Readings in Cointegration,* Oxford, Oxford University Press.

GRANGER, C. W. J. (1981), 'Some properties of time series data and their use in econometric model specification', *Journal of Econometrics*, Vol. 16, pp.121–30.

GREGORY, A.W. (1991), 'Testing for Cointegration in Linear Quadratic Models', mimeo, Dept of Economics, Queen's University, Ontario.

GREGORY, A. W., A. R. PAGAN and G. W. SMITH (1990), 'Estimating Linear Quadratic Models with Integrated Processes', Rochester Center for Economic Research Working Paper No. 247, University of Rochester.

HANSEN, B. E. (1990), 'A Powerful Simple Test for Cointegration using Cochrane-Orcutt', Working Paper No. 260, University of Rochester.

—— (1991), 'Tests for Parameter Instability in Regression with I(1) Processes', mimeo, University of Rochester.

—— (1992), 'Consistent Covariance matrix Estimation for Dependent Heterogenous Processes', *Econometrica,* Vol. 60, No. 4, pp. 967–72.

HANSEN, B. E. and P. C. B. PHILLIPS (1990), 'Estimation and Inference in Models of Cointegration: a Simulation Study', pp. 225–48 in *Advances in Econometrics,* Vol. 8, JAI Press.

HARGREAVES, C. P. (ed.) (1992), *Macroeconomic Modelling of the Long Run*, Edward Elgar Publishing Ltd, Aldershot, UK.

—— (1993), 'Estimation of Multidimensional Cointegrating Spaces by OLS', mimeo, submitted for publication.

HENDRY, D. F. and J.-F. RICHARD (1982), 'On the formulation of empirical models in dynamic econometrics', *Journal of Econometrics,* Vol. 20, pp. 3–33.

—— (1983), 'The Econometric Analysis of Economic Time-Series (with discussion)', *International Statistical Review*, Vol. 51, pp. 111–63.

HILLIER, G. H. (1988), 'On the Interpretation of Exact Results for Structural Equation Estimators', Discussion Paper No. 6188, Monash University.

HSIAO, C. (1983), 'Identification', Ch. 4, pp. 223–83, in *Handbook of Econometrics*, Vol. 1, edited by Z. Griliches and M.D. Intriligator, North Holland, Amsterdam.

JOHANSEN, S. (1988), 'Statistical Analysis of Cointegrating Vectors', *Journal of Economic Dynamics and Control,* Vol. 12, pp. 231–54.

—— (1991), 'Estimation and Hypothesis Testing of Cointegration Vectors in Gaussian Vector Autoregressive Models', *Econometrica,* Vol. 59, pp. 1551–81.

—— (1992), 'An I(2) Cointegration Analysis of the Purchasing Power Parity between Australia and the United States', Chapter 9, pp. 229–28, in Hargreaves (1992).

JUSELIUS, K. and C. P. HARGREAVES (1992), 'Long-Run Relations in Australian Monetary Data', Chapter 10, pp. 249–86, in Hargreaves (1992).

NEWEY, W. K. and K. D. WEST (1987), 'A Simple, Positive Semi-definite, Heteroskedasticity and Autocorrelation Consistent Covariance Matrix', *Econometrica*, Vol. 55, No. 3, pp.703–8.

OSTERWALD-LENUM, M. (1992), 'A Note with Fractiles of the Asymptotic Distribution of the Likelihood Cointegration Rank Test statistics: Four Cases', *Oxford Bulletin of Economics and Statistics*.

PARK, J. Y. (1989), 'Canonical cointegrating regressions', *Econometrica,* Vol. 60, pp.119–43.

PARK, J. Y. and M. OGAKI (1991), 'Inference in Cointegrated Model using VAR Prewhitening to Estimate Short Run Dynamics', Rochester Center for Economic Research Working Paper No. 281.

PARK, J. Y., S. OULIARIS and B. CHOI (1988), 'Spurious Regressions and Tests for Cointegration', mimeo, Cornell University.

PARK, J. Y. and P. C. B. PHILLIPS (1988), 'Statistical Inference in Regressions with Integrated Processes: Part 1', *Econometric Theory*, Vol. 4, pp. 468–97.

—— (1989), 'Statistical Inference in Regressions with Integrated Processes: Part 2', *Econometric Theory*, Vol. 5, pp. 95–131.

PHILLIPS, P. C. B. (1983), 'Exact Small Sample Theory in the Simultaneous Equations Model', Ch. 8, pp. 449–516, in M. D. Intrilligator and Z. Griliches (eds.), *Handbook of Econometrics*, Amsterdam, North-Holland.

—— (1987), 'Time Series Regression with a Unit Root', *Econometrica*, Vol. 55, No. 2, pp. 277–301.

—— (1990), 'T.W. Anderson's Contributions to the Study of Structural Equation Estimation', pp. 1643–9, in G. Styan (ed.), *The Collected Works of T.W. Anderson,* NY: Wiley.

—— (1991*a*), 'Optimal Inference in Cointegrated Systems', *Econometrica,* Vol. 59, pp. 283–306.

—— (1991*b*), 'Spectral Regression for Cointegrated Time Series', in W. Barnett, J. Powell and G. Tauchen (eds.), *Nonparametric and Semiparametric Methods in Economics and Statistics,* NY: Cambridge University Press.

—— (1992), 'The Tail Behaviour of Maximum Likelihood Estimators of Cointegrating Coefficients in Error Correction Models', mimeo, Cowles Foundation for Research in Economics, Yale University.

PHILLIPS, P. C. B. and S. N. DURLAUF (1986), 'Multiple Time Series Regression with Integrated Processes', *Review of Economic Studies,* Vol. 53, pp. 473–95.

PHILLIPS, P. C. B. and B. E. HANSEN (1990), 'Statistical Inference in Instrumental Variables Regression with I(1) Processes', *Review of Economic Studies,* Vol. 57, pp. 99–125.

PHILLIPS, P. C. B. and M. LORETAN (1991), 'Estimating Long-Run Eonomic Equilibria', *Review of Economic Studies,* Vol. 59, pp. 407–36.

PHILLIPS, P. C. B. and S. OULIARIS (1988), 'Testing for Cointegration using Principal Components Methods', *Journal of Economic Dynamics and Control,* Vol. 12, pp. 105–30.

PHILLIPS, P. C. B. and J. Y. PARK (1992), 'Unidentified Components in Reduced Rank Regression Estimation of ECM's', mimeo, Cowles Foundation for Research in Economics, Yale University.

PHILLIPS, P. C. B. and P. PERRON (1988), 'Testing for a Unit Root in Time Series Regression', *Biometrika,* Vol. 75, pp. 335–46.

SAID, S. E. and D. A. DICKEY (1984), 'Testing for Unit Roots in Autoregressive Moving Average Models of Unknown Orders', *Biometrika,* Vol. 71, pp. 599–607.

SAIKONNEN, P. (1991), 'Asymptotically Efficient Estimation of Cointegration Regressions', *Econometric Theory,* Vol. 7, pp. 1–21.

STOCK, J. H. (1987), 'Asymptotic properties of least squares estimators of cointegrating vectors', *Econometrica,* Vol. 55, pp. 1035–56.

STOCK, J. H. and M. WATSON (1989), 'A simple MLE of cointegrating vectors in higher order integrated systems', Technical Working Paper No. 83, National Bureau of Economic Research.

WICKENS, M. R. and T. S. BRESUCH (1988), 'Dynamic Specification, the Long Run and the Estimation of Transformed Regression Models', *Economic Journal,* Vol. 98 (Conference 1988), pp. 189–205.

5

A test of the null hypothesis of cointegration

David Harris and Brett Inder

The cointegration tests of Engle and Granger (1987) test the null hypothesis of no cointegration. We extend the unit root testing framework of Kwiatkowski *et al.* (1992) to testing the null hypothesis of cointegration. A test is developed which is asymptotically equivalent to the locally best invariant (LBI) test and is applicable to a wide range of nuisance parameters, and is dependent only on the number of regressors in the cointegrating regression. We tabulate asymptotic critical values for the test based on this distribution and report on a small power comparison with the Dickey-Fuller test.

1. INTRODUCTION

In the literature on cointegrated time series, hypothesis tests developed for testing for cointegration have a null hypothesis of no cointegration. Such tests can be found in Engle and Granger (1987), Phillips and Ouliaris (1990) and Johansen (1988). However, there would seem to be some merit in constructing a test of the null hypothesis of cointegration. In fact, Engle is quoted in Phillips and Ouliaris (1990) as writing 'a null hypothesis of cointegration would be far more useful in empirical research than the natural null of non-cointegration'. Phillips and Ouliaris proceed to suggest two tests of this hypothesis, but then show that these tests are inconsistent. They leave the problem unsolved.

The merit in testing the null hypothesis of cointegration can be seen if we were building a model where the variables were believed, *a priori*, to be cointegrated. The classic example of aggregate consumption and income could be one such case. If we were to use the standard Dickey-Fuller tests of the null hypothesis of cointegration, then we are implicitly saying that we believe the variables are not cointegrated unless the data can convincingly demonstrate otherwise. Instead, we suggest a test of the null hypothesis of cointegration, so that we will believe the variables to be cointegrated unless the data can strongly convince us otherwise. Such an approach may prove to be useful.

Not all applications of cointegration tests may involve some *a priori* beliefs on the presence or absence of cointegration. In such cases, one procedure could make use of tests of both null hypotheses. If we accept no cointegration and reject cointegration, then this is strong evidence for no cointegration. Similarly, if we reject no cointegration and accept cointegration then there is strong evidence for cointegration. There is also the possibility of two inconclusive results. If both null hypotheses are rejected, then a type I error may have occurred in one of the tests! If both null hypotheses are accepted, then at least one of the tests is the victim of a lack of power against the particular data generating process.

This paper can be considered a generalisation of Kwiatkowski *et al.* (1992) (referred to as KPSS), who proposed a unit root test with the null hypothesis of stationarity. We extend their model to include explanatory variables, so that we can estimate a cointegrating regression and test the null hypothesis of stationarity of the residuals. On the issue of model specification, we find that the KPSS model is not a completely general data generating process, and we suggest a more general model. However, in the context of the hypothesis testing problem, this difference is found to be unimportant. Locally best invariant (LBI) tests are constructed for quite restrictive models, but we also construct an asymptotically equivalent test which is applicable to very general data generating processes.

In the process of deriving a test that is asymptotically equivalent to the LBI test, we find that the asymptotic distribution of the test is a function of independent standard Brownian motions, and is not dependent on any nuisance parameters. The only variable in the distribution is the number of explanatory variables in the cointegrating regression. Hence we can construct tables of percentage points of the distribution under different numbers of regressors for use as critical values.

In Section 2 we discuss two models of nonstationary variables, and in Section 3 we derive tests of the null hypothesis of cointegration for these models, which turn out to be identical. To make the tests of Section 3 more generally applicable, an asymptotically equivalent test is constructed in Section 4, and consistency is proved in Section 5. Methods for finding critical values for the test are discussed

---

[1] Another possibility is that the potential error correction term is fractionally integrated but this is not explored in this paper.

in Section 6, and Section 7 provides a brief summary of the procedure for our test. Section 8 presents a simple Monte Carlo comparison of the new test with the Dickey-Fuller test for cointegration. An application of our test to the Fisher effect in Australia, following Inder and Silvapulle (1992), is given for illustrative purposes in Section 9.

2. THE MODELS

In this section we suggest two models from which tests for cointegration can be derived. Both are basically extensions of the idea of the cointegrating regression as introduced by Engle and Granger (1987) to allow for the possibility of nonstationarity in the error term.

2.1 Model 1: Independent random walk

Model 1 is a simple extension of the model used by KPSS for testing for unit roots, and is specified as

$$y_t = x_t'\beta_0 + \mu_t + u_t \tag{1}$$

$$x_t = x_{t-1} + v_t \tag{2}$$

$$\mu_t = \mu_{t-1} + w_t, \tag{3}$$

where y_t is the dependent variable, x_t is a vector of k nonstationary explanatory variables and μ_t is a random walk in the error term of the cointegrating regression (1). If we define $\xi_t = [u_t, v_t', w_t]'$ to be a k+2 dimensional process, then we assume that ξ_t is serially independent and $\xi_t \sim \mathrm{N}(0, \Sigma)$ where

$$\Sigma = \begin{bmatrix} \sigma_1^2 & 0 & 0 \\ 0 & \Sigma_{22} & 0 \\ 0 & 0 & \sigma_3^2 \end{bmatrix}.$$

With this set of assumptions on ξ_t, we can regard x_t as a set of exogenous variables. Also the random walk component of the error term is independent of the white noise component. If model 1 is the true data generating process then y_t and x_t are not cointegrated, due to the presence of the random walk μ_t in equation (1). However, if we could impose the restriction $\sigma_3^2 = 0$ then μ_t would collapse to μ_0 for all t. That is, equation (1) would be a cointegrating regression with a constant and a stationary error term. Hence, testing the null hypothesis of $\sigma_3^2 = 0$ against the alternative $\sigma_3^2 > 0$ will test the null hypothesis of cointegration against the alternative of no cointegration.

Repeated back substitution for μ_{t-1} in equation (3) leads to the following representation for (1):

$$y_t = \mu_0 + x_t'\beta_0 + \sum_{k=1}^{t} w_k + u_t. \tag{4}$$

If we defined $z_t = \sum_{k=1}^{t} w_k + u_t$, then $\mathrm{E}(z_t) = 0$ and

$$\begin{aligned} \mathrm{E}(z_i z_j) &= \min(i,j)\sigma_3^2 + \sigma_1^2 \quad i = j \\ &= \min(i,j)\sigma_3^2 \qquad\quad i \neq j. \end{aligned}$$

Now equation (1) can be expressed as a linear regression in matrix form,

$$y = \mathbf{X}\beta + z, \quad z \sim \mathrm{N}\left[0, \sigma_1^2 \mathbf{Q}_1(\lambda)\right], \tag{5}$$

where $\lambda = \sigma_3^2/\sigma_1^2$, $\mathbf{Q}_1(\lambda) = \mathbf{A}\lambda + \mathbf{I}$, and $\mathbf{A}$ is the matrix with ij^{th} element $\min(i,j)$. The null hypothesis of $\sigma_3^2 = 0$ is equivalent to testing $\lambda = 0$ in equation (5) against $\lambda > 0$. If $\lambda = 0$ then $z_t = u_t$ and is stationary, whereas if $\lambda > 0$ then z_t is nonstationary. The advantage of formulating the model in this way is that this is the exact hypothesis testing problem considered by King and Hillier (1985) for constructing LBI tests. This is considered further in Section 3.

One possible objection to the specification of model 1 arises from the independence of u_t and w_t. This places a restriction on the data generating process for z_t when $\sigma_3^2 > 0$, which can be seen by writing down the first difference of z_t as $\Delta z_t = w_t + u_t - u_{t-1}$. This imposes a negative first order autocorrelation on the moving average process generating Δz_t, since $\mathrm{E}(\Delta z_t \Delta z_{t-1}) = -\sigma_1^2$. This restriction is expressed in terms of the the spectrum of the process by Watson (1986). In this sense, model 1 is not a completely general data generating process. In the next section, we suggest a model which does not impose this restriction, although there is no difference between the models under the null hypothesis.

2.2 Model 2: Perfectly correlated random walk

In model 1, the random walk and stationary components of the error term in equation (1) were driven by independent processes. Following the idea of Snyder (1985), we let the processes driving the two components be perfectly correlated. As we shall see, this allows for an unrestricted data generating process. Model 2 is

$$y_t = x_t'\beta_0 + \mu_t + u_t \tag{6}$$

$$x_t = x_{t-1} + v_t \tag{7}$$

$$\mu_t = \mu_{t-1} + \theta u_t. \tag{8}$$

If $\zeta_t = [u_t, v_t']'$ then we assume that ζ_t is a serially independent process with distribution $N(0, \Sigma)$ where

$$\Sigma = \begin{bmatrix} \sigma_1^2 & 0 \\ 0 & \Sigma_{22} \end{bmatrix}.$$

Under these assumptions, x_t is a set of exogenous explanatory variables. At first glance, imposing perfect correlation between the processes driving the random walk and stationary components of the error of (6) may seem restrictive, but as discussed below, this is not the case. The coefficient θ is a scale parameter which reflects the relative size of the effects of a shock on the random walk and stationary components. We do not restrict θ to be positive, so that the direction of the effect of a shock may be different for the random walk and stationary components. Clearly, if $\theta = 0$ then the random walk μ_t collapses to a constant μ_0 and it follows that y_t and x_t are cointegrated. Thus we are interested in the two sided testing problem of $\theta = 0$ against $\theta \neq 0$.

Back substituting in equation (8) leads to an equivalent representation for model 2 as given in equation (4) for model 1:

$$y_t = \mu_0 + x_t'\beta_0 + \theta \sum_{k=1}^{t} u_k + u_t. \tag{9}$$

We can define the compound error term $z_t = \theta\left(\sum_{k=1}^{t} u_k\right) + u_t$, and we note that $E(z_t) = 0$ and

$$\begin{aligned} E(z_i z_j) &= \min(i,j)\theta^2\sigma_1^2 + 2\theta\sigma_1^2 + \sigma_1^2 & i = j \\ &= \min(i,j)\theta^2\sigma_1^2 + \theta\sigma_1^2 & i \neq j. \end{aligned}$$

Equation (6) can now be expressed in matrix form as follows

$$y = \mathbf{X}\beta + z, \quad z \sim N\left[0, \sigma_1^2 \mathbf{Q}_2(\theta)\right], \tag{10}$$

Where $\mathbf{Q}_2(\theta) = \mathbf{A}\theta^2 + (\mathbf{J} + \mathbf{I})\theta + \mathbf{I}$ and $\mathbf{J}$ is a $T \times T$ matrix with ij^{th} element equal to 1. Thus we have expressed the testing problem in the same form as for model 1 in equation (5), and the theory of locally best unbiased invariant (LBUI) tests in King and Hillier (1985) is applicable.

We showed in Section 2.1 that the data generating process for Δz_t was constrained to have a negative first order autocorrelation. For model 2, we find $\Delta z_t = (1+\theta)u_t - u_{t-1}$ and $E(\Delta z_t \Delta z_{t-1}) = -(1+\theta)\sigma_1^2$. Thus the data generating process for Δz_t is positively autocorrelated if $\theta < -1$ and negatively correlated for $\theta > -1$. Watson (1986) showed that this model is completely general (in terms of the Wold representation), in constrast to model 1. More discussion on this point can be found in Snyder (1985) and Watson (1986).

3. TESTS FOR COINTEGRATION

3.1 LBI test for model 1

We want to test $\lambda = 0$ against $\lambda > 0$ in equation (5) with $\mathbf{Q}_1(\lambda) = \mathbf{A}\lambda + \mathbf{I}$. Since $\mathbf{Q}_1(0) = \mathbf{I}$, we can directly apply the results of King and Hillier (1985) for the form of the LBI test. The form of the LBI test in the general case is given by

$$s_1 = \frac{\hat{z}'\mathbf{A}_0\hat{z}}{\hat{z}'\hat{z}} < c_0,$$

where $\mathbf{A}_0 = -\partial\mathbf{Q}_1(\lambda)/\partial\lambda|_{\lambda=0}$ and $\hat{z}$ are the residuals from the estimation of equation (5) by OLS. Evaluating the partial derivative gives the following LBI test

$$s_1 = \frac{\hat{z}'\mathbf{A}\hat{z}}{\hat{z}'\hat{z}} > c_0, \tag{11}$$

which KPSS show is equivalent to

$$s_1 = \frac{\sum_{t=1}^{T} K_t^2}{\hat{\sigma}_z^2} \tag{12}$$

up to a power of T, where $K_t = (\sum_{i=1}^{t} \hat{z}_i)$ and $\hat{\sigma}_z^2 = T^{-1}\hat{z}'\hat{z}$ is the estimator of the error variance. The required normalisation by a power of T is explored further in Section 4.

3.2 LBUI test for model 2

We consider the test for cointegration in model 2 in more detail. The equation of interest is (10) with $\mathbf{Q}_2(\theta) = \mathbf{A}\theta^2 + (\mathbf{J} + \mathbf{I})\theta + \mathbf{I}$ and since $\mathbf{Q}_2(0) = \mathbf{I}$ it is straightforward to follow King and Hillier (1985) to derive locally best tests. The maximal invariant for this hypothesis testing problem is defined to be

$$w = \frac{\mathbf{P}\hat{z}}{\hat{z}'\hat{z}}$$

where $\mathbf{M} = \mathbf{I} - \mathbf{X}(\mathbf{X}'\mathbf{X})^{-1}\mathbf{X}'$, $\hat{z} = \mathbf{M}y$, $\mathbf{P}'\mathbf{P} = \mathbf{M}$ and $\mathbf{P}\mathbf{P}' = \mathbf{I}_m$, and $m = T - k$. The probability density function of the maximal invariant is given by

$$f(w|\theta) = \frac{1}{2}\Gamma\left(\frac{m}{2}\right)\pi^{-m/2}\left|\mathbf{P}\mathbf{Q}_2(\theta)\mathbf{P}'\right|^{-1/2}\left(w'(\mathbf{P}\mathbf{Q}_2(\theta)\mathbf{P}')^{-1}w\right)^{-m/2} \tag{13}$$

and the log of the likelihood ratio test for $\theta = 0$ against $\theta = \delta$ is

$$\log r(\theta)|_{\theta=0} = \log f(w|\theta+\delta)|_{\theta=0} - \log f(w|\theta)|_{\theta=0} > c_1 \tag{14}$$

for some value of δ. At this stage, we are following the steps to construct a one sided test, and direction of the alternative depends on the sign of δ. In what follows, we will show that the critical region for (14) depends on δ^2 (and not δ as in a standard LBI test), and hence that the critical region is unaffected by the choice of alternative hypothesis. Expanding $\log f(w|\theta+\delta)$ in a Taylor series about $\delta=0$ gives

$$\log f(w|\theta+\delta)=\log f(w|\theta)+\delta\frac{\partial\log f(w|\theta)}{\partial\theta}+\frac{\delta^2}{2}\frac{\partial^2\log f(w|\theta)}{\partial\theta^2}+\mathrm{O}\left(\delta^3\right).$$

Substituting this expansion into equation (14) gives the following test:

$$\log r(\theta)\Big|_{\theta=0}=\delta\frac{\partial\log f(w|\theta)}{\partial\theta}\bigg|_{\theta=0}+\frac{\delta^2}{2}\frac{\partial^2\log f(w|\theta)}{\partial\theta^2}\bigg|_{\theta=0}+\mathrm{O}\left(\delta^3\right)>c_1. \quad (15)$$

Evaluating the first derivative gives

$$\frac{\partial\log f(w|\theta)}{\partial\theta}\bigg|_{\theta=0}=\text{constant }+\frac{m}{2}\frac{\hat{z}'(\mathbf{J}+\mathbf{I})\hat{z}}{\hat{z}'\hat{z}}=\text{constant }+\frac{m}{2},$$

where the constant term is constant with respect to $\hat{z}$. Clearly the first derivative is constant with respect to $\hat{z}$ and can be transferred to the right hand side of (15) and combined with c_1. In this way, the dependence of the test on δ disappears, and we are left only with δ^2. Evaluation of the second derivative gives

$$\frac{\partial^2\log f(w|\theta)}{\partial\theta^2}\bigg|_{\theta=0}=\text{constant }+\frac{m}{2}\frac{\hat{z}'\mathbf{A}\hat{z}}{\hat{z}'\hat{z}}.$$

Now dividing (15) (without the first term) by δ^2, and letting δ approach zero to maximise power local to the null hypothesis, we obtain the following test:

$$s_2=\frac{\hat{z}'\mathbf{A}\hat{z}}{\hat{z}'\hat{z}}>c_2. \quad (16)$$

Thus the test statistic is the same for models 1 and 2.

4. RELAXING SOME ASSUMPTIONS

The tests developed in Section 3 are finite sample tests, in the sense that the optimal local power properties hold for finite samples. However, the assumptions of the models from which the tests are derived are quite restrictive. In particular, imposing exogeneity on the explanatory variables and serial independence on the error process of the cointegrating regression is quite unrealistic in practice. In this section, we relax these assumptions and derive a transformed test statistic that has the same asymptotic distribution as the locally best test statistic. Hence we greatly improve the generality of the model, but at the expense of relying on asymptotic theory.

4.1 A generalised model

Since the tests for models 1 and 2 are identical, we concentrate on model 2. Using $\zeta_t = [u_t, v_t']'$ and defining $S_t = [S_{ut}, S_{vt}']' = \sum_{i=1}^{t} \zeta_i$, we assume that ζ_t satisfies the multivariate invariance principle as set out in Theorem 2.1 of Phillips and Durlauf (1986), so that

$$T^{-1/2} S_{[Tr]} \Rightarrow \mathrm{B}(r) \text{ as } T \to \infty$$

where B(r) is a k+1 dimensional Brownian motion defined on $r \in [0,1]$ with covariance matrix Ω. The Brownian motion can be partitioned conformably with ζ_t as $\mathrm{B}(r) = [B_1(r), \mathrm{B}_2(r)']'$, and similarly the covariance matrix is

$$\Omega = \lim_{T\to\infty} T^{-1}\mathrm{E}(S_T S_T') = \begin{bmatrix} \omega_1^2 & \omega_{12} \\ \omega_{21} & \Omega_{22} \end{bmatrix}.$$

We can regard Ω as the long-run covariance matrix of ζ_t in the sense that it captures both contemporaneous variances and covariances as well as covariances at all lags. It can be expressed as a sum of the contemporaneous and lagged components by writing $\Omega = \Sigma + \Lambda + \Lambda'$ where

$$\Sigma = \mathrm{E}(\zeta_t \zeta_t') = \begin{bmatrix} \sigma_1^2 & \sigma_{12} \\ \sigma_{21} & \Sigma_{22} \end{bmatrix}$$

and

$$\Lambda = \sum_{j=1}^{\infty} \mathrm{E}(\zeta_t \zeta_{t-j}') = \begin{bmatrix} \lambda_{11} & \lambda_{12} \\ \lambda_{21} & \Lambda_{22} \end{bmatrix}.$$

The conditions imposed by Phillips and Durlauf (1986) on ζ_t allow for some nonstationarity, but in defining the preceding covariance matrices we have added the assumption of stationarity. This is necessary so that we can interpret u_t as a stationary error term in the cointegrating regression. These conditions on ζ_t generalise model 2 considerably, since x_t is now endogenous, and the error terms u_t and v_t may be any of a wide range of stationary processes, including all stationary ARMA processes. The LBUI test from Section 2.2 is no longer exactly LBUI for this generalised model, but we will show that the only difference to its asymptotic distribution is the introduction of correlation between $B_1(r)$ and $\mathrm{B}_2(r)$, which can be removed with a nonparametric correction due to Phillips and Hansen (1990). This is described in Section 4.3.

4.2 Asymptotic distribution of the test

In this section, we derive the asymptotic distribution of the test given by (16) when applied to model 2 (that is, not including the generalisations described above). Since u_t and v_t are mutually independent i.i.d. processes in model 2, ζ_t is a special case of the general class of processes described above. Thus the convergence of $T^{-1/2}S_{[Tr]}$ to a Brownian motion is assured. The lagged covariance matrix Λ is zero, and the covariance matrices Σ and Ω are block diagonal and equal, so the two components of the Brownian motion B(r) are uncorrelated. To reflect this in the notation, we write the partition of B(r) as $\mathrm{B}(r) = [B_{1.2}(r), \mathrm{B}_2(r)]$, so $\mathrm{B}_{1.2}(r)$ is independent of $\mathrm{B}_2(r)$.

The asymptotic distribution is found by considering the form of the test statistic given by equation (12). Under the null hypothesis, we know that $z_t = u_t$ and so

$$\hat{z}_t = u_i - [1 \quad x_i'](\mathbf{X}'\mathbf{X})^{-1}\mathbf{X}'u, \tag{17}$$

where

$$\mathbf{X}'\mathbf{X} = \begin{bmatrix} T & \sum_{t=1}^{T} x_t' \\ \sum_{t=1}^{T} x_t & \sum_{t=1}^{T} x_t x_t' \end{bmatrix} = \begin{bmatrix} \mathbf{E} & \mathbf{F} \\ \mathbf{G} & \mathbf{H} \end{bmatrix}$$

and

$$\mathbf{X}'u = \begin{bmatrix} \sum_{t=1}^{T} u_t \\ \sum_{t=1}^{T} x_t u_t \end{bmatrix} = \begin{bmatrix} \mathbf{B} \\ \mathbf{C} \end{bmatrix}.$$

Then

$$(\mathbf{X}'\mathbf{X})^{-1} = \begin{bmatrix} \mathbf{D}^{-1} & -\mathbf{D}^{-1}\mathbf{F}\mathbf{H}^{-1} \\ -\mathbf{H}^{-1}\mathbf{G}\mathbf{D}^{-1} & \mathbf{H}^{-1} + \mathbf{H}^{-1}\mathbf{G}\mathbf{D}^{-1}\mathbf{F}\mathbf{H}^{-1} \end{bmatrix}$$

Where $\mathbf{D} = \mathbf{E} - \mathbf{F}\mathbf{H}^{-1}\mathbf{G}$. This leads to the following expression for $K_t = \sum_{i=1}^{t} \hat{z}_i$:

$$K_t = S_{ut} - \mathbf{D}^{-1}(\mathbf{B} - \mathbf{F}\mathbf{H}^{-1}\mathbf{C})t - \left(\sum_{i=1}^{t} x_i'\right)\left[\mathbf{H}^{-1}\mathbf{C} - \mathbf{H}^{-1}\mathbf{G}\mathbf{D}^{-1}(\mathbf{B} - \mathbf{F}\mathbf{H}^{-1}\mathbf{C})\right] \tag{18}$$

by summing (17). We can now consider the behaviour of each of these terms as $T \to \infty$.

The notation will be simplified in the following results by assuming the integrals are taken from 0 to 1 and with respect to r unless stated otherwise, and by omitting the argument r from the Brownian motions. Lemma 2.1 of Park and Phillips (1988) provides a number of useful intermediate results.

$$T^{-3/2}\mathbf{F} \Rightarrow \int B_2'; \quad T^{-3/2}\mathbf{G} \Rightarrow \int B_2; \quad T^{-2}\mathbf{H} \Rightarrow \int B_2 B_2';$$

$$T^{-1/2}\mathbf{B} \Rightarrow B_{1.2}(1); \quad T^{-1}\mathbf{C} \Rightarrow \int B_2 dB_{1.2}$$

We will also make use of the following two results:

$$T^{-1}t \Rightarrow r; \quad T^{-3/2}\sum_{i=1}^{t} x_i' \Rightarrow \int_0^r B_2(s)'ds.$$

Now we can obtain the limiting behaviour of each of the terms in equation (18).

$$T^{-1/2}S_{ut} \Rightarrow B_{1.2}(r)$$

$$T^{-1}\mathbf{D} \Rightarrow 1 - \int B_2'\left(\int B_2 B_2'\right)^{-1}\int B_2$$

$$T^{-1/2}\left(\mathbf{B} - \mathbf{F}\mathbf{H}^{-1}\mathbf{C}\right) \Rightarrow B_{1.2}(1)' - \int B_2'\left(\int B_2 B_2'\right)^{-1}\int B_2 dB_{1.2}$$

$$T^{-1/2}\mathbf{H}^{-1}\mathbf{G} \Rightarrow \left(\int B_2 B_2'\right)^{-1}\int B_2$$

$$T\mathbf{H}^{-1}\mathbf{C} \Rightarrow \left(\int B_2 B_2'\right)^{-1}\int B_2 dB_{1.2}$$

These limits can be combined to obtain the limiting distribution of K_t. However, these limits contain implicit nuisance parameters in the form of the variances of the Brownian motions $B_{1.2}$ and B_2 given by $\omega_{1.2}^2$ and Ω_{22} respectively. To show this, we define $W(r) = \Omega^{-1/2}B(r)$ to be standard Brownian motion, since its covariance matrix is the identity. Since Ω is block diagonal for this case, we can also define the elements of the partitioned standard Brownian motion to be $W_{1.2}(r) = \omega_1^{-1}B_{1.2}(r)$ and $W_2(r) = \Omega_{22}^{-1/2}B_2(r)$.

For convenience, we define the following functions of standard Brownian motions:

$$P = 1 - \int W_2'\left(\int W_2 W_2'\right)^{-1}\int W_2$$

$$\mathbf{Q} = W_{1.2}(1)' - \int W_2'\left(\int W_2 W_2'\right)^{-1}\int W_2 dW_{1.2}$$

$$\mathbf{R} = \left(\int W_2 W_2'\right)^{-1}\int W_2$$

$$\mathbf{S} = \left(\int W_2 W_2'\right)^{-1}\int W_2 dW_{1.2}$$

Combining the limits from the previous paragraph, and replacing Brownian motions by standard Brownian motions gives the following limit for K_t:

$$T^{-1/2}K_t \Rightarrow \omega_1 V(r), \tag{19}$$

where

$$V(r) = W_{1.2}(r) - P^{-1}\mathbf{Q}r - \left(\int_0^r \mathrm{W}_2(s)'ds\right)\left(\mathbf{S} - \mathbf{R}P^{-1}\mathbf{Q}\right).$$

Now we can evaluate the limit of the numerator of (18) as follows:

$$\begin{aligned} T^{-2}\sum_{t=1}^{T} K_t^2 &= T^{-1}\int_0^1 K_{[Tr]}^2 dr \\ &\Rightarrow \omega_1^2 \int_0^1 V(r)^2\, dr. \end{aligned} \quad (20)$$

The bottom line of (18) is the estimate of the error variance

$$\hat{\sigma}_z^2 = T^{-1}\sum_{t=1}^{T} \hat{z}_t^2,$$

which will be a consistent estimator of σ_1^2 since $z_t = u_t$ under the null hypothesis. Also $\omega_1^2 = \sigma_1^2$ since u_t is i.i.d., so we can say

$$\hat{\sigma}_z^2 \Rightarrow \omega_1^2. \quad (21)$$

Now the asymptotic distribution of the test statistic can be found by the ratio of (20) and (21).

$$s_2 = \frac{T^{-2}\sum_{t=1}^{T} K_t^2}{\hat{\sigma}_z^2} \Rightarrow \int V(r)^2\, dr. \quad (22)$$

4.3 Correction for endogenous regressors in the generalised model

When applied to the generalised model described in Section 4.1, the LBUI test no longer has the same finite sample optimal power properties. In this section, we consider the asymptotic behaviour of the test under the more general assumptions that u_t and v_t are correlated stationary processes.

Following the derivation in Section 4.2, equation (18) is unchanged as the expression for K_t. Since the elements of the Brownian motion $\mathrm{B}(r)=[B_1(r),\mathrm{B}_2(r)']'$ are no longer independent, we have new limits for **B** and **C**.

$$T^{-1/2}\mathbf{B} \Rightarrow B_1(1); \quad T^{-1}\mathbf{C} \Rightarrow \int \mathrm{B}_2 dB_1 + \delta_{21},$$

where $\Delta = \Sigma + \Lambda$ is a matrix partitioned conformably with $\mathbf{\Omega}$, Σ, Δ. Thus we have two differences in the limits introduced by the generalised model — correlation between the Brownian motions and a related nuisance parameter δ_{21}. To remove these effects, we make use of the method of fully-modified OLS suggested by Phillips and Hansen (1990). Notice that both of the problems are due to the endogeneity of x_t, not to the serial correlation in u_t and v_t. The effect of the serial correlation will be discussed later in the section.

To deal with the endogenous regressors, we consider the modified error process $u_t^+ = u_t - \omega_{12}\Omega_{22}^{-1}v_t$. If $\zeta_t^+ = [u_t^+, v_t]$ and $S_t^+ = \sum_{i=1}^{t}\zeta_t^+$ then the long-run covariance matrix of ζ_t^+ is

$$\Omega^+ = \lim_{T\to\infty} T^{-1}\mathrm{E}\left(S_T^+ S_T^{+'}\right) = \begin{bmatrix} \omega_{1.2}^2 & 0 \\ 0 & \Omega_{22} \end{bmatrix},$$

where $\omega_{1.2}^2 = \omega_1^2 - \omega_{12}\Omega_{22}^{-1}\omega_{21}$. Now $T^{-1/2}S_{[Tr]} \Rightarrow \mathrm{B}(r)$ where $\mathrm{B}(r)=[B_{1.2}(r),\mathrm{B}_2(r)']'$. We use $B_{1.2}(r)$ since the covariance matrix of the Brownian motion $\mathrm{B}(r)$ is block diagonal. Also, the following limits containing u_t^+ hold.

$$T^{-1/2}\sum_{t=1}^{T} u_t^+ \Rightarrow B_{1.2}(1)$$

and

$$T^{-1}\sum_{t=1}^{T} x_t u_t^+ \Rightarrow \int \mathrm{B}_2 dB_{1.2} + \delta_{21}^+,$$

where $\delta_{21}^+ = \delta_{21} - \Delta_{22}\Omega_{22}^{-1}\omega_{21}$.

Even with endogenous regressors, OLS estimation of the cointegrating regression (under the null hypothesis) is consistent. Taking the estimated residuals, $\hat{z}_t$ from this regression, we can form $\zeta_t = [\hat{z}_t, \Delta x_t']'$ and obtain consistent estimates of Ω and Δ by

$$\hat{\Omega} = T^{-1}\sum_{t=1}^{T}\hat{\zeta}_t\hat{\zeta}_t' + T^{-1}\sum_{k=1}^{l} w(k,l)\sum_{t=k+1}^{T}\left(\hat{\zeta}_{t-k}\hat{\zeta}' + \hat{\zeta}_t\hat{\zeta}_{t-k}'\right)$$

$$\hat{\Delta} = T^{-1}\sum_{k=0}^{l}\sum_{t=k+1}^{T}\hat{\zeta}_t\hat{\zeta}_{t-k}',$$

where $w(k,l)$ are a set of weights to ensure the positive-definiteness of the estimated matrix (see Newey and West (1987) for example), and the truncation parameter l increases at a slower rate than T (say $o(T^{1/4})$). Then if we define the transformed variable $y_t^+ = y_t - \hat{\omega}_{12}\hat{\Omega}_{22}^{-1}\Delta x_t$, we can write the cointegrating regression as

$$y_t^+ = \mu_0 + x_t'\beta_0 + \left(u_t - \hat{\omega}_{12}\hat{\Omega}_{22}^{-1}v_t\right), \tag{23}$$

and the error term in this regression is asymptotically equivalent to u_t^+. The Phillips-Hansen fully-modified OLS estimator is then defined to be

$$\hat{\beta}^+ = (\mathrm{X}'\mathrm{X})^{-1}\left(\mathrm{X}'y^+ - e_k T\hat{\delta}_{21}^+\right), \tag{24}$$

where $e_k = [0, \mathrm{I}_k]'$ and $\hat{\delta}_{21}^{+} = \hat{\delta}_{21} - \hat{\Delta}_{22}\hat{\Omega}_{22}^{-1}\hat{\omega}_{21}$. The extra term involving $\hat{\delta}_{21}^{+}$ is included to remove the nuisance parameter δ_{21}^{+} from the asymptotic distribution of $\hat{\beta}^{+}$. This correction will also remove the nuisance parameter from the asymptotic distribution of our test statistic.

We can now derive the asymptotic distribution of the test in equation (12) using the residuals obtained from the fully modified OLS regression. The estimated residuals will be given by

$$\hat{u}^{+} = u^{+} - \mathbf{X}(\mathbf{X}'\mathbf{X})^{-1}\left(\mathbf{X}'u^{+} - e_k T\hat{\delta}_{21}^{+}\right), \tag{25}$$

where

$$\mathbf{X}'u^{+} - e_k T\hat{\delta}_{21}^{+}\begin{bmatrix} \sum_{t=1}^{T} u_t^{+} \\ \sum_{t=1}^{T} x_t u_t^{+} - T\hat{\delta}_{21}^{+} \end{bmatrix} = \begin{bmatrix} \mathbf{B}^{+} \\ \mathbf{C}^{+} \end{bmatrix}.$$

The limits of these terms are easily found and they are not dependent of the nuisance parameter δ_{21}^{+} and involve only independent Brownian motions.

$$T^{-1/2}\mathbf{B}^{+} \Rightarrow B_{1.2}(1)$$

$$T^{-1}\mathbf{C}^{+} = T^{-1}\sum_{t=1}^{T} x_t u_t - \hat{\delta}_{21}^{+} \Rightarrow \int \mathrm{B}_2 dB_{1.2}$$

Notice that these terms have the same limiting distributions as **B** and **C** for model 2 (see Section 4.2). Now we can define $K_t^{+} = \sum_{i=1}^{t}\hat{u}_t^{+}$, and following the steps leading to equation (18) we obtain

$$K_t^{+} = S_{ut}^{+} - \mathbf{D}^{-1}\left(\mathbf{B}^{+} - \mathbf{F}\mathbf{H}^{-1}\mathbf{C}^{+}\right)t - \left(\sum_{i=1}^{t} x_i'\right)\left[\mathbf{H}^{-1}\mathbf{C}^{+} - \mathbf{H}^{-1}\mathbf{G}\mathbf{D}^{-1}\left(\mathbf{B}^{+} - \mathbf{F}\mathbf{H}^{-1}\mathbf{C}^{+}\right)\right]. \tag{26}$$

Those terms without a '+' superscript are left unchanged by fully modified OLS, since they involve only functions of x_t. We can find the limits of each term in (26), and those of interest are

$$T^{-1/2}S_{ut}^{+} \Rightarrow B_{1.2}(r)$$

$$T^{-1/2}\left(\mathrm{B}^{+} - \mathbf{F}\mathrm{H}^{-1}\mathbf{C}^{+}\right) \Rightarrow B_{1.2}(r) - \int \mathrm{B}_2'\left(\int \mathrm{B}_2\mathrm{B}_2'\right)^{-1}\left(\int \mathrm{B}_2 dB_{1.2}\right)$$

$$T\mathrm{H}^{-1}\mathbf{C}^{+} \Rightarrow \left(\int \mathrm{B}_2\mathrm{B}_2'\right)^{-1}\left(\int \mathrm{B}_2 dB_{1.2}\right).$$

Hence the limit of each term of K_t^+ is the same as the limit of each term of K_t when constructed from the restricted model 2. Thus the limit of the top line of (12) using K_t^+ will be the same as given in equation (20) for model 2.

$$T^{-2}\sum_{t=1}^{T} K_t^{+^2} \Rightarrow \omega_{1.2}^2 \int_0^1 V(r)^2 dr. \tag{27}$$

The one difference between (27) and (20) is that, in equation (27), $\omega_{1.2}^2$ is not equal to $\sigma_{1.2}^2$ since u_t^+ is not i.i.d. in the general case. If, however, we define the transformed test statistic

$$s_2^+ = \frac{T^{-2}\sum_{t=1}^{T} K_t^{+^2}}{\hat{\omega}_{1.2}^2}, \tag{28}$$

where $\hat{\omega}_{1.2}^2$ is a consistent estimator of $\omega_{1.2}^2$, then the asymptotic distribution of s_2^+ is

$$s_2^+ \Rightarrow \int_0^1 V(r)^2 dr. \tag{29}$$

That is, s_2^+ for the general model has the same asymptotic distribution as s_2 for model 2. Hence we have a test which is asymptotically equivalent to the LBUI test, and is applicable to a wide range of data generating processes.

5. CONSISTENCY

In this section, we consider the asymptotic behaviour of the test under the alternative hypothesis. We refer to equation (9) as the model under the alternative hypothesis, where $\theta \neq 0$. Defining $u_t^a = \theta S_{ut} + u_t$ we obtain the following regression model

$$y = \mathbf{X}\beta + u^a, \tag{30}$$

from which the test statistic s_2 can be constructed. The OLS residuals from equation (30) are

$$\hat{u}^a = u^a - \mathbf{X}(\mathbf{X}'\mathbf{X})^{-1}\mathbf{X}'u^a, \tag{31}$$

where

$$\mathbf{X}'u^a = \begin{bmatrix} \theta\sum_{t=1}^{T} S_{ut} + S_{uT} \\ \theta\sum_{t=1}^{T} x_t S_{ut} + \sum_{t=1}^{T} x_t u_t \end{bmatrix} = \begin{bmatrix} \mathbf{K}+\mathbf{B} \\ \mathbf{L}+\mathbf{C} \end{bmatrix}.$$

If we define $K_t^a = \sum_{i=1}^{t} \hat{u}_t^a$ to be the partial sum process under the alternative hypothesis, then

$$\begin{aligned} K_t^a &= S_{ut} - \mathbf{D}^{-1}\left(\mathbf{B}-\mathbf{FH}^{-1}\mathbf{C}\right)t - \left(\sum_{i=1}^{t} x_i'\right)\left[\mathbf{H}^{-1}\mathbf{C}-\mathbf{H}^{-1}\mathbf{GD}^{-1}\left(\mathbf{B}-\mathbf{FH}^{-1}\mathbf{C}\right)\right] \\ &\quad +\theta\left[\sum_{i=1}^{t} S_{ui} - \mathbf{D}^{-1}\left(\mathbf{K}-\mathbf{FH}^{-1}\mathbf{L}\right)t - \left(\sum_{i=1}^{t} x_i'\right)\left[\mathbf{H}^{-1}\mathbf{L}-\mathbf{H}^{-1}\mathbf{GD}^{-1}\left(\mathbf{K}-\mathbf{FH}^{-1}\mathbf{L}\right)\right]\right] \\ &= K_t + \theta M_t. \end{aligned}$$

From Section 4, we know that K_t is $O_p(T^{-1/2})$. Thus we are only concerned with finding the order of probability of M_t. The orders of the terms of M_t can be found by noting that **K**, **F**, **G** and $\sum_{i=1}^{t} x_i$ are $O_p(T^{3/2})$ and **D** and t are $O_p(T)$. Then $\sum_{i=1}^{t} S_{ui}$, $\mathbf{D}^{-1}(\mathbf{K}-\mathbf{FH}^{-1}\mathbf{L})t$, and $(\sum_{i=1}^{t} x_i')[\mathbf{H}^{-1}\mathbf{L}-\mathbf{H}^{-1}\mathbf{GD}^{-1}(\mathbf{K}-\mathbf{FH}^{-1}\mathbf{L})]$ are $O_p(T^{3/2})$. Thus M_t, and hence K_t^a, are $O_p(T^{3/2})$. The top line of s_2 under the null hypothesis is $T^{-2}\sum_{t=1}^{T} K_t^{a\,2}$ which is $O_p(T^2)$. From Phillips (1987) or KPSS, we know that the long-run variance estimator of $\hat{\omega}_{1.2}^2$ is $O_p(lT)$, and so under the alternative hypothesis s_2 is $O_p(T/l)$. Since $l \to \infty$ at a lesser rate than $T \to \infty$ (typically $l = O(T^{1/4})$), we deduce that $s_2 \to \infty$ under the alternative hypothesis, and hence the test is consistent.

6. CRITICAL VALUES

As can be seen from equation (29), the asymptotic distribution of the test statistic s_2^+ is a function of $V(r)$ only, which in turn is dependent only on two independent standard Brownian motions. In particular, we have been able to produce a test which does not depend on nuisance parameters such as $\omega_{1.2}, \Omega_{22}$ or δ_{21}. The only way in which the asymptotic distribution depends on the model specification is that the number of explanatory variables in the cointegrating regression determines the dimensions of $W_2(r)$. Hence we can obtain sets of asymptotic critical values for various numbers of regressors. These are found by simulation of (29) using a GAUSS program with a sample size of 4000 and 50000 replications for 1 to 5 explanatory variables. The results of this simulation are reported in Table 1.

Table 1. Asymptotic Critical Values

Number of Regressors (excluding constant)	10%	5%	1%
1	0.2335	0.3202	0.5497
2	0.1617	0.2177	0.3727
3	0.1203	0.1590	0.2756
4	0.0929	0.1204	0.1983
5	0.0764	0.0972	0.1560

7. SUMMARY

In this section we give a summary of the testing procedure that has been derived in the paper. First apply OLS to the regression

$$y_t = \mu_0 + x_t'\beta_0 + z_t, \tag{32}$$

and obtain the estimated residuals $\hat{z}_t$. Form $\hat{\zeta}_t = [\hat{z}_t, \Delta x_t']'$ and calculate the estimated covariance matrices

$$\hat{\Omega} = \begin{bmatrix} \hat{\omega}_1^2 & \hat{\omega}_{12} \\ \hat{\omega}_{21} & \hat{\Omega}_{22} \end{bmatrix} = T^{-1}\sum_{t=1}^{T} \hat{\zeta}_t\hat{\zeta}_t' + T^{-1}\sum_{k=1}^{l} w(k,l) \sum_{t=k+1}^{T} \left(\hat{\zeta}_{t-k}\hat{\zeta}_t' + \hat{\zeta}_t\hat{\zeta}_{t-k}'\right) \tag{33}$$

$$\hat{\Delta} = \begin{bmatrix} \hat{\delta}_{11} & \hat{\delta}_{12} \\ \hat{\delta}_{21} & \Delta_{22} \end{bmatrix} = T^{-1}\sum_{k=0}^{l}\sum_{t=k+1}^{T} \hat{\zeta}_t\hat{\zeta}_{t-k}', \tag{34}$$

where $w(k,l) = 1 - k/(l+1)$. Then calculate

$$y_t^+ = y_t - \hat{\omega}_{12}\hat{\Omega}_{22}^{-1}\Delta x_t$$

and

$$\hat{\delta}_t^+ = \hat{\delta}_{21} - \hat{\Delta}_{22}\hat{\Omega}_{22}^{-1}\hat{\omega}_{21}$$

and re-estimate the cointegrating regression using the fully modified OLS estimator

$$\hat{\beta}^+ = (\mathbf{X}'\mathbf{X})^{-1}\left(\mathbf{X}'y^+ - e_k T\hat{\delta}_{21}^+\right), \tag{35}$$

where $e_k = [0, I_k]'$. Obtain the estimated residuals $\hat{u}_t^+$ from this regression and construct the test statistic

$$s_2^+ = \frac{T^{-2}\sum_{t=1}^{T} K_t^{+^2}}{\hat{\omega}_{1.2}^2}, \tag{36}$$

where $K_t^+ = \sum_{i=1}^{t} \hat{u}_i^+$ and $\hat{\omega}_{1.2}^2 = \hat{\omega}_{1.2}^2 - \hat{\omega}_{12}\hat{\Omega}_{22}^{-1}\hat{\omega}_{21}$.

Critical values for this test statistic are provided in Table 1 above. This procedure provides a test for the null hypothesis of cointegration which is consistent and is asymptotically equivalent to the locally best invariant test.

8. MONTE CARLO COMPARISON

In this section we examine briefly the power properties of our test by comparing its ability to distinguish between null and alternative hypotheses with that of the Dickey-Fuller test. A power comparison between these two tests is difficult, as they are derived from different assumed data generating processes (DGPs), and have opposite null and alternative hypotheses. To resolve these difficulties, the Monte Carlo study is undertaken as follows.

We first specify a DGP as in equations (6) to (8), as designed for the s_2^+ test, with $T = 100$ and $k = 1$. A 10% critical value for the s_2^+ test is taken from Table 1 (0.2335). For the Dickey-Fuller test, we generate 10000 samples under the hypothesis of cointegration ($\theta = 0$), and compute Dickey-Fuller t statistics for cointegration. A 'critical value' of –8.89 is found where 10% of statistics greater than this value (leading to a conclusion of no cointegration). Both tests thus have a 10% chance of concluding that there is no cointegration when $\theta = 0$. 'Powers' are then calculated for a range of values of θ under the alternative, with 10000 replications. These results are given in the first half of Table 2.

To ensure a fair opportunity is given to each test, further results were obtained with a DGP for which the Dickey-Fuller test is designed, namely

$$y_t = x'\beta_0 + \mu_t \tag{37}$$

$$\mu_t = \rho\mu_{t-1} + u_t. \tag{38}$$

For the Dickey-Fuller test, a 10% critical value for testing the null of no cointegration is found in Phillips and Ouliaris (1990) (–3.0657), and for the s_2^+ test we use 10000 replications under the hypothesis of no cointegration ($\rho = 1$) to find a 'critical value' of 0.7543 where 10% of the s_2^+ statistics are less than this value (leading to a conclusion of cointegration). Both tests thus have a 10% chance of

Table 2. Powers of Tests[a]

DGP: Equations (6), (7), (8); Null Hypothesis: Cointegration

	$\theta=0$	$\theta=0.05$	$\theta=0.1$	$\theta=0.15$	$\theta=0.2$	$\theta=0.25$
s_2^+	0.103	0.281	0.519	0.673	0.764	0.839
Dickey-Fuller	0.100	0.164	0.344	0.567	0.718	0.837

DGP: Equations (32), (32); Null Hypothesis: No Cointegration

	$\rho=1$	$\rho=0.95$	$\rho=0.9$	$\rho=0.85$	$\rho=0.8$	$\rho=0.75$
s_2^+	0.100	0.235	0.386	0.528	0.643	0.720
Dickey-Fuller	0.105	0.181	0.375	0.642	0.866	0.966

[a] In both DGP's the innovations u_t are i.i.d. N(0,1), so the Dickey-Fuller procedure is not augmented by any lags of the differenced residuals, and the s_2^+ test has a lag truncation parameter (l) of zero.

chance of concluding there is cointegration when $\rho=1$. 'Powers' are then calculated for values of ρ under the alternative, with 10000 replications. These results are given in the second half of Table 2.

It is difficult to draw definitive conclusions from such a limited Monte Carlo study, but Table 2 does suggest a couple of points worthy of note. Firstly, it is not surprising that each test performs at its best when the DGP follows that for which the test is designed. The s_2^+ test shows clear domination over Dickey-Fuller in the error components DGP, particularly for small values of θ. Likewise, the autoregressive DGP favours the Dickey-Fuller test, especially as ρ moves further from one. The s_2^+ test actually dominates for ρ close to one, but this is reversed with smaller ρ.

Given that we do not know which DGP is more realistic, it is inappropriate to 'choose' between the two tests on the basis of power. What we can say, though, is that deriving a test which reverses the null and alternative hypotheses has not obviously hurt the capacity to distinguish between cointegration and no cointegration.

9. APPLICATION

Inder and Silvapulle (1992) used cointegration methods to test if the Fisher effect applies in Australia. One question related to this is whether nominal interest rates and inflation are cointegrated. Consider the equation

$$i_t = \beta_0 + \beta_1 \pi_t + u_t \tag{39}$$

where i_t is the nominal interest rate and π_t is the inflation rate. Inder and Silvapulle (1992) treat (39) as a long-run relationship which implies that the interest rate and the inflation rate are cointegrated. They use the augmented Dickey-Fuller test to test for cointegration between the series as represented by the change in the log of CPI for inflation and the bank accepted bill rate for interest rates. The estimated parameters for equation (37) were $\hat{\beta}_0 = 7.2275$ and $\hat{\beta}_1 = -0.4387$, and the Dickey-Fuller t statistic was –2.5921. Compared with a 10% critical value of –2.84, it can be seen that the null hypothesis of no cointegration could not be rejected.

However, it is possible that the failure to reject the null hypothesis was due only to lack of power of the Dickey-Fuller test, rather than true non-cointegration. To check this, we computed s_2^+ to test the null hypothesis of cointegration. We chose the lag length l by following Inder (1991), who suggested choosing the largest significant lag in the autocorrelation function of the OLS residuals. For this application, this gave a choice of $l = 4$. We obtained $s_2^+ = 1.6674$, which when compared to the 1% critical value of 0.5497 allows rejection of the null hypothesis of cointegration. This provides further evidence for the results of Inder and Silvapulle (1992) rejecting the cointegration hypothesis.

REFERENCES

ENGLE, R. F. and C. W. J. GRANGER (1987), 'Cointegration and Error Correction: Representation, Estimation and Testing', *Econometrica*, 55, pp. 251–76.

INDER, B. (1991), 'Some Strong Evidence for a Unit Root in Economic Time Series', Mimeo, Monash University.

INDER, B. and P. SILVAPULLE (1992), 'Does the Fisher Effect Apply in Australia?', Working Paper No. 5/92, Monash University.

JOHANSEN, S. (1988), 'Statistical Analysis of Cointegration Vectors', *Journal of Economic Dynamics and Control*, 12, pp. 231–54.

KING, M. L. and G. H. HILLIER (1985), 'Locally Best Invariant Tests of the Error Covariance Matrix of the Linear Regression Model', *Journal of the Royal Statistical Society B*, 47, pp. 98–102.

KWIATKOWSKI, D., P. C. B. PHILLIPS, P. SCHMIDT and Y. SHIN (1992), 'Testing the Null Hypothesis of Stationarity Against the Alternative of a Unit Root: How Sure are We that Economic Time Series Have a Unit Root?', *Journal of Econometrics*, 54, pp. 159–78.

PARK, J. Y. and P. C. B. PHILLIPS (1988), 'Statistical Inference in Regressions with Integrated Processes: Part 1', *Econometric Theory*, 4, pp. 468–97.

PHILLIPS, P. C. B. and S. N. DURLAUF (1986), 'Multiple Time Series Regression with Integrated Processes', *Review of Economic Studies*, 53, pp. 473–95.

PHILLIPS, P. C. B. and B. E. HANSEN (1990), 'Statistical Inference in Instrumental Variables Regression with I(1) Processes', *Review of Economic Studies*, 57, pp. 99–125.

PHILLIPS, P. C. B. and S. OULIARIS (1990), 'Asymptotic Properties of Residual Based Tests for Cointegration', *Econometrica*, 58, pp. 165–93.

SNYDER, R. D. (1985), 'Recursive Estimation of Dynamic Linear Statistical Models', *Journal of the Royal Statistical Society B*, 47, pp. 272–76.

WATSON, M. W. (1986), 'Univariate Detrending Methods with Stochastic Trends', *Journal of Monetary Economics*, 18, pp. 49–75.

6

Modelling seasonal variation

Svend Hylleberg[*]

The realisation among econometricians and applied economists that seasonal variation in many economic time series is often larger and less regular than often hitherto acknowledged, has led to an increased interest in seriously modelling seasonality. The relative size of the seasonal variation also means that such modelling is of major economic interest.

In the last few years important developments in modelling seasonality have occurred. The use of model based procedures and periodic models, the extension of integration and cointegration to the seasonal frequencies and the development of economic theories of seasonality are some of the most promising areas of research. However, the procedures applied by the official data-producing agencies have not been affected. In this paper we discuss the implications of these developments on the treatment of seasonality within the data-producing agencies.

1. INTRODUCTION

Modelling seasonal variations has recently been subject to a number of studies of both econometricians and economics theorists. A major cause for this development is the realization that the seasonal components account for the major part of the variations in many economic time series, that the seasonal variation is far from regular but varying and changing, and that economic considerations are playing an important part in the analysis. Hence, the attitude of many economists expressed in the following quotation by Jevons is yielding at last. Jevons (1862, p. 4) wrote

* Financial support from the U.K. Economic and Social Research Council under grant R000233447 is gratefully acknowledged by both authors. We are indebted to Neil Ericsson for helpful comments.

yearly, must be detected and exhibited not only as a subject of study in itself, but because we must ascertain and eliminate such periodic variations, before we can correctly exhibit those which are irregular or non-periodic, and probably of more interest and importance.'

In the following we will present some of the main features of seasonal macroeconomic time series, and discuss a definition of seasonality in economics. In Section 3 we present some formal tests of a varying seasonal pattern while Section 4 presents the theory of seasonal cointegration. The final section presents some alternative models and try to look into the future development of the area.

2. THE DEFINITION OF SEASONALITY AND SOME CHARACTERISTIC FEATURES OF SEASONAL ECONOMIC TIME SERIES

The complexities and variation of seasonal economic time series can be illustrated just by graphically studying a few macroeconomic time series. In Figure 1 the log of an index of the quarterly industrial production in the Netherlands from 1960.1 to 1986.4 is depicted. From the first panel, Figure 1.1, which shows the data series it is clear that the seasonal pattern is more varying in the latter part of the period and especially that the third quarter through is deeper. This can also be seen from the first difference of the series in Figure 1.2.

In Figure 1.4 the first quarter series etc. is depicted after subtraction of the calendar year average. This is done to remove the trend and thereby clarify the message of the figure. In case the seasonal pattern is totally regular and non-changing the four lines in Figure 1.4 should be parallel straight lines, but this is obviously not the case.[1]

From the tradition of Box and Jenkins (1970) the fourth differences transformation $y_{4t} = (1-B^4)y_t$ is used to remove the seasonal components. This series is shown in Figure 1.3 and the seasonal pattern seems to be removed. In Hylleberg, Engle, Granger and Yoo (1990) this fourth difference transformation is decomposed into $1-B^4 = (1-B)(1+B)(1+B^2)$ where a series which contains the first difference component is called integrated at the long frequency zero as the pseudo spectrum has the form $f(\omega) = \text{constant}/(\theta - \omega)^2$ for the frequency ω near $\theta = 0$. The component $1+B$ corresponds to the series being integrated at the semiannual frequency $\theta = ½$ of 2π, while $1+B^2 = (i-B)(-i-B)$ corresponds[2] to the series being

[1] The graph in Figure 1.4 is a variation of a graph advocated by Franses (1991).
[2] $i^2 = -1$

Figure 1. Logarithm of the Quarterly Industrial Production in the Netherlands, 1960.1 to 1986.4

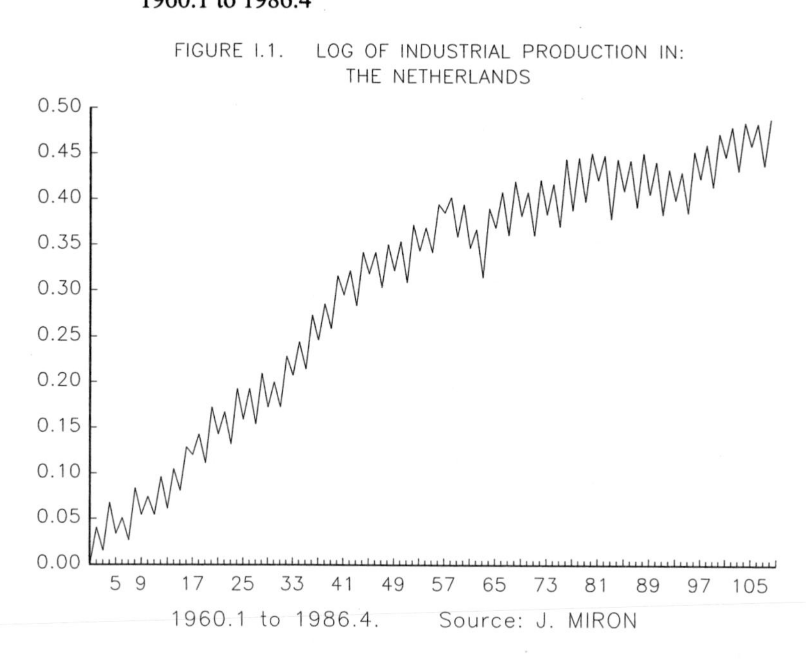

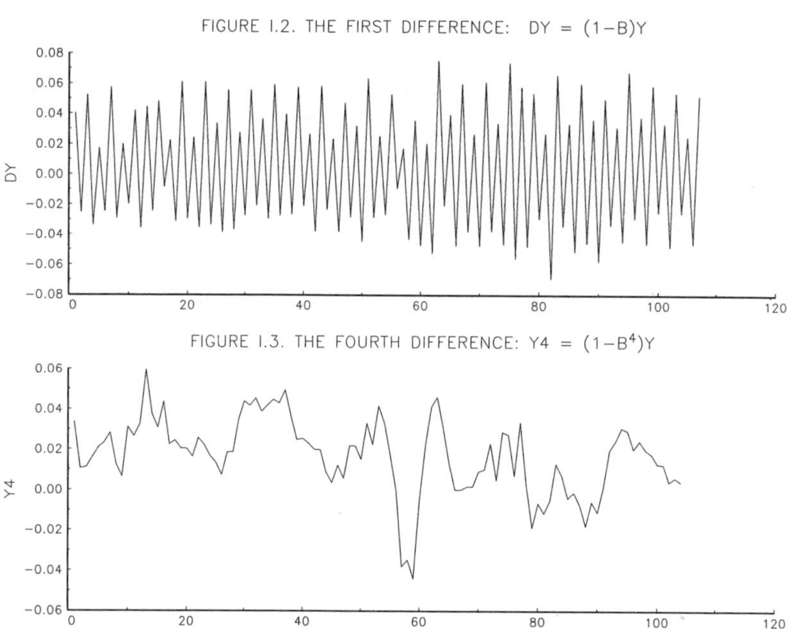

Figure 1. (continued)

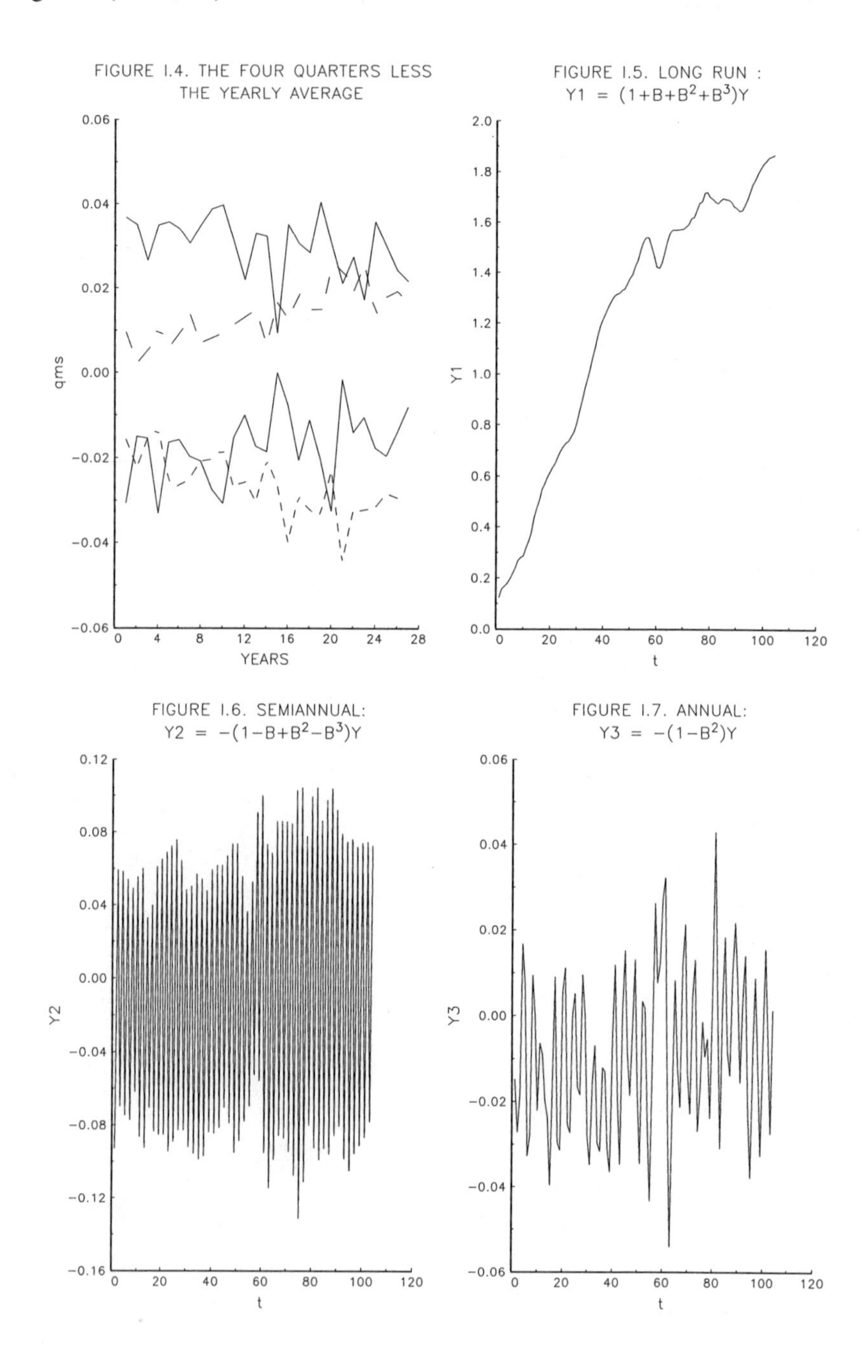

Figure 2. Logarithm of the Quarterly Production in Australia, 1963(1) to 1986(4)

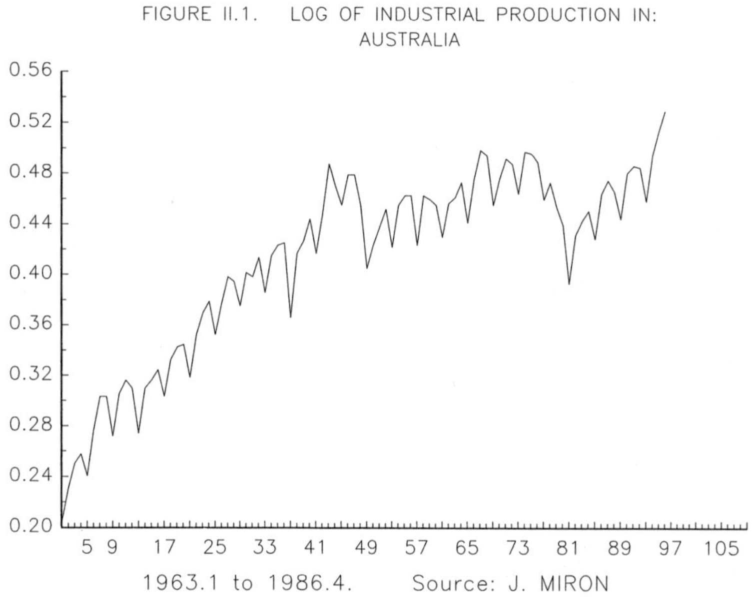

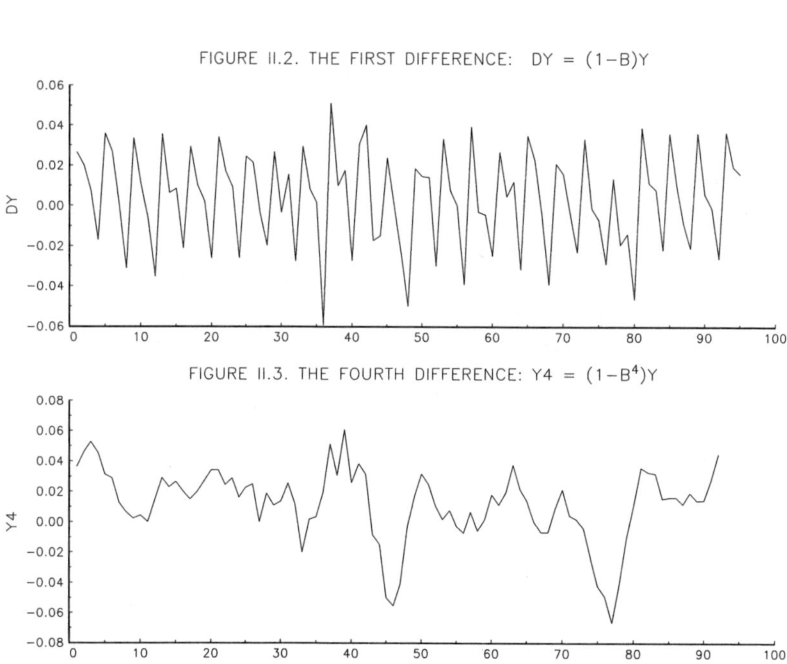

Figure 2. (continued)

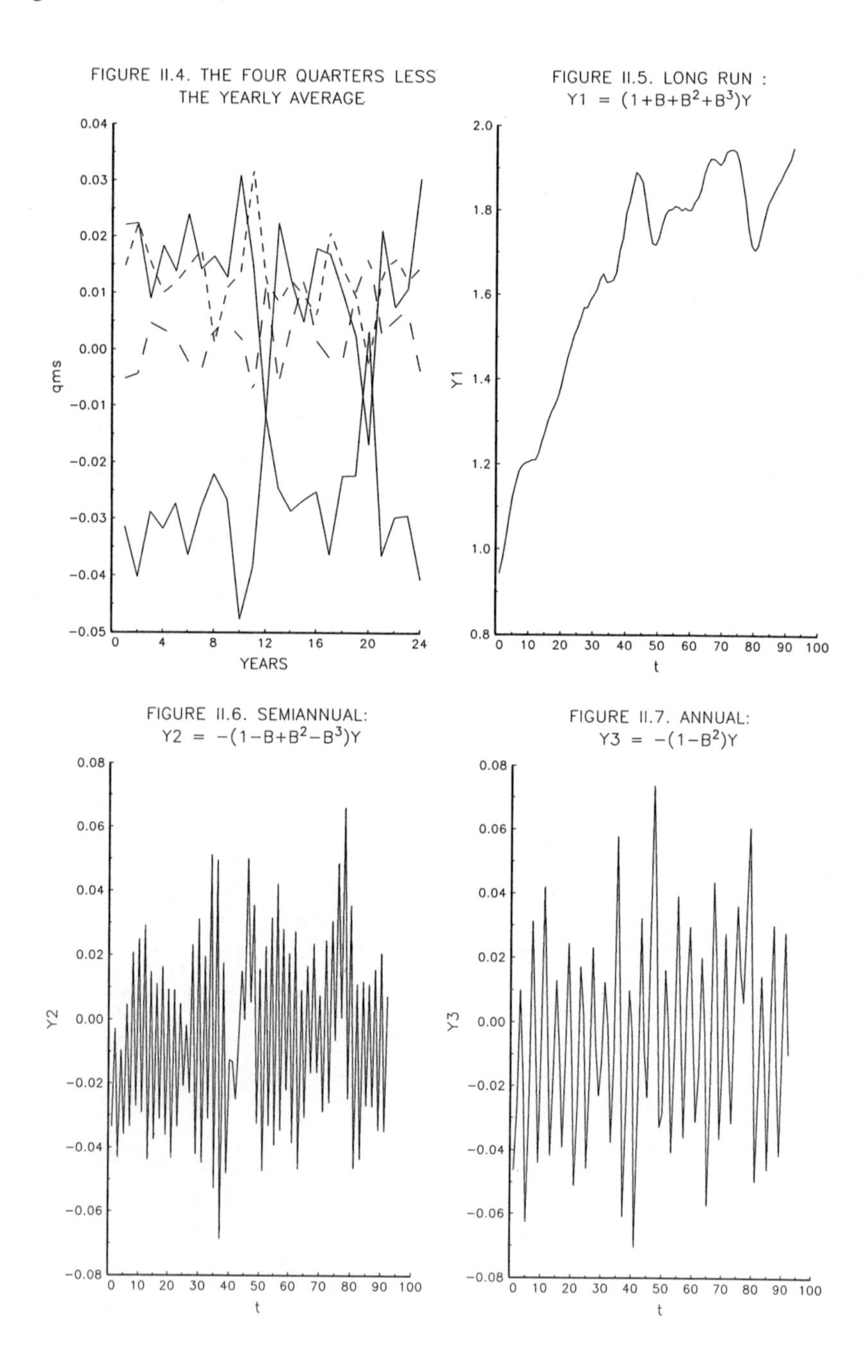

Figure 3. Logarithm of the Monthly Industrial Production in Canada, 1960(1) to 1986(12)

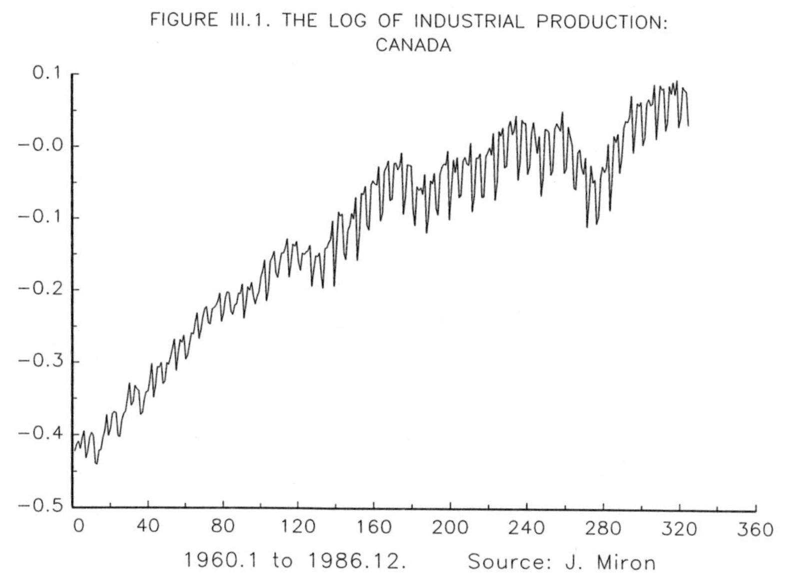

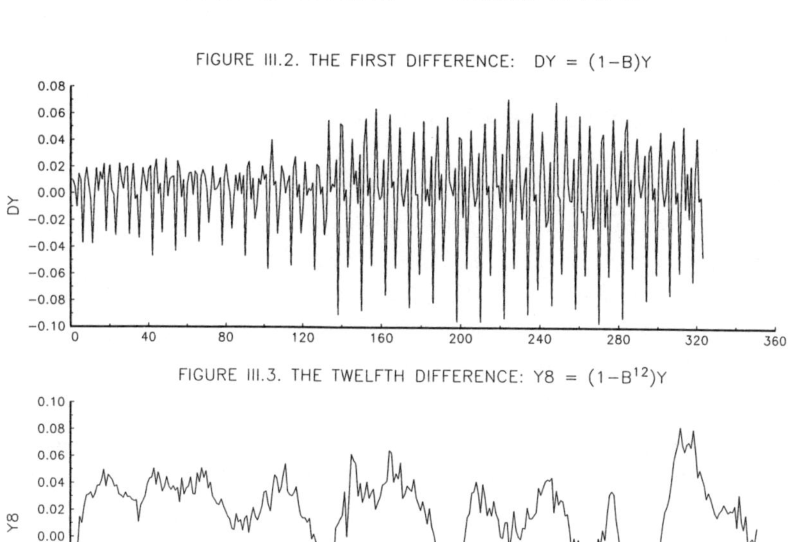

Figure 3. (continued)

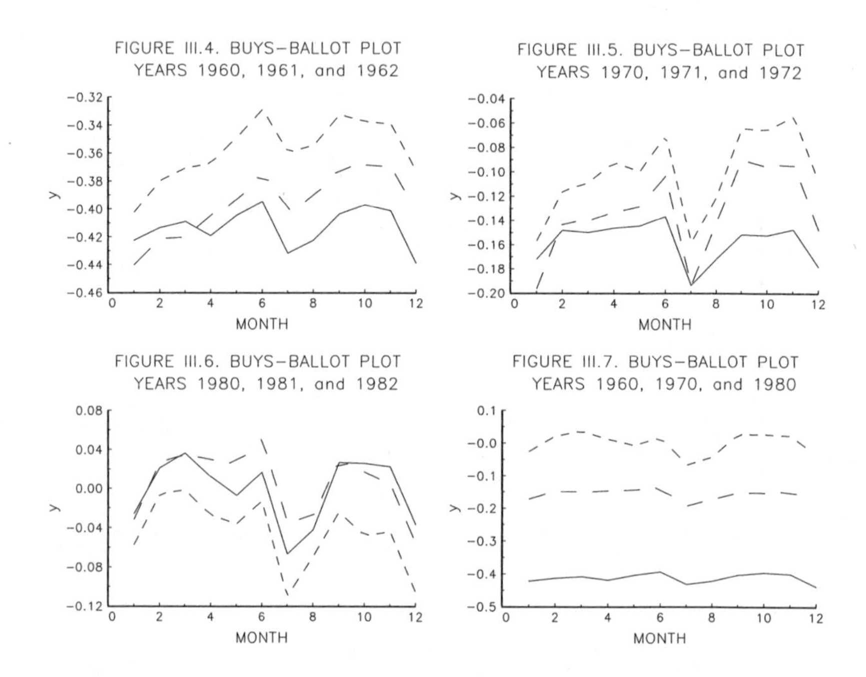

series being integrated at the annual frequency 1/4 and its complex conjugate partner 3/4 of 2π.

In case a series is integrated of order 1 at the frequency $\theta = 0$, 1/4, 1/2, 3/4, a fourth difference will render the series stationary.[3] If this is the case for the log of the industrial production in the Netherlands the transformations $y_{1t} = (1+B)(1+B^2)y_t = -(1-B+B^2-B^3)y_t$, and $y_{3t} = -(1-B)(1+B)y_t = -(1-B^2)y_t$, where the minus sign is just for convenience, draw forward the semiannual $(1 + B)$ and the annual $(1+B^2)$ components, respectively. The graphs of these transformations are shown in Figures 1.6 and 1.7.

In case the seasonal components of a time series is deterministic, i.e. very regular, the transformation y_{2t} and/or y_{3t} will be very regular too, but in case of a varying and changing seasonal pattern y_{2t} and/or y_{3t} will show some variation. However, if no seasonality is present both y_{2t} and y_{3t} will lose the sinusoidal pattern totally. In case of the Dutch industrial production the variations seem too large to be explained for instance by a deterministic seasonal dummy plus stationary noise.

[3] A series may be integrated of higher order than 1 and also of different order at each frequency.

A common criticism of the concept of seasonal integration is that the existence of seasonal unit roots in the data generating process implies that the seasonal pattern is varying too much and that summer may become winter. However, here we take the view that univariate models like the ones described above are approximations which can tell you about the variation in the seasonal pattern, and therefore be a possible useful tool in a proper modelling strategy.

Another quarterly series is the log of the industrial production in Australia from 1963.1 to 1986.4 and again the panels of Figure 2 indicate some variation in the seasonal pattern.

In Figure 3 the monthly industrial production of Canada for the period 1960.1 to 1986.12 is shown. From Figures 3.1 and 3.2 it is clear already that the seasonal pattern is changing both in form and variability through the sample period. Obviously transformation similar to the unit root transformation for the quarterly case could be depicted as well and also graphs corresponding to Figure 1.4 showing the series for the first month, the second month etc.[4] However, another useful graph is the so-called Buys Ballot plot, Fig. 3.4–7, see Hylleberg (1986, 1992), and Nerlove, Grether and Carvalho (1979). Here the series is graphed against the 12 months and a regular deterministic pattern is indicated by parallel series.

From the illustrations above, the results of Hylleberg, Jørgensen and Sørensen (1993) and of Canova and Hansen (1991), but not Beaulieu and Miron (1990, 1992, 1993), it is clear that the seasonal pattern of many economic time series cannot possibly be described by deterministic seasonal dummies, i.e. by a model which assumes that the seasonal component is regular and non-changing.

The reasons for this claim may be found in the causes for the seasonal variation as for instance the climate where both precipitation, temperature and number of sunny hours vary considerably over the years. Combined with endogenous changes in production techniques due to irrigation, storage possibilities, biological innovations etc. this may often cause the seasonal pattern to change. For instance, the experience of a few very dry and hot summers in Northern Europe in the mid seventies has implied a massive investment in irrigation, which has changed the whole farm technology which again has implied

---

[4] The transformations are $y_{1t} = (1+B)(1+B^2)(1+B^4+B^8)y_t = (1+B+B^2+...+B^{11})y_t$, i.e. a transformation which removes the seasonal unit roots and preserves the long run or zero frequency unit root, $y_{2t} = -(1-B)(1+B^2)(1+B^4+B^8)y_t$, which preserves the frequency 6/12 corresponding to a six month period, $y_{3t} = -(1-B^2)(1+B^4+B^8)y_t$ retains the frequency 3/12 (9/12) corresponding to a four month period, $y_{4t} = -(1-B^4)(1-\sqrt{3}B+B^2)(1+B^2+B^4)y_t$ retains the frequency 5/12 (7/12), while $y_{5t} = -(1-B^4)(1+\sqrt{3}B+B^2)(1+B^2+B^4)y_t$, $y_{6t} = -(1-B^4)(1-B+B^2)(1-B^2+B^4)y_t$, and $y_{7t} = -(1-B^4)(1+B+B^2)(1-B^2+B^4)y_t$ retain the frequencies 1/12 (11/12), 4/12 (8/12), and 2/12 (10/12). Finally $y_{8t} = (1-B^{12})y_t$.

Figure 4. Logarithm of Quarterly Real Disposable Income in Japan, 1961(1) to 1987(4).

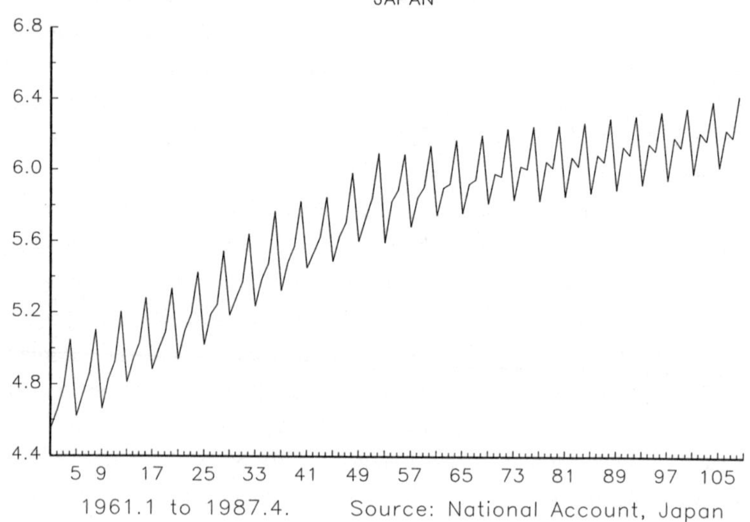

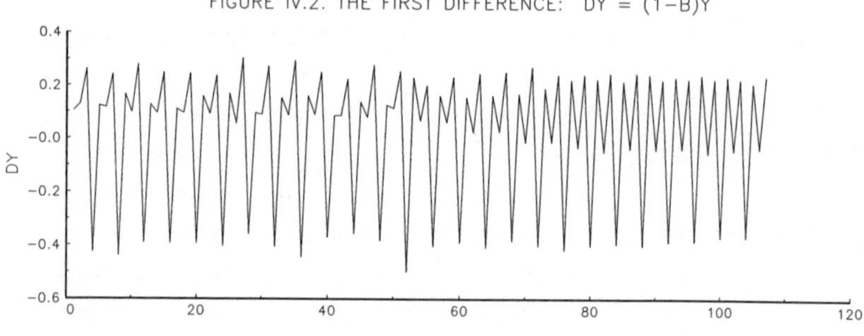

FIGURE IV.3. THE FOURTH DIFFERENCE: Y4 = $(1-B^4)Y$

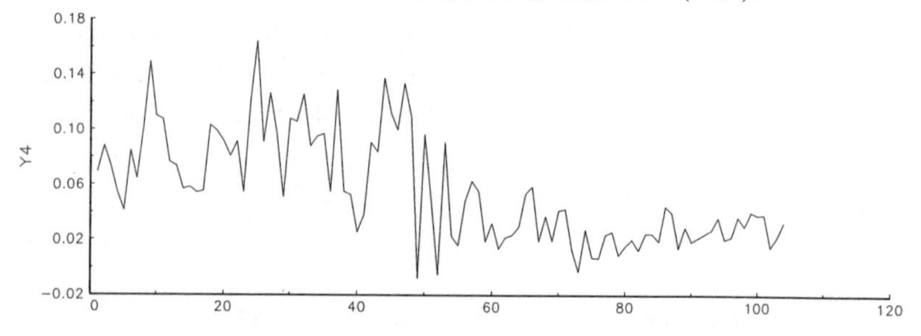

Figure 4. (continued)

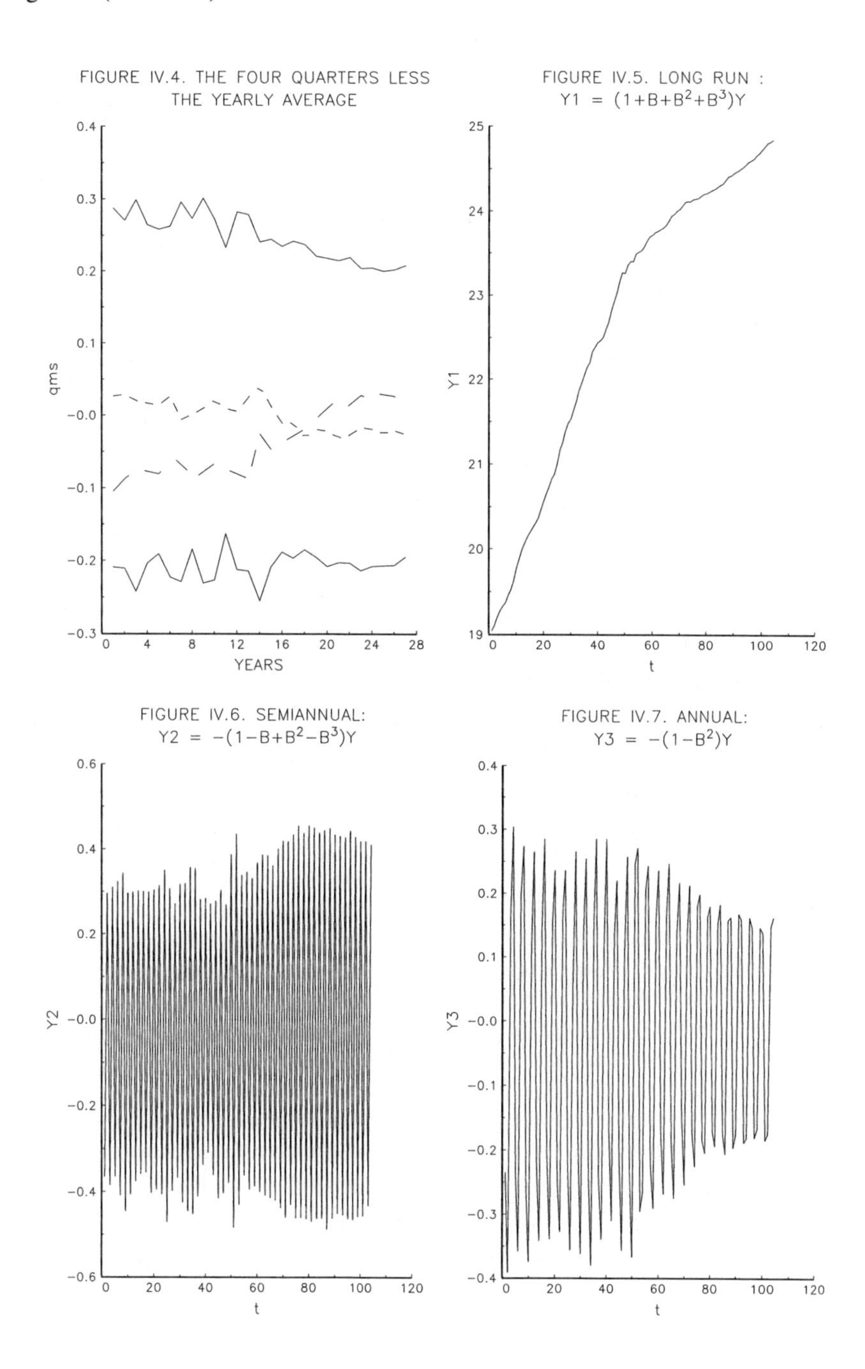

less dependence on the weather. The drought in 1992 in the same area will, undoubtedly, push this process even further.

Demand factors may change as well. Since the 1960s the tourist streams have been directed towards the sunny beaches of for instance the Mediterranean but in the last 10 years the streams have partly changed direction to the snowy areas of the Alps and Scandinavia. This in turn has changed the timing of the vacations. In addition, institutional changes such as changes in the length and in the timing of school vacations, payment of taxes, wages, bonuses etc. are also factors that may imply a shift in the seasonal pattern. In Figure 4 which depicts the log of real disposable income in Japan from 1961.1 to 1987.4 the high fourth quarter peak is caused by massive bonus payments from the Japanese firms, while the emergence of the second quarter peak is caused by an increasing number of firms paying out bonuses here as well.

Obviously, these basic causes of seasonal variability may have direct effects on economic variables, but they may often be changed and filtered through the decision of economic agents. An economic agent may choose the degree of smoothness of his consumption and production over the year in accordance with his expectations, preferences, costs, credit constraints, and physical possibilities. Factors that may change over the years as well.

In the light of this the definition of seasonality proposed in Hylleberg (1992)[5] p. 4 is 'Seasonality is the systematic, although not necessarily regular, intra-year movement caused by the changes of the weather, the calendar, and timing of decision, directly or indirectly through the production and consumption decisions made by the agents of the economy. These decisions are influenced by the endowments, the expectations and preference of the agents, and the production techniques available in the economy.'

3. TESTS FOR A STOCHASTIC SEASONAL PATTERN

Interpreting the existence of seasonal unit roots as indication of a varying and changing seasonal pattern there is a need for a formal test of the existence of such roots. Hylleberg *et al.* (1990) propose a test based on ideas similar to those of the well-known Dickey-Fuller test for a unit root at the zero frequency and the tests proposed by Dickey, Hasza and Fuller (1984) for testing for a unit root, $\rho = -1$, in models such as $y_t = \rho y_{t-s} + \varepsilon_t$, where $s = 2, 4, 12$.

The test proposed by Hylleberg *et al.* (1990), HEGY, is based on the autoregressive model $\phi(B)y_t = \varepsilon_t$, where $\phi(B)$ is a pth lag polynomium with roots

[5] This definition is only slightly different from the one proposed in Hylleberg (1986).

on or outside the unit circle. $\phi(B)$ can be written as

$$\phi(B)=\sum_{k=1}^{p}\frac{\delta_k\Delta(B)[1-\delta_k(B)]}{\delta_k(B)}+\Delta(B)\phi^*(B) \quad (1)$$

where $\delta_k=1-\frac{1}{\theta_k}B$ and $\Delta(B)=\prod_{k=1}^{p}\delta_k(B)$ and in case $\{\theta_k,\ k = 1,2,...,p\}$ contains all the possible roots in $\phi(B)$ of modules 1, $\phi^*(B)$ is stationary. In the quarterly case where $\theta_k = \pm 1, \pm \mathrm{i}$ and $\Delta(B) = 1-B^4$, $\phi(B)y_t=\varepsilon_t$ can be written as

$$y_{4,t}=\left(1-B^4\right)y_t=\pi_1 y_{1,t-1}+\pi_2 y_{2,t-1}+\pi_3 y_{3,t-2}+\pi_4 y_{4,t-1}+\sum_k \phi_k y_{4,t-k}+\varepsilon_t \quad (2)$$

where y_{it}, $i = 2, 3$ is defined earlier as the transformation retaining the unit roots at the zero frequency, the semiannual frequency, and the annual frequency if they exist. The coefficients π_i, $i = 1, 2, 3, 4$, are defined by $\pi_1 = -\lambda_1$, $\pi_2 = -\lambda_2$, $-\pi_3 + i\pi_4 = 2\lambda_3$, and $-\pi_3 - i\pi_4 = 2\lambda_4$, and the test of a zero frequency unit root against a stationary alternative is a test of $\pi_1 = 0$ against $\pi_1 < 0$ with the t-value distributed as the usual Dickey-Fuller t-value. Likewise a test of a seasonal unit root at the semiannual frequency, 1/2 of 2π, is a test of $\pi_2 = 0$ against $\pi_2 < 0$, with the t-value distributed as in the zero frequency case.[6]

The test of an annual unit root is slightly more complicated due to the complex conjugate roots, but, under the null of a unit root at the frequency 1/4 of 2π (and 3/4), $\pi_3 \cap \pi_4 = 0$ and the test proposed is based on the F-value with the critical value supplied by Hylleberg *et al.* (1990).[7] Alternatively, a two step procedure is available based on a test of $\pi_4 = 0$ against $\pi_4 \neq 0$ with a twosided test. The distribution of this t-value is also given by Hylleberg *et al.* (1990). In case $\pi_4 = 0$ a test of $\pi_3 = 0$ against $\pi_3 < 0$ can be based on the t-value of π_3 and the critical values are supplied by Dickey, Hasza and Fuller (1984) for the case $y_t = \rho y_{t-2} + \varepsilon_t$, $\rho = -1$.

An analogous transformation can be found for the monthly case as shown by Beaulieu and Miron (1993) and Franses (1990)[8]. The auxiliary regression is

$$\begin{aligned}y_{8,t}=\left(1-B^{12}\right)y_t &= \pi_1 y_{1,t-1}+\pi_2 y_{2,t-1}+\pi_3 y_{3,t-1}+\pi_4 y_{3,t-2}\\ &+\pi_5 y_{4,t-1}+\pi_6 y_{4,t-2}+\pi_7 y_{5,t-1}+\pi_8 y_{5,t-2}\\ &+\pi_9 y_{6,t-1}+\pi_{10} y_{6,t-2}+\pi_{11} y_{7,t-1}+\pi_{12} y_{7,t-2}\\ &+\sum_k \phi_k y_{8,t-k}+\varepsilon_t.\end{aligned} \quad (3)$$

[6] The use of the minus in the definition of $y_{2,t}$ is made in order to be able to use the Dickey-Fuller distribution and not its mirror image.

[7] Engle, Granger, Hylleberg and Lee (1993) provide the asymptotic results.

[8] See the definitions in footnote 4.

The critical values of the appropriate t- and F-values are provided by Franses (1990) and Beaulieu and Miron (1993).

Both the quarterly auxiliary regression and the monthly can be augmented by deterministic terms as intercept, I, seasonal dummies, SD, and trends, Tr. In order for the test to have reasonable power and the stipulated size the lag polynomial, ϕ_k, $k = 1, 2,...$, must be chosen with care. In case too many ϕ_k's are used, the power is low and if too few are applied, the size is far above the chosen level of significance.

Dickey, Hasza and Fuller (1984) have suggested to test $\rho = 1$ in the quarterly model $y_t = \rho y_{t-4} + \varepsilon_t$ or in the monthly case in $y_t = \rho y_{t-12} + \varepsilon_t$. The null hypothesis is here what compares to the joint hypotheses $\pi_1 = \pi_2 = \pi_3 = \pi_4 = 0$ and the alternative is that all roots are stationary with the same modules. The alternative may be loosened, however, by the augmentation of lagged dependent variables. The shortcoming of the test is that we cannot test the roots at the individual frequencies, see Ghysels, Lee and Noh (1991) for a Monte Carlo study comparing the HEGY test with the DHF test and finding that the former is preferable in small samples.

A test almost as the HEGY test is proposed by Osborn, Chui, Smith and Birchenhall (1988) while Franses (1991) proposed a test based on the stacking of the quarters or months and a cointegration test based on the ML approach of Johansen (1988). In case cointegration is rejected it is an evidence against seasonal unit roots. However, the stacking implies a possible large vector autoregression which may easily ruin the practical use of the test, especially in the monthly case.

Yet another test is proposed by Canova and Hansen (1991). This test is based on an auxiliary regression of the series on seasonal dummies and a subsequent test of whether the residuals from that regression contain components of the form $\sum_{j=0}^{t-1}(-1)^j\varepsilon_{t-j}$, which corresponds to a semiannual unit root component $(1+B)y_t = \varepsilon_t$, and/or of the form $\sum_{j=0}^{\text{Int}[(t-1)/2]}(-1)^j\varepsilon_{t-2j}$ corresponding to an annual component $(1+B^2)y_t = \varepsilon_t$. However, the size and power properties of this test depend crucially on there not being unit roots at other frequencies, see Hylleberg (1992).

The results of applying the HEGY test described above to the quarterly series presented in Section 2 are given in Table 1.[9] Table 1 also contains tests for seasonal unit roots in the log of real Japanese total consumption, a series which is depicted in Figure 5.

[9] In Hylleberg, Jørgensen and Sørensen (1993) seasonal unit roots were found in the log of monthly Canadian industrial production at the frequencies of 0/12, 3/12, 4/12, 5/12 and 6/12 of 2π.

Figure 5. Logarithm of Real Total Consumption in Japan, 1961(1) to 1987(4)

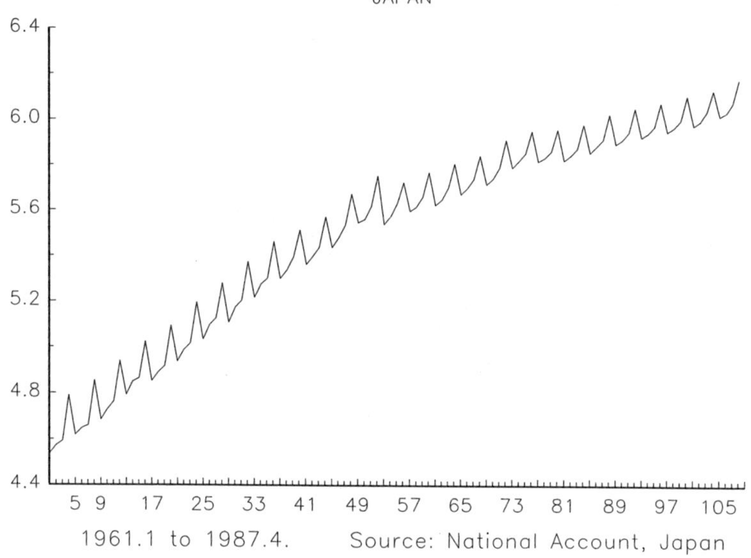

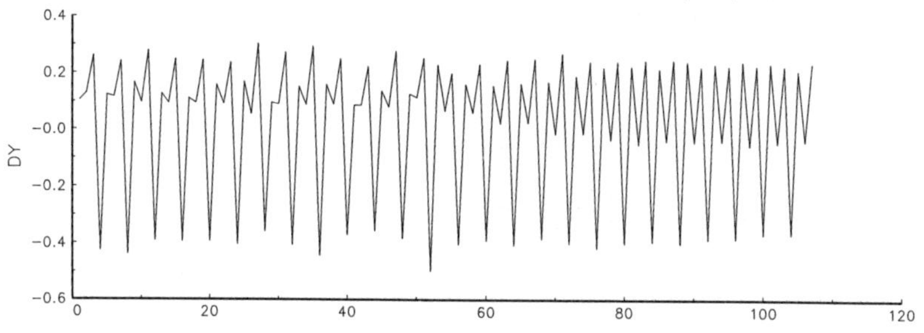

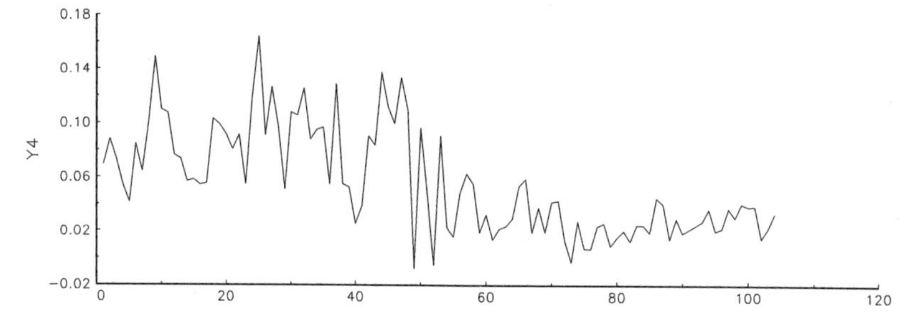

Figure 5. (continued)

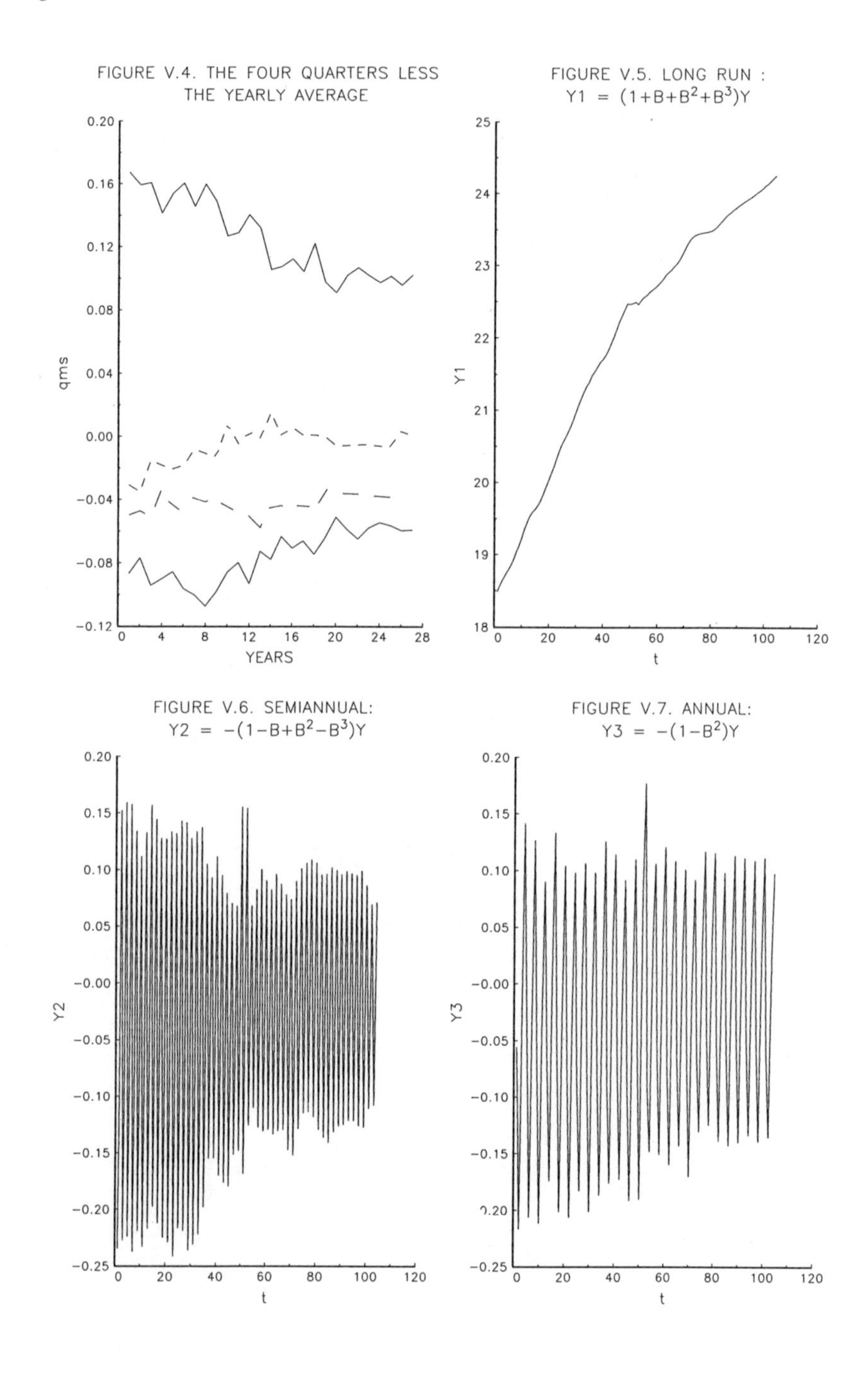

From the results in Table 1 it is clear that the seasonal pattern of the Japanese disposable income is varying as seasonal unit roots cannot be rejected at neither the frequency 1/2 nor the frequency 1/4. The total consumption contains a more regular seasonal pattern but a seasonal unit root cannot be rejected at frequency 1/2 but maybe at frequency 1/4, where the test rejects if the auxiliary regression contains seasonal dummies. The results for the two quarterly industrial production series are that the seasonal pattern of the Australian figure seems to be so regular that seasonal dummies may do the job while the series for the Netherlands have a varying seasonal pattern. However, the augmentation is difficult as white noise errors are not obtained.

4. SEASONAL COINTEGRATION

In the macroeconomic theory, consumption is related to income through the effect of the quarterly income stream on the expectation of the future income or permanent income. This usually implies that consumption is predicted to be much smoother than income unless consumers are for instance credit rationed. From the general theory of consumption under perfect capital markets one would expect a non parallel seasonal pattern in income and consumption, while a parallel seasonal pattern or seasonal cointegration imply that at least parts of the assumption behind the permanent income theory are violated. From Figures 4.1 and 5.1 it is evident that the seasonal pattern of Japanese consumption and income has the major peak in common although the emergence of the second quarter peak is not evident in the consumption series. In addition, the seasonal variation accounts for a smaller part of the total variation in the consumption series than in the income series.

In Hylleberg *et al.* (1990) and Engle *et al.* (1993) the theory of seasonal cointegration is developed following Engle and Granger (1987) for the zero frequency case. In the quarterly case cointegration between n series integrated at the semiannual frequency is said to exist if there is at least one linear combination of the series that is stationary at that frequency.

The annual frequency case is a little more complicated due to the existence of a complex conjugate pair of unit roots. Here cointegration is said to exist if there is at least one linear combination of the series, all integrated at the annual frequency and the series lagged one quarter which is stationary at that particular frequency.[10]

Still, confining ourselves to the quarterly case tests of seasonal cointegration can be performed similarly to the test proposed in Engle and Granger (1987), see

[10] In other words, we need the notion of a cointegrating polynomial vector, see Yoo (1987).

Table 1. HEGY tests of quarterly series.

Series	Period	Auxiliary regression		t_{π_1}	t_{π_2}	t_{π_3}	t_{π_4}	$F_{\pi_3 \cap \pi_4}$	Comments
		Determi-nistic	Lags of y_{4t}						
Log of industrial production: The Netherlands	1960.1–1986.4	I,SD,Tr	1,4–6,8	–1.06	–1.52	–3.67[a]	–1.33	7.04[a]	Unit roots: Semiannual but maybe not annual
Log of industrial production: Australia	1963.1–1986.4	I,SD,Tr	1–11	–1.98	–4.80[b]	–2.25	–1.98	4.08	Unit roots: Annual but not semiannual.
Log of real disposable income in Japan[c]	1961.1–1987.4	I,SD,Tr I,Tr	1,3,4,7,8 1,3,4,7,8	–1.11 –1.14	–0.67 0.62	–2.00 –1.58	–0.55 –0.46	2.24 1.48	Unit roots: Semiannual and annual
Log of real total consumption in Japan[c]	1961.1–1987.4	I,SD,Tr I,Tr	1,4–6,8,10 1,2,4–6,8,10	–1.91 –1.49	–0.94 –1.71	–4.23[b] –1.67	1.03 0.94	10.11[b] 1.81	Unit roots: Semiannual but maybe not annual

[a] Significant at a 5% level but not at a 2.5% level. [b] Significant at a 1% level. [c] Source: Engle, Granger, Hylleberg and Lee (1993).

also Engle and Yoo (1987).[11] This implies that the test for seasonal non cointegration at a particular frequency is based on a test for a unit root at that frequency in the residuals from a first step regression. The first step regression is a regression of one of the series on the other, but after proper transformations so that no unit roots exist at other frequencies. In the Japanese case presented above and provided both the consumption and income series are integrated at the frequencies $\theta = 0$, 1/4, 1/2 a test of non cointegration at the long-run frequency is a test for a unit root at the zero frequency in the residuals, u_t, from a regression of $c_{1,t}$ on $y_{1,t}$ where $c_{1,t}$ is $(1+B+B^2+B^3)c_t$, i.e. the sum of four consecutive values of the log of the consumption series while $y_{1,t}$ is defined analogously for the log of the disposable income series.

Likewise, a test of non cointegration at the semiannual frequency is a test of there being a unit root at that frequency in the residuals, v_t, from a regression of $c_{2,t} = -(1-B+B^2-B^3)c_t$ on $y_{2,t} = -(1-B+B^2-B^3)y_t$. The appropriate critical values can be found in Engle and Yoo (1987). For the annual frequency the first step regression is $c_{3,t} = -(1-B^2)c_t$ on $y_{3,t} = -(1-B^2)y_t$ and $y_{3,t-1}$ and the test for a unit root at the annual frequency in the residuals w_t is based on the F-value for $\pi_3 = \pi_4 = 0$ in the regression $(w_t = w_{t-2}) = \pi_3(-w_{t-2}) + \pi_3(-w_{t-1})$ + augmentation etc. The critical values and the asymptotics of this are found in Engle *et al.* (1992).

They also provide the actual test result for the Japanese case and they find that the consumption and income series are not cointegrated at the long-run and semiannual frequency but maybe at the annual frequency where the residuals from the first step regression are shown in Figure 6. The test of the residuals rejects a unit root at the 10% but not quite at the 5% level implying that Japanese consumption and income may be cointegrated at the frequency.

5. ALTERNATIVE MODELS AND SOME REFLECTIONS

The main objective of the treatment of sesonality has been to help in historical studies of the business cycles, in the appraising of current economic conditions, and in econometrically oriented studies, see Hylleberg (1986). Nowadays it is widely accepted among researchers working within the field that

(A) the seasonal variation often accounts for the major part of the variation in economic time series, see Miron (1994).

(B) the seasonal and the non-seasonal component are dependent and that economic considerations are important in explaining both components; see

[11] In Lee (1992) the ML approach due to Johansen (1988) is extended to the seasonal case.

the collection of articles and the introduction in Hylleberg (1992) and the invited Barcelona papers of Miron (1994) and Ghysels (1994) to be published in Sims (1994).

Currently, it is disputed

(C) whether the seasonal component is very regular and strikingly similar all over the globe as argued by Miron and associates; see Barsky and Miron (1989), Beaulieu and Miron (1990, 1992, 1993) and Miron (1994) or varying and changing, see Hylleberg, Jørgensen and Sørensen (1993), Hylleberg (1992, 1994*a*, 1994*b*), Ghysels (1994), and Canova and Hansen(1991) .

This controversy has no effect on the agreement with respect to (A) and (B), but it has a major impact on the modelling strategy and on the specific conclusion one may draw, see the discussion of Miron (1994) by Hylleberg (1992).

But where do these findings leave the most common treatment of seasonality, namely

(i) Official Adjustment by X-11, see Hylleberg (1986 ch. 5 or, 1992 part 3) (or soon X-12) or the

(ii) Do it yourself adjustment by seasonal dummies or band spectrum regression or the like, see Engle (1974) and Hylleberg (1986)?

The X-11 adjustments are all based on the view that seasonality is noise contaminating the data and that the noise (read seasonality) in a specific variable can be treated in isolation from any other variable and from the context in which it is used. Obviously, the views set out above are very much at odds with this, and they lead to the conclusion that X-11 should be used when only a quick glance at the data series is needed. Any serious economic analysis must take the dependence of the components and their changing and varying nature into account.

However, being realistic about this, X-11 seasonal adjusted data may be used for econometric analysis in the future as well, at least because such data may be the only data available. It is therefore of major interest to study the properties of the X-11 (X-12) filters and to persuade the data collecting agencies to publish the data without them being seasonally adjusted[12].

---

[12] Wallis (1974) and Ghysels (1984, 1994) show that X-11 may seriously distort the dynamics of econometric models, while Ericsson, Hendry and Tran (1992) show that the seasonally adjusted and unadjusted data are cointegrated at the zero frequency, but that both the exogeneity status and the dynamic adjustment are changed. A popular and fruitful way to analyse the X-11 method is to analyse linear approximations to the filter as in Burridge and Wallis (1984).

Figure 6. The Residuals from the Cointegrating Regression $c_{3,t}$ on I, SD, $y_{3,t}$ and $y_{3,t-1}$.

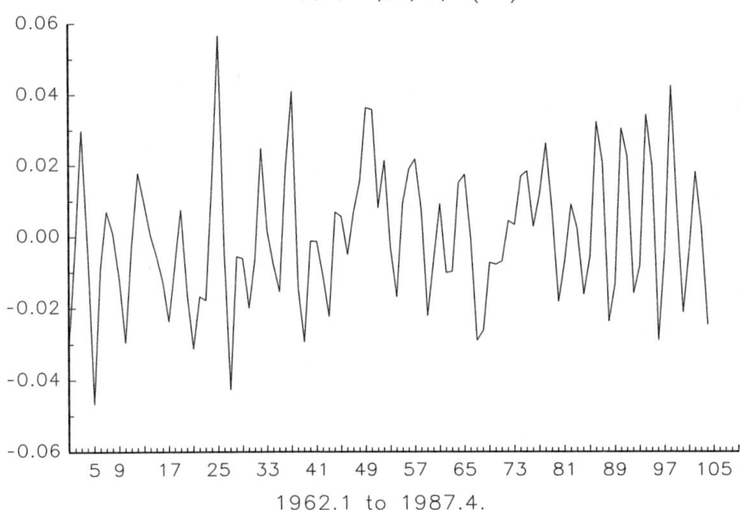

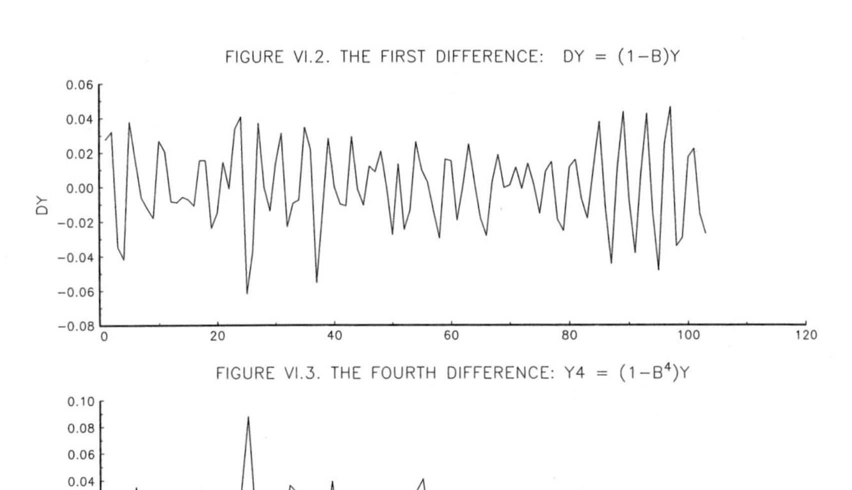

Figure 6. (continued)

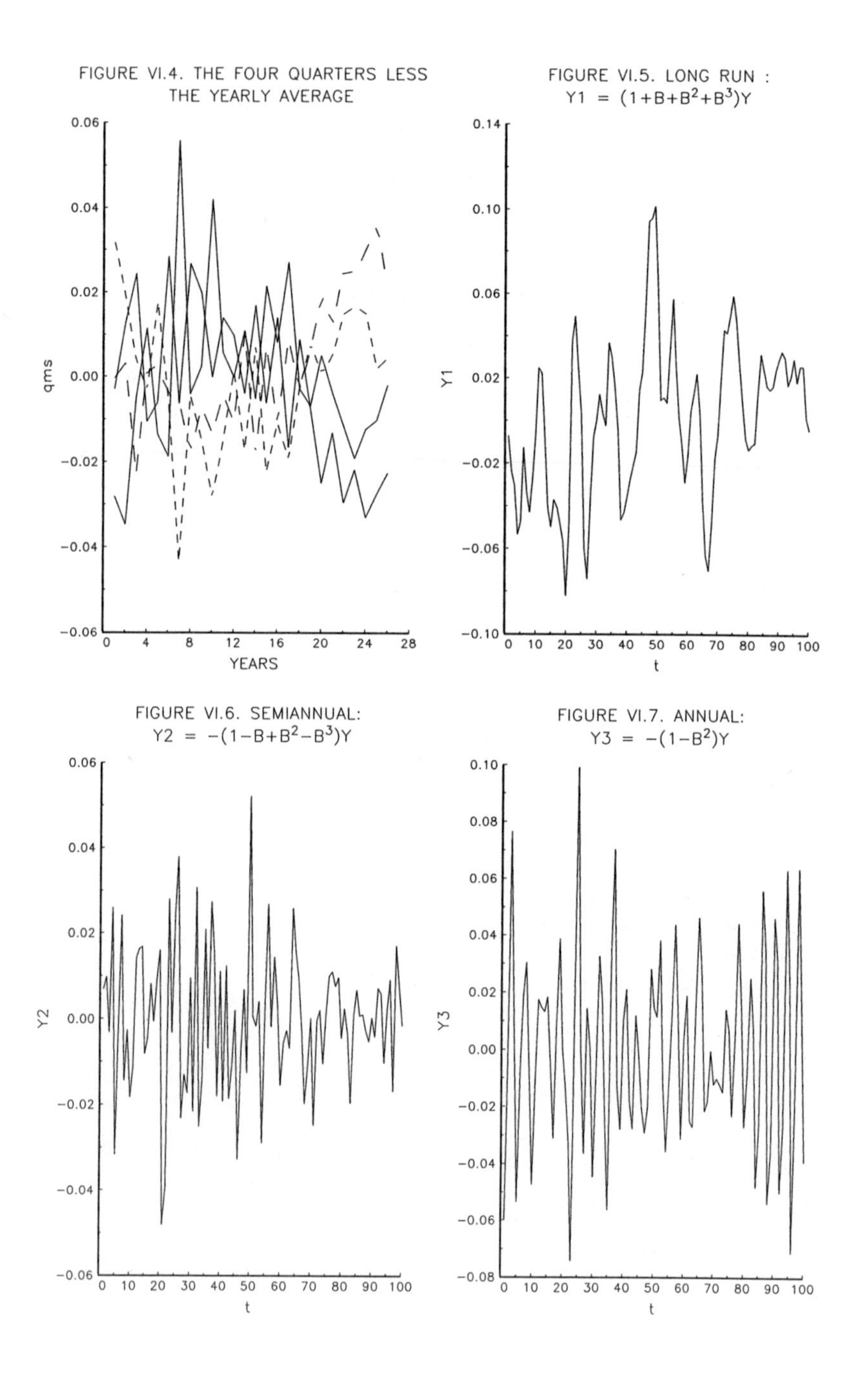

A similar argument is of course applicable against the use of the do it yourself method of including seasonal dummies in a regression, or the slightly more flexible band spectrum regression method where bands of seasonal frequencies are excluded before the regression. But here at least it is the choice of the user and in addition it is possible to test, whether the same models apply to all frequency bands, see Engle (1974) and Hylleberg (1986).

The so-called model based procedures, i.e. modelling procedures based on time series models like the Box-Jenkins model, see Box and Jenkins (1970) and Hillmer and Tiao (1982) or the unobserved components models, which may be considered a parsimonious version of the Box-Jenkins model, see Engle (1974), Harvey and Todd (1983) and Maravall and Pierce (1987), must in the light of the views expressed earlier best be seen as techniques available for the preliminary analysis of economic data. Techniques that may enable you to understand the characteritics of the series analysed. This is especially true for the univariate modelling, while the multivariate model could be used as an intermediate step towards a more satisfactory structural model.

This is also the case for the extension of the error correction model to include error correction terms for each frequency component, see Hylleberg *et al.* (1990) and Engle *et al.* (1993). As an example consider the Japanese consumption income model discussed in Engle *et al.*(1993). In its original form the error correction model looks like

$$c_{4,t} = \sum_{j=0}^{q} \beta_j c_{4,t-j} + \sum_{i=1}^{p} \delta_i y_{4,t-j} + \gamma_{11}\left(c_{1,t-1} - \alpha_{12} y_{1,t-1}\right) + \gamma_{12}\left(c_{2,t-1} - \alpha_{22} y_{2,t-1}\right) \\ - \left(\gamma_{13} + \gamma_{14} B\right)\left(c_{3,t-2} - \alpha_{32} y_{3,t-2} - \alpha_{42} y_{3,t-3}\right) + \varepsilon_t \tag{4}$$

where c_{it} and y_{it} , i = 1, 2, 3, 4 are the transformations of the consumption and income presented earlier. The three brackets define the error correction terms at the long-run, semiannual, and annual frequency respectively.

Another possibility is to apply so-called periodic models where the coefficients and parameters variations etc. are allowed to be different for different seasons, see Gersowitz and MacKinnon (1978), Hylleberg (1986), Miron (1986), Miron and Zeldes (1988), Burridge and Wallis (1990), Osborn (1988), Franses and Boswijk (1991), Franses and Kloek (1991) among others. A severe difficulty with this model is, however, that the numbers of parameters often increase dramatically, and that it may be difficult to connect the different parts of the model. But there seems to be a fertile research field here.

In concluding I am quite optimistic about the current development with the area of seasonal modelling. What needs to be done now is to combine good economic theory with adequate statistical models, but that this is a formidable task is well known. However, a promising beginning is Crutchfield and Zellner (1962), Ghysels (1984, 1988), Osborn (1988), Barsky and Miron (1989), Todd (1990) and Hansen and Sargent (1990, 1991).

REFERENCES

BARSKY, R. B. and J. A. MIRON (1989), 'The Seasonal Cycle and the Business Cycle', *Journal of Political Economy*, Vol. 97, No. 3, pp. 503–34, reprinted in HYLLEBERG (1992).

BEAULIEU, J. J. and J. A. MIRON (1990), 'The Seasonal Cycle in U.S. Manufacturing', NBER Working Paper No. 3450.

BEAULIEU, J. J. and J. A. MIRON (1992), 'A Cross Country Comparison of Seasonal Cycles and Business Cycles', *Economic Journal*, 102, pp. 772–88.

—— (1993), 'Seasonal Unit Roots in Aggregate U.S. Data', *Journal of Econometrics*, 55, pp. 305–28.

BOX, G. E. P. and G. M. JENKINS (1970), *Time Series Analysis: Forecasting and Control*, San Francisco: Holden-Day.

BURRIDGE, P. and K. F. WALLIS (1984), 'Unobserved-Components Models for Seasonal Adjustment Filters', *Journal of Business and Economic Statistics*, Vol. 2, No. 4, pp. 350–60, reprinted in HYLLEBERG (1992).

—— (1990), 'Seasonal adjustment and Kalman filtering: extension to periodic variances'. *Journal of Business and Economic Statistics*, 2, pp. 350–9, reprinted in HYLLEBERG (1992).

CANOVA, F. and B. E. HANSEN (1991), 'Are Seasonal Patterns Constant Over Time: A Test for Seasonal Stability', Working Paper, Brown University and University of Rochester.

CRUTCHFIELD, J. and A. ZELLNER (1962), 'Analysis of port pricing of halibut: theoretical and empirical results', chs. 6 and 7 in *Economic Aspects of Halibut Fishery*, United States Department of the Interior, Washington DC. Reprinted in HYLLEBERG (1992).

DICKEY, D. A., D. P. HASZA and W. A. FULLER (1984), 'Testing for Unit Roots in Seasonal Time Series', *Journal of the American Statistical Association*, Vol. 79, No. 386, pp. 355–68, reprinted in HYLLEBERG (1992).

ENGLE, R. F. (1974), 'Band Spectrum Regression', *International Economic Review*, Vol. 15, No. 1, pp. 1–12.

ENGLE, R. F. and C. W. J. GRANGER (1987), 'Co-Integration and Error Correction: Representation, Estimation and Testing', *Econometrica*, Vol. 55, No. 2, pp. 251–77.

ENGLE, R. F. and B. S. YOO (1987), 'Forecasting and testing in co-integrated systems', *Journal of Econometrics*, Vol. 35, No. 1, pp. 143–61.

ENGLE, R. F., C. W. J. GRANGER, S. HYLLEBERG and H. LEE (1993), 'Seasonal Cointegration: The Japanese Consumption Function', *Journal of Econometrics*, 55, pp. 275–98.

ERICSSON, N., D. F. HENDRY and H. TRAN (1992), 'Cointegration, Seasonality, Encompassing and the Demand for Money in the United Kingdom', Chapter 6 in this book.

FRANSES, P. H. (1990), 'Testing for Seasonal Unit Roots in Monthly Data', Research Memorandum Series No. TI-1990/30, Tinbergen Institute, Erasmus University. Presented at ESEM, Barcelona.

—— (1991), 'A Multivariate Approach to Modeling Univariate Seasonal Time Series', Working Paper, Erasmus University Rotterdam. Forthcoming, *Journal of Econometrics*.

FRANSES, P. H. and H. P. BOSWIJK (1991), 'Testing for Periodic Cointegration', Paper presented at 'Structure and Dynamics in Econometrics II', Tinbergen Institute.

FRANSES, P. H. and T. KLOEK (1991), 'A Periodic Cointegration Model of Quarterly Consumption in Austria, Japan and the United Kingdom', Working Paper, Erasmus University Rotterdam.

GERSOWITZ, M. and J. G. MACKINNON (1978), 'Seasonality in regression: an application of smoothness priors', *Journal of the American Statistical Association*, 73, pp. 264–73.

GHYSELS, E. (1984), 'The economics of seasonality: the case of the money supply', unpublished, Ph.D. dissertation, Northwestern University, Department of Managerial Economics and Decision Science.

—— (1988), 'A study towards a dynamic theory of seasonality for economic time series', *Journal of the American Statistical Association*, 83, pp. 168–72. Reprinted in HYLLEBERG (1992).

—— (1994), 'On the Economics and Econometrics of Seasonality' in SIMS (1994).

GHYSELS, E., H. S. LEE and J. NOH (1991), 'Testing for Unit Roots in Seasonal Time Series. Some Theoretical Extensions and a Monte Carlo Investigation', Working Paper, University of Montreal, Tulane University and University of California, San Diego. Forthcoming, *Journal of Econometrics*.

HANSEN, L. P. and T. J. SARGENT (1990), 'Recursive linear models of dynamic economies', Manuscript, Hoover Institution, Stanford University.

HANSEN, L. P. and T. J. SARGENT (1991), 'Seasonality and approximation errors in rational expectations models', Manuscript, Hoover Institution, Stanford University.

HARVEY, A. C. and P. H. J. TODD (1983), 'Forecasting Economic Time Series with Structural and Box-Jenkins Models: A Case Study', *Journal of Business and Economic Statistics*, Vol. 1, No. 4, pp. 299–307.

HILLMER, S. C. and G. C. TIAO (1982), 'An ARIMA-Model-Based Approach to Seasonal Adjustment', *Journal of the American Statistical Association*, Vol. 77, No. 377, pp. 63–71, reprinted in HYLLEBERG (1992).

HYLLEBERG, S. (1986), *Seasonality in Regression,* Academic Press, Orlando.

—— (ed.), (1992), *Modelling Seasonality*, Oxford University Press.

—— (1994*a*), 'Tests for Seasonal Unit Roots. General to Specific or Specific to General', Forthcoming, *Journal of Econometrics*.

—— (1994*b*), 'The Economics of Seasonal Cycles. A Comment', in SIMS (1994).

HYLLEBERG, S., R. F. ENGLE, C. W. J. GRANGER and S. YOO (1990), 'Seasonal Integration and Cointegration', *Journal of Econometrics*, 44, pp. 215–38, reprinted in HYLLEBERG (1992).

HYLLEBERG, S., C. JØRGENSEN and N. K. SØRENSEN (1993), 'Seasonality in Macroeconomic Time Series', *Empirical Economics*, Vol 18, pp. 321–35.

JEVONS, W. S. (1862), 'On the study of periodic commercial fluctuations', in *Investigations in Currency and Finance*, pp. 2–11. MacMillan, London, 1884.

JOHANSEN, S. (1988), 'Statistical Analysis of Cointegration Vectors', *Journal of Economic Dynamics and Control*, 12, pp. 231–54.

LEE, H. S. (1992), 'Maximum Likelihood Inference on Cointegration and Seasonal Cointegration', *Journal of Econometrics*, 54, pp. 1–47.

MARAVALL, A. and D. A. PIERCE (1987), 'A Prototypical Seasonal Adjustment Model', *Journal of Time Series Analysis*, Vol. 8, No. 2, pp. 177–95, reprinted in HYLLEBERG (1992).

MIRON, J. A. (1986), 'Seasonal fluctuations and the life cycle-permanent income model of consumption', *Journal of Political Economy*, 94, pp. 1258–79.

—— (1994), 'The Economics of Seasonal Cycles', in SIMS (1994).

MIRON, J. A. and S. P. ZELDES (1988), 'Seasonality, cost shocks, and the production smoothing model of inventories', *Econometrica*, 56, pp. 877–908. Reprinted in HYLLEBERG (1992).

NERLOVE, M., D. M. GRETHER and J. L. CARVALHO, (1979), *Analysis of Economic Time Series, A Synthesis*, Academic Press, New York.

OSBORN, D. R. (1988), 'Seasonality and Habit Persistence in a Life Cycle Model of Consumption', *Journal of Applied Econometrics*, Vol. 3, No. 4, pp. 255–67. Reprinted in HYLLEBERG (1992).

OSBORN, D. R., A. P. L. CHUI, J. P. SMITH and C. R. BIRCHENHALL (1988), 'Seasonality and the Order of Integration for Consumption', *Oxford Bulletin of Economics and Statistics*, Vol. 50, No. 4, pp. 361–79, reprinted in HYLLEBERG (1992).

SIMS, C. A. (1994), *Advances in Econometrics. 6th World Congress*, Cambridge University Press.

TODD, R. (1990), 'Periodic linear-quadratic methods for modelling seasonality', *Journal of Economic Dynamics and Control*, 14, pp. 763–95.

WALLIS, K. F. (1974), 'Seasonal adjustment and relations between variables', *Journal of the American Statistical Association*, 69, pp. 18–31. Reprinted in HYLLEBERG (1992).

YOO, S. (1987), 'Co-integrated time-series: Structure, forecasting and testing', Ph.D. dissertation, University of California, San Diego.

7

Cointegration, seasonality, encompassing, and the demand for money in the UK

Neil R. Ericsson, David F. Hendry and Hong-Anh Tran[*]

Virtually all previous narrow money demand studies for the United Kingdom have used seasonally adjusted data for money, prices and expenditure. This paper develops a constant, data-coherent M_1 demand equation for the United Kingdom with seasonally *unadjusted* data. For that model, we address issues of cointegra-tion, error correction, general-to-specific modelling, dynamic specification, model evaluation and testing, parameter constancy and exogeneity. We also establish theoretical and empirical relationships between seasonally adjusted and unadjusted data, and so between models using those data. Finally, we derive and implement encompassing tests for comparing models using adjusted data with models using unadjusted data. Unlike the 'standard' encompassing framework, variance dominance is *not* always a necessary condition for encompassing.

1. INTRODUCTION

Wallis (1974) and Sims (1974) examine the effects of seasonal adjustment when estimating econometric relationships. Wallis considers the implications of estimation with seasonally adjusted data when the underlying economic relation involves the unadjusted data. Sims investigates the converse situation, in which the

[*] Financial support from the U.K. Economic and Social Research Council under grant R000233447 is gratefully acknowledged by both authors. We are indebted to Neil Ericsson for helpful comments.

relation is in terms of the non-seasonal components of the economic variables, but the econometrician uses the unadjusted seasonal data. In each case, the empirical model is misspecified and coefficient estimates generally are inconsistent.

Using two recently developed concepts, cointegration and encompassing, this paper sheds new light on the use of adjusted and unadjusted data in econometric modeling.[1] First, under mild assumptions about the seasonal adjustment procedure, the adjusted and unadjusted series of a given variable are cointegrated, with cointegrating vector (+1 : −1). Second, the cointegrating vector for a set of variables is invariant to the choice of adjusted or unadjusted data. Even so, conducting inference may be problematic with adjusted data. In particular, dynamic adjustments and exogeneity status are usually altered. Third, parameter-encompassing statistics can be constructed to compare a model using unadjusted data with one using adjusted data. These statistics aim to test empirically the dissimilar premises of Sims and Wallis. Contrasting with the classical encompassing framework, variance dominance is not always a necessary condition for parameter encompassing.

Section 2 constructs an analogue model of seasonal adjustment and obtains the cointegration results for the adjusted and unadjusted series. Section 3 modifies an existing parameter-encompassing statistic to compare 'adjusted' with 'unadjusted' models. Sections 4 and 5 illustrate the cointegration propositions and the encompassing test via a substantive empirical study of narrow money demand in the United Kingdom. Section 4 reports Johansen's system-based cointegration tests and estimates for adjusted and unadjusted series, variable by variable, and for sets of variables, whether adjusted or unadjusted. The theoretical and empirical results match closely. Section 5 develops a conditional money demand model with the unadjusted data, summarizes Hendry and Ericsson's (1991b) model with adjusted data, and applies the encompassing test of Section 3 to these two models. Surprisingly, the model with adjusted data does not encompass the model with unadjusted data, even though the former has a long and favorable track record. Section 6 concludes. Appendix A documents data sources, and Appendix B describes a sequential reduction for obtaining the conditional model of money demand using the unadjusted data. Ericsson, Hendry, and Tran (1993, Appendix

[1] A recent burgeoning literature discusses integration and cointegration at seasonal frequencies; cf. Dickey, Hasza, and Fuller (1984), Hylleberg *et al.* (1990), Bell and Wilcox (1990), Ghysels, Lee, and Noh (1991), Hylleberg, Jørgensen, and Sørensen (1991), Lee (1992), and Beaulieu and Miron (1993) *inter alia.* Osborn (1988, 1991), Birchenhall *et al.* (1989), and Franses and Kloek (1991) consider seasonally varying slope coefficients for cointegrated processes. Our paper focuses on cointegration at the zero frequency only, both for seasonally unadjusted data and for seasonally adjusted data.

C) compare the results of cointegration analysis for several variants on the money demand system.

2. RELATIONSHIPS BETWEEN SEASONALLY ADJUSTED AND UNADJUSTED DATA

This section establishes several theoretical results on the cointegration relationships between adjusted and unadjusted data, where seasonal adjustment is approximated by a certain two-sided linear filter. As a preliminary, Section 2.1 discusses linear filters, defines an analogue model of X–11 seasonal adjustment, establishes a notation for cointegration, and defines mean equality between series. Section 2.2 considers conditions under which the original and filtered series are cointegrated and under which their mean difference is zero, and relates these conditions to the properties of seasonal adjustment filters, based on the analogue model. Section 2.3 extends the analysis to vector processes and matrix filters, specializing these for seasonal adjustment filters. If the unadjusted series are cointegrated, the adjusted series are also cointegrated, and with the same cointegrating vector(s) as for the unadjusted series. Section 2.4 comments on the approximations made in using the analogue model of seasonal adjustment. For comparison with these analytical results, Section 4 provides empirical evidence on the relationships between actual adjusted and unadjusted data. Below, the abbreviation NSA means 'not seasonally adjusted', and SA means 'seasonally adjusted' or 'seasonal adjustment', depending upon the context.

2.1 Preliminaries

After briefly reviewing some properties of linear filters, this subsection formulates a linear analogue model of the X–11 SA procedure and derives implications of that analogue model for the NSA and SA data.

Linear filters. Consider a single variable x, and denote the filtered and unfiltered series of that variable as $\{x_t^f\}$ and $\{x_t\}$ respectively, where t is the time subscript. Throughout, $\{x_t\}$ and $\{x_t^f\}$ have the following relationship:

$$x_t^f = f(L)x_t, \tag{1}$$

where $f(L)$ is a finite-order, two-sided linear filter in the lag operator L:

$$f(L) = \sum_{i=-n}^{n} f_i L^i, \tag{2}$$

with half-length n and fixed, finite weights $\{f_i\}$. Some f_i could be zero, so $f(L)$ could be a one-sided filter in practice. Only *finite*-order filters are considered. Generalization to an infinite-order filter is feasible, but is of limited interest in the context of actual SA procedures.

When examining the properties of $f(L)$ in the context of SA filters, it will prove useful to re-express $f(L)$ as:

$$\begin{aligned} f(L) &= f(1) + f^*(L)\Delta \\ &= f(1) + f^*(1)\Delta + f^{**}(L)\Delta^2, \end{aligned} \tag{3}$$

where Δ is the difference operator, $(1-L)$; and $f^*(L)$ and $f^{**}(L)$ are themselves finite-order, fixed-weight, two-sided linear filters with polynomial coefficients denoted $\{f_i^*\}$ and $\{f_i^{**}\}$. Here and below, superscript asterisks * and ** denote polynomials obtained as in (3). While (3) finishes with a polynomial in Δ^2, the recursion can be repeated to any order. The sum of coefficients in each successive lag polynomial can be obtained recursively, noting that:

$$f^*(1) = -\left.\frac{\partial f(L)}{\partial L}\right|_{L=1}, \quad f^{**}(1) = -\left.\frac{\partial f^*(L)}{\partial L}\right|_{L=1}, \quad \text{etc.} \tag{4}$$

Consequently, the sums $f^*(1)$ and $f^{**}(1)$ are $-\sum_{i=-n}^{n} i\cdot f_i$ and $-\sum_{i=-n}^{n-1} i\cdot f_i^*$.

An analogue model of seasonal adjustment. In the analogue model of SA, x_t^f and x_t correspond to SA and NSA series respectively. As the analogue SA filter, $f(L)$ satisfies three assumptions:

(i) the weights $\{f_i\}$ sum to unity,
(ii) $f(L)$ is symmetric in L, and
(iii) $f(L)$ eliminates deterministic seasonals.

While this analogue model only approximates the highly complex X–11 SA procedure, the approximation appears a good one for linear properties; cf. Nerlove (1964), Wallis (1974, 1983), Cleveland and Tiao (1976), and Bell (1992). See Wallis (1974, 1978) and Sims (1974) on the possible econometric consequences of using SA (or NSA) data. See Lovell (1963), Grether and Nerlove (1970), Granger (1978), Kenny and Durbin (1982), Wallis (1982, 1983), Burridge and Wallis (1984), Hylleberg (1986), and Maravall and Pierce (1987) on properties of existing and more 'optimal' SA procedures.

The nature of the three assumptions for the analogue model is now discussed.

ASSUMPTION 1. *The sum* $f(1)$ *is unity.*

Under Assumption 1, the first equality in (3) is:

$$f(L) = 1 + f^{*}(L)\Delta. \tag{5}$$

By restricting the sum of coefficients in the scalar polynomial $f(L)$ to unity, Assumption 1 ensures that x_t^f and x_t are in the same units. Assumption 1 is with loss of generality by excluding unit roots in $f(L) = 0$, but otherwise represents a normalization of $f(L)$. The normalized polynomial is sufficiently general for current purposes since the focus is on seasonal adjustment. Further, if $f(1)$ were zero rather than unity, x_t^f would be a finite-weight, finite-order distributed lag of Δx_t, and so would be integrated of an order different from that of x_t. The restriction $f(1) = 0$ thus affects long-run properties of the data whereas seasonal adjustment is meant to leave those properties intact, so $f(1) \neq 0$ is a relatively innocuous assumption.

ASSUMPTION 2. *The polynomial $f(L)$ is symmetric in the lag operator.*

Assumption 2 means that $f(L) = f(L^{-1})$ or, equivalently, $f_i = f_{-i}$, $i = 1,\ldots,n$. From (3) and (4), Assumption 2 implies $f^{*}(1) = 0$, and so $f(L)$ can be written as:

$$f(L) = f(1) + f^{**}(L)\Delta^2. \tag{6}$$

That $f^{*}(1)$ is zero can be seen by solving for the coefficients $\{f_i^{*}\}$ in terms of the $\{f_i\}$. In general, the polynomial $f^{*}(L)$ is:

$$f^{*}(L) = \sum_{i=-n}^{n-1} f_i^{*} L^i = \sum_{i=1}^{n} f_{-i}^{*} L^{-i} + \sum_{i=1}^{n} f_{i-1}^{*} L^{i-1}. \tag{7}$$

From (3), the coefficients f_i^{*} are:

$$f_i^{*} = \begin{cases} \sum_{j=|i|}^{n} f_{-j} & for \quad -n \leq i < 0 \\ -\sum_{j=i+1}^{n} f_j & for \quad 0 \leq i < n. \end{cases} \tag{8}$$

Under symmetry, $f_i = f_{-i}$ $(0 < i \leq n)$, so $f_{-i}^{*} = -f_{i-1}^{*}$ $(0 < i \leq n)$. Substituting (8) into (7), $f^{*}(1)$ is zero. Symmetry is sufficient for $f^{*}(1) = 0$, but is not necessary. For example, a unit root in $f^{*}(L)$ ensures $f^{*}(1) = 0$, but does not imply symmetry. Extensions of this result appear in Osborn (1993) and Wallis (1993).

ASSUMPTION 3. *The polynomial $f(L)$ has a factor $v(L)$, where*

$$v(L) = s^{-1}\sum_{i=0}^{s-1} L^i, \text{ and } s \text{ is the periodicity of seasonality.}$$

Assumption 3 ensures that the SA filter eliminates any fixed seasonal pattern. Other assumptions about $f(L)$ will achieve the same result, but Assumption 3 is one of the simplest and most intuitive: $v(L)$ averages the data over the seasonal interval. Even so, Assumption 3 is not innocuous. An SA filter with a factor of $v(L)$ will eliminate seasonal unit roots in x_t if they are present. If they are not present, application of the SA filter will be similar to over-differencing; see also Maravall (1993).

A fixed seasonal pattern can be represented by $s(L)S_{1t}$, where $s(L)$ is an $(s-1)^{\text{th}}$-order polynomial and S_{it} is a mean-adjusted dummy for the ith season. Hence $f(L)$ should annihilate $s(L)S_{1t}$:

$$f(L)s(L)S_{1t} = 0. \tag{9}$$

Under Assumption 3, $f(L) = f^o(L)v(L)$, where $f^o(L)$ is a fixed-weight, finite-order polynomial; and the superscript ^{o} indicates that $v(L)$ has been factored from the polynomial. Because S_{it} is mean-adjusted, it follows that:

$$f(L)s(L)S_{1t} = f^o(L)v(L)s(L)S_{1t} = f^o(L)s(L)[v(L)S_{1t}] = 0. \tag{10}$$

Thus, Assumption 3 ensures that $f(L)$ annihilates the seasonal dummies.

To summarize, under Assumptions 1–3, $f(L)$ may be written as:

$$f(L) = 1 + f^{**}(L)\Delta^2 = f^o(L)v(L). \tag{11}$$

In discussing filtered series satisfying Assumptions 1–3, x_t^f will be referred to as x_t^a ('*a*' for analogue or adjusted), the weights $\{f_i\}$ as $\{a_i\}$, and so the filter $f(L)$ as the analogue SA filter $a(L)$.

Cointegration. Cointegration is discussed at length elsewhere; cf. Engle and Granger (1987), Hendry (1986), Johansen (1988, 1991), Johansen and Juselius (1990), Phillips (1991), Ericsson (1992*a*), and Banerjee *et al.* (1993). Here, the interest is in cointegration of the pair of integrated series $(x_t^f : x_t)'$. If x_t is I(d) and $x_t^f - x_t$ is (at most) I(k) for integers d and k such that $d > k \geq 0$, then x_t^f and x_t cointegrate from I(d) to I(k) with unit coefficients, denoted CI1(d,k). In the context of SA, filters for which x_t^f and x_t are CI1(d,0) are of particular interest, yet the order of integration d may be unknown in practice.

Mean equality. Even for a filter ensuring CI1(d,0), the expectation of $x_t^f - x_t$ need not be zero. (Throughout, this expectation is assumed to exist.) Yet, it may be desirable to have an SA procedure such that the adjusted and unadjusted series are equal 'on average'. When the SA procedure does so, the series satisfy 'mean equality'. Because integrated processes need not have finite means and because fixed seasonal patterns are non-ergodic, mean equality is defined as $E[\Sigma_s(x_t^f - x_t)/s] = E[v(L)(x_t^f - x_t)] = 0$, where the summation is over the seasonal interval, the expectation E[·] is over possible realizations, and x_t^f and x_t are CI1(d,0).

Implications for the SA filter $a(L)$ follow directly. Let Δ_s [$=(1-L^s)$] be the annual change operator, and note that $\Delta v(L) = s^{-1}\Delta_s$. Since $x_t^a = a(L)x_t$ and $a(L) = 1 + a^{**}(L)\Delta^2$ from (11), then:

$$E\left[\sum_s (x_t^a - x_t)/s\right] = E\left[v(L)\{a^{**}(L)\Delta^2 x_t\}\right] = E\left[a^{**}(L)\Delta^2 v(L)x_t\right] \\ = E\left[a^{**}(L)\Delta\Delta_s x_t\right]/s. \quad (12)$$

Thus, mean equality requires a lag polynomial $a^{**}(L)$ for which $a^{**}(L)\Delta\Delta_s x_t$ is zero on average. Suitable conditions are discussed in the next subsection.

2.2 Relationships between series

This subsection considers the relationships between x_t^f and x_t under Assumptions 1–3 about the polynomial $f(L)$, applying those assumptions only as necessary.

The following result establishes the relationship between unit cointegrationand the properties of the linear filter $f(L)$. Without loss of generality, suppose:

$$f(L) = f(1) + h(L)\Delta^q, \quad (13)$$

where $h(L)$ is a finite-order, fixed-weight linear filter; $h(1)$ is finite; and q is positive. Under Assumption 1, x_t^f and x_t are CI1(d,0) if and only if a q exists such that $q \geq d$. From (1),

$$x_t^f - x_t = \left[1 + h(L)\Delta^q\right]x_t - x_t = h(L)\Delta^q x_t, \quad (14)$$

which is I(0) if and only if $q \geq d$. For autoregressive processes of x_t as in (19) below, mean equality requires an extra order of differencing to ensure that $E[h(L)\Delta^q x_t]$ is not a function of (e.g., for $d = 1$) the variable's average growth rate.

Under Assumption 1 alone, x_t^f and x_t are CI1(d, d–1) from (5). Thus, if $d = 1$, the filtered and raw data are cointegrated in the usual sense. If the filter is symmetric as well (Assumption 2), $q = 2$ by construction from (6). Hence, for

$d = 1$, x_t^f and x_t will also be mean equal. As a corollary, x_t and the series from the analogue SA process x_t^a are CI1(d, d–2), and are CI1(d, d–1) with mean equality (provided the expectation exists). From the general properties of cointegrated series, Granger causality must run in at least one direction between x_t^f and x_t.

2.3 Implications for sets of series

For a given type of data (filtered or unfiltered), several series may themselves be cointegrated. This subsection shows that such cointegration implies that the corresponding series of the 'other' type are cointegrated with the same cointegrating vector(s).

Consider p variables and denote the filtered and unfiltered series of the jth variable as x_{jt}^f and x_{jt} respectively ($j = 1,...,p$). The associated $p \times 1$ vectors of data are denoted x_t^f and x_t, with their context below clarifying that they are vectors (rather than scalars as in Sections 2.1 and 2.2). The vectors x_t^f and x_t have the relationship:

$$x_t^f = F(L)x_t, \tag{15}$$

where

$$F(L) = \sum_{i=-n}^{n} F_i L^i = F(1) + \Delta \sum_{i=-n}^{n-1} F_i^* L^i = F(1) + F^*(L)\Delta, \tag{16}$$

and the F_i are $p \times p$ matrices. For the time, assume $F(L)$ is the diagonal matrix of scalar polynomials, diag[$f_1(L)$,...,$f_p(L)$], where $f_j(L)$ is the filter generating x_{jt}^f from x_{jt} (j = 1,...,p). That is, each filtered series is a weighted moving average of the corresponding unweighted series, and that unweighted series alone. Filters for different series need not be the same. Under Assumption 1, $F(1) = I_p$.

Cointegration within x_t. If, under Assumption 1, β is a cointegrating vector for the integrated vector process x_t, then β is also a cointegrating vector for x_t^f. The converse also applies. Intuitively, seasonal adjustment should affect only the dynamics of a process, and not its long-run properties. A general proof for equivalent cointegrating vectors follows directly from the definition of $F(L)$ under Assumption 1. The special case where x_t can be represented as a finite-order Gaussian vector autoregression (VAR) illustrates this invariance to data type.

A Proof. Under Assumption 1, premultiplying (15) by β' yields:

$$\beta' x_t^f = \beta' x_t + \beta' F^*(L)\Delta x_t. \tag{17}$$

If x_t is I(1), $\beta' x_t$ and Δx_t are both I(0) in (17), and so $\beta' x_t^f$ must be I(0). That is, β is a cointegrating vector for x_t^f. More generally, if x_t is CI(d,k) with cointegrating vector β, x_t^f is CI(d,d–1) with cointegrating vector β, and $\beta' x_t$ and $\beta' x_t^f$ differ by a term that is I(d–1). For x_t and x_t^f interchanged, the proof is immediate. A corollary also follows directly: if x_t has r cointegrating vectors, then so does x_t^f, and vice versa.

Under Assumptions 1–3, equation (17) becomes:

$$\beta' x_t^a = \beta' x_t + \beta' A^{**}(L)\Delta^2 x_t, \tag{18}$$

in an obvious notation. The disequilibrium measures $\beta' x_t^a$ and $\beta' x_t$ differ by the term $\beta' A^{**}(L)\Delta^2 x_t$, which is two orders of integration less than the order of integration of x_t. If x_t is I(1), then $\beta' x_t^a$ and $\beta' x_t$ satisfy mean equality. Figure 8 in Section 4.3 below shows how close the empirically estimated $\beta' x_t^a$ and $\beta' x_t$ can be.

Importantly, the cointegration results both for pairs of series and for sets of series require assumptions about the seasonal filter only. *No* assumption is made about the actual seasonality (or lack thereof) in the data, other than implicitly (and rather weakly) through the assumed order of integration of the data.

If Assumption 1 is *not* satisfied, the cointegrating vector(s) for x_t and those for x_t^f need not be the same. Thus, unless implemented carefully, multiple-series SA procedures run the risk of affecting long-run as well as short-run relationships between the series; cf. Bartelsman and Cleveland (1993).

VARs. In the cointegration literature, Johansen (1988) and Johansen and Juselius (1990) have stimulated interest in finite-order Gaussian VARs. If x_t follows such a process, then inference in a VAR for the SA data x_t^a generally is affected, even though β and the number of cointegrating vectors r are invariant to the transformation from x_t to x_t^a. Also, as discussed in Hendry and Mizon (1978) and Davidson *et al.* (1978), series may be filtered, but not relationships. In practice, the relationship for the SA data is obtained by filtering each NSA series individually (potentially using different filters) and then combining the filtered (SA) data. Thus, the remainder of this subsection derives the VAR for x_t^a from a VAR for x_t by using the relationship between the NSA and SA data in (15) and discusses empirical implications for analyzing x_t^a rather than x_t.

Suppose x_t has the representation:

$$x_t = \mu + \sum_{i=1}^{\ell} \pi_i x_{t-i} + \Phi S_t + \varepsilon_t, \quad \varepsilon_t \sim \mathrm{IN}(0,\Omega), \tag{19}$$

where μ is a $p \times 1$ vector of constants, $\{\pi_i\}$ are $p \times p$ matrices of autoregressive coefficients, ℓ is the maximal lag length, S_t is a vector of seasonal dummies

$(S_{1t},\ldots,S_{st})'$, Φ is the corresponding matrix of coefficients, and ε_t is a mean zero Gaussian innovation with covariance matrix Ω. By adding and subtracting various lags of x_t, (19) may be rewritten as:

$$\Delta x_t = \mu + \pi x_{t-1} + \sum_{i=1}^{\ell-1} \Gamma_i \Delta x_{t-i} + \Phi S_t + \varepsilon_t, \tag{20}$$

where the $\{\Gamma_i\}$ and π are:

$$\Gamma_i = -(\pi_{i+1} + \cdots + \pi_\ell), \quad i = 1, \ldots, \ell - 1, \tag{21}$$

$$\pi = \left(\sum_{i=1}^{\ell} \pi_i \right) - I_p. \tag{22}$$

For convenience below, (20) can be rewritten as a polynomial in Δx_t:

$$\Gamma(L)\Delta x_t = \mu + \pi x_{t-1} + \Phi S_t + \varepsilon_t, \tag{23}$$

where $\Gamma(L) = I_p - \Sigma_{i=1}^{\ell-1}\Gamma_i L^i$. Also, for simplicity and ease of exposition, assume that x_t is I(1).

The matrix π contains the 'impact' coefficients of the lagged level x_{t-1}. Defining r as rank(π), π can be expressed as the outer product of two (full column rank) $p \times r$ matrices α and β:

$$\pi = \alpha\beta', \tag{24}$$

for $0 < r \leq p$. The matrix β' is the set of cointegrating vectors, α is the matrix of 'weighting elements', and r is the number of cointegrating vectors. If $r = 0$, then $\pi = 0$, in which case Δx_t is solely a function of its lags, μ, S_t, and ε_t, and there is no cointegration. If $r > 0$, the representation in (24) is unique only up to nonsingular $r \times r$ linear transformations, since $\alpha\beta' = (\alpha Q)(Q^{-1}\beta') = \alpha^\dagger\beta^{\dagger\prime}$ (say) for a nonsingular $r \times r$ matrix Q. It is assumed that there are sufficient a priori restrictions on α and β to identify them uniquely.

To derive the VAR for x_t^a, two operations are useful: pre-multiplication of (23) by $F(L)$ and substitution of x_t by $x_t^a + (x_t - x_t^a)$. In combination with Assumptions 1 and 3, these operations obtain the VAR for x_t^a. From the first operation, (23) becomes:

$$F(L)\Gamma(L)\Delta x_t = F(L)\mu + F(L)\pi x_{t-1} + F(L)\Phi S_t + F(L)\varepsilon_t. \tag{25}$$

The terms in (25) are transformed as follows. Applying the second operation, the left-hand side of (25) is:

$$F(L)\Gamma(L)\Delta x_t = F(L)\Gamma(L)\Delta x_t^a + F(L)\Gamma(L)\Delta\left(x_t - x_t^a\right). \tag{26}$$

Using Assumption 1, $F(L)\mu = \mu$. Applying the second operation and using Assumption 1, the second term on the right-hand side of (25) is:

$$\begin{aligned} F(L)\pi x_{t-1} &= F(L)\pi x_{t-1}^a + F(L)\pi\left(x_{t-1} - x_{t-1}^a\right) \\ &= \pi x_{t-1}^a + F^*(L)\pi\Delta x_{t-1}^a + F(L)\pi\left(x_{t-1} - x_{t-1}^a\right). \end{aligned} \tag{27}$$

Under Assumption 3, the seasonal dummies vanish in (25):

$$F(L)\Phi S_t = F^o(L)v(L)\Phi S_t = F^o(L)\Phi v(L)S_t = 0, \tag{28}$$

in an obvious notation. Thus, by substitution and re-arrangement of terms, (25) may be rewritten as:

$$G(L)\Delta x_t^a = \mu + \pi x_{t-1}^a + \eta_t, \tag{29}$$

where:

$$G(L) = F(L)\Gamma(L) - F^*(L)\pi L, \tag{30}$$

$$\eta_t = -F(L)\pi(L)\left(x_t - x_t^a\right) + F(L)\varepsilon_t, \tag{31}$$

and $\pi(L)$ is defined as $I_p - \sum_{i=1}^{\ell} \pi_i L^i$ (equivalently, $\pi(L) = \Gamma(L)\Delta - \pi L$).

While (29) is formally similar to (23), several important differences exist. First, (23) is a conditional model, whereas (29) generally is not: $G(L)$ is in general a two-sided polynomial. Second, no seasonal dummies appear in (29) because $F(L)$ annihilates them. Third, ε_t in (23) is an innovation whereas η_t in (29) is not. This can be problematic when conducting inference on (29); cf. Ghysels and Perron (1993) for the univariate case. Fourth, as noted by Wallis (1974, p. 21), use of SA data may induce residual autocorrelation at seasonal lags: note the presence of the seasonal factor $x_t - x_t^a$ in the error η_t.

From (17), the number of cointegrating vectors r and the cointegrating vectors β are invariant to the type of data (NSA or SA). From (23) and (29), the impact matrix π also appears invariant in a conditional VAR, but it is not. In (29), πx_{t-1}^a is $\alpha\beta' x_{t-1}^a$; $\beta' x_{t-1}^a$ is I(0); and in general $\beta' x_{t-1}^a$ is correlated with η_t. Consequently, the weighting matrix α is not invariant in conditional VAR models, so neither is the product $\alpha\beta'$. Thus, whether or not a set of variables is weakly exogenous may depend upon which type of data is used. Additionally, because $G(L)$ is in general a two-sided non-diagonal polynomial matrix, the conditional representation of (29) confounds dynamics from future Δx^a with $\beta' x_{t-1}^a$, and dynamics in η_t with lagged dynamics in Δx^a. Both affect α, and so weak exogeneity for β. Even with the invariance of cointegration itself, the empirical power of cointegration tests on SA and on NSA data may differ, as illustrated in Lee and Siklos (1991, 1993).

2.4 Comments

Wallis (1974) finds that the linear filter (2) approximates the actual X–11 procedure well. However, as Wallis (1974, p. 20) notes, the linear filter ignores several features of X–11: graduation of extreme values, constraints on calendar-year totals, corrections at the ends of series, and multiplicative models of SA. For integrated data, the first three of the ignored features affect only the short-run dynamics for a wide class of x_t processes, so the results above should still hold. However, see Granger and Hallman (1991) and Ermini and Granger (1993) on nonlinear transformations of integrated processes, as arise in multiplicative models.

3. ENCOMPASSING TESTS IN THEORY

Consider two empirical models, one developed on NSA data and the other on SA data. The following description of parameter encompassing suggests how to compare and evaluate these two models. Suppose one model (Model 1) has an estimate $\hat{\theta}$ for its parameter θ, and the other model (Model 2) implies that that parameter should be $\tilde{\theta}$. Then test the closeness of $\tilde{\theta}$ to $\hat{\theta}$, accounting for the uncertainty from estimation. If $\tilde{\theta}$ is 'statistically close' to $\hat{\theta}$, Model 2 parameter-encompasses Model 1. That is, Model 2 explains why Model 1 obtains the estimate that it does. See Mizon and Richard (1986) and Hendry and Richard (1989) for extensive discussions of encompassing.

Numerous forms of encompassing have been proposed, including those of: the error variance [Cox (1961, 1962), Mizon and Richard (1986)], parameters [Hendry (1983), Mizon and Richard (1986)], the reduced form [Ericsson (1983), Hendry and Mizon (1993)], exogeneity [Hendry (1988)], and forecasts [Chong and Hendry (1986), Ericsson (1992*b*)]. However, none of these tests are directly applicable because the dependent variables of the SA and NSA models are inherently different. This section modifies the parameter-encompassing test to address this problem, using the relation between the adjusted and unadjusted data from the SA filter. Because of a fundamental asymmetry in the two models, this section first considers whether or not the SA model encompasses the NSA model (Section 3.1), and then the converse (Section 3.2).

3.1 Does an SA model encompass an NSA model?

For ease of exposition, this section switches to and modifies a notation common to the literature on encompassing and non-nested hypothesis tests, even while this new notation does conflict with that in Section 2. Denote the conditional (regression) models on the SA and NSA data as:

$$M_{SA}: \; y_t^a = x_t^{a\prime}\beta + u_{0t}, \qquad u_{0t} \sim \mathrm{IN}\left(0, \sigma_0^2\right) \tag{32}$$

$$M_{NSA}: \; y_t = z_t'\gamma + u_{1t}, \qquad u_{1t} \sim \mathrm{IN}\left(0, \sigma_1^2\right), \tag{33}$$

where y_t^a and y_t are the SA and NSA observations on the dependent variable at time t, x_t^a and z_t are vectors of SA and NSA regressors for the models M_{SA} and M_{NSA}, and β and γ are the corresponding coefficients. The error u_{0t} (u_{1t}) is assumed to be independently and normally distributed with mean zero and variance σ_0^2 (σ_1^2) under the hypothesis that M_{SA} (M_{NSA}) is correctly specified. Independence and normality are chosen for expositional simplicity. The regressors in M_{NSA} are z_t rather than x_t so as to allow for different dynamic structures in the two models. Even so, x_t^a and z_t may well derive from the same set of basic variables.

In the standard framework, the F-statistic for the significance of z_t in M_{SA} would be used to test whether or not M_{SA} parameter-encompasses M_{NSA}. Under the hypothesis M_{SA}, this F-statistic is distributed as an F-ratio asymptotically, and possibly in finite samples as well. However, the dependent variables in (32) and (33) differ, so some modification of the statistic is necessary. One possibility is to use the regression:

$$y_t^a = x_t^{a\prime}\beta + z_t'c + \left(y_t - y_t^a\right)d_0 + e_{0t}, \tag{34}$$

with regression error e_{0t}, and test for the significance of z_t and the seasonal factor $(y_t - y_t^a)$. Under the hypothesis M_{SA}, $c = 0$ and $d_0 = 0$, and both z_t and $(y_t - y_t^a)$ are valid conditioning variables.[2] As usual, variance dominance ($\sigma_0^2 < \sigma_1^2$) is a necessary condition for encompassing. Under M_{NSA}, $\beta = 0$, $c = \gamma$, and $d_0 = -1$ although, as implied by Section 3.2 below, the precise power of the test may be difficult to derive.

The encompassing test proposed in (34) is easy to calculate. Several variants of the test are possible, depending upon which variables are added to (32): z_t and

---

[2] Lovell (1963, p. 995) discusses the desirability of an SA procedure being orthogonal, i.e., where $\Sigma_t(y_t - y_t^a)y_t^a = 0$. Even so, some SA procedures are not orthogonal, in which case the resulting SA data contain a seasonal component. Unless seasonal adjustment is orthogonal, $(y_t - y_t^a)$ and the dependent variable in (34) will be correlated.

$(y_t - y_t^a)$ [as proposed], z_t only, x_t and $(y_t - y_t^a)$, and x_t.[3] While only the first implies a nesting model for M_{SA} and M_{NSA}, the others may prove useful as general diagnostic procedures.

3.2 Does an NSA model encompass an SA model?

In light of (34), the obvious procedure for testing whether or not M_{NSA} encompasses M_{SA} is to estimate the regression:

$$y_t = z_t'\gamma + x_t^{a\prime}b + \left(y_t - y_t^a\right)d_1 + e_{1t}, \tag{35}$$

with regression error e_{1t}, and test for the significance of x_t^a and $(y_t - y_t^a)$, i.e., test $b = 0$ and $d_1 = 0$. Unfortunately, this procedure is invalid: in general, x_t^a and $(y_t - y_t^a)$ are not valid conditioning variables under M_{NSA} because they include future values of y and x. The problem is most obvious when x_t^a contains the SA lagged dependent variable y_{t-1}^a. From the two-sided nature of the SA filter, y_{t-1}^a includes y_t, y_{t+1}, and so on. Under M_{NSA}, these are not valid 'explanatory' variables.

An encompassing test is feasible, but it requires analysis of y_t, z_t, and x_t jointly; and it is inherently more difficult to calculate. The procedure is as follows, where $x_t = z_t$ (assumed purely for ease of exposition). Suppose y_t and z_t are both modeled, as in a VAR or in a conditional/marginal factorization. From that model, the moments of $(y_t : z_t')'$ can be calculated. The matrix SA filter $F(L)$ transforms $(y_t : z_t')'$ to $(y_t^a : z_t^{a\prime})'$; and conditioning y_t^a on z_t^a generates the regression coefficient on z_t^a in terms of the moments of $(y_t^a : z_t^{a\prime})'$. Thus, from the model for $(y_t : z_t')'$, an implied coefficient from regressing y_t^a on z_t^a can be calculated via $F(L)$. The encompassing statistic compares that implied coefficient with the estimated coefficient.[4]

This procedure has several difficulties. First, analysis of y_t conditional on z_t may have been chosen precisely because modeling $(y_t : z_t')'$ is more difficult. E.g., in Sections 4–5, a congruent, constant-parameter, parsimonious, economically interpretable, conditional money demand model was (relatively) easy to obtain.

---

[3] Other possible inclusions are: $(x_t - x_t^a)$ and $(y_t - y_t^a)$; $(x_t - x_t^a)$; $(z_t - z_t^a)$ and $(y_t - y_t^a)$; $(z_t - z_t^a)$; z_t, $(z_t - z_t^a)$, and $(y_t - y_t^a)$; and z_t and $(z_t - z_t^a)$. These emphasize the seasonal discrepancies between the adjusted and unadjusted data. The first two are equivalent to two of the main variants.

[4] In light of Campos, Ericsson, and Hendry (1990), this procedure parallels what would be required to test whether or not Hendry and Ericsson's (1991*a*) U.K. money demand equation on annual data parameter-encompasses Friedman and Schwartz's (1982) money demand equation on phase-average data. The empirical *lack* of parameter encompassing in the reverse direction follows directly from those equations' standard errors, where variance dominance is necessary for parameter encompassing.

As an alternative encompassing approach, foreshadowed by Wallis (1974), premultiply both sides of (33) by $f_y(L)$ to obtain the same dependent variable as in (32). Then apply standard encompassing tests. This approach is complicated by the induced two-sided moving averages of the error and z_t in the modified (33).

Finding an economically and statistically acceptable model of interest rates, inflation, and total final expenditure is much more difficult. Second, the mapping by $F(L)$ only approximates X–11. Third, the actual numerical calculations are substantial, and nontrivial to program. Thus, Section 5 does not calculate this statistic, but reports statistics from regressions like (35), recognizing that those regressions may include invalid conditioning variables.

Equations (34) and (35) are algebraically identical, with $d_1 = 1 + d_0$, $\gamma = c$, and $b = \beta$, as follows from adding $(y_t - y_t^a)$ to both sides of (34). That is, the coefficients on y_t and y_t^a sum to unity. A parallel structure appears in Ericsson's (1992*b*) modification to Chong and Hendry's (1986) forecast-encompassing test statistic. The validity of (34) [or (35)] as a maintained hypothesis depends upon the null hypothesis, whether (32) or (33). Specifically, the validity of conditioning on the 'additional' variables in (34) [or (35)] is at issue.

Surprisingly, variance dominance is *not* a necessary condition for M_{NSA} to encompass M_{SA}. Under M_{NSA}, the model M_{SA} (possibly) conditions inappropriately on x_t^a, giving M_{SA} an 'artificially low' error variance. Parallel situations arise in comparing least squares and instrumental variables error variances in a simultaneous equations framework, and in testing conditional versus expectational models; cf. Hendry (1988) and Favero and Hendry (1992) on the latter. Also, M_{SA} may have a smaller error variance than M_{NSA} even if conditioning is not an issue: by averaging the dependent variable, the SA filter may reduce the dependent variable's variability.

4. EMPIRICAL RESULTS ON COINTEGRATION

This section analyses UK data on money demand with the system-based cointegration procedures in Johansen (1988) and Johansen and Juselius (1990). Section 4.1 examines cointegration between the adjusted and unadjusted series, variable by variable. Sections 4.2 and 4.3 analyse the unadjusted data and adjusted data as separate sets. The remainder of this introduction summarizes the economic theory of money demand and describes the data. The notation is one common to the money demand literature, albeit conflicting occasionally with the notation above.

The standard theory of money demand posits:

$$m^d - p = \delta_0 \cdot y + \delta' R, \tag{36}$$

where M^d is nominal money demanded, P is the price level, Y is a scale variable ('income'), R is a vector of interest rates, and variables in lower case are in logarithms. Equation (36) assumes log-linearity in money, prices, and incomes and linearity in interest rates, a common functional form. The income elasticity δ_0 is one half in Baumol's (1952) and Tobin's (1956) transactions demand theory and unity in Friedman's (1956) quantity theory. The elements in δ are semi-elasticities for interest rates, and a given element is negative (positive) if the associated asset is excluded from (included in) the selected monetary aggregate. See Laidler (1985) and Goldfeld and Sichel (1990) for general discussions of money demand, Goodhart (1984, 1989) on UK financial institutions, and Miller and Orr (1966), Milbourne (1983), and Smith (1986) for additional developments on the theory of money demand.

In the empirical analysis below, M, Y, and P are nominal M_1, real total final expenditure (TFE) at 1985 prices, and the TFE deflator. There are two interest rates, the three-month local authority interest rate ($R3$) and the M_1 retail sight-deposit interest rate (Rr). The first is the dominant short-term interest rate in the secondary market and measures the return on (some) assets outside M_1. The second is the interest rate on checkable interest-bearing accounts at commercial banks and is a return on an asset within M_1. Three derived variables are of interest: the inflation rate (Δp), the learning-adjusted retail sight-deposit interest rate (Rra; see Baba, Hendry, and Starr (1992), Hendry and Ericsson (1991*b*), and Appendix A for details), and the net interest rate or opportunity cost (defined as $R3$–Rra and denoted R^*). Money and expenditure are in £ millions, the deflator is unity for 1985, and interest rates are in fractions. The data are quarterly, 1963(1)–1989(2). Allowing for lags and transformations, estimation is over 1964(3)–1989(2), which is 100 observations ($T = 100$). For details on the data, see Appendix A.

Data description begins with six pairs of graphs, where the first of each pair (e.g., Figure 1a) plots the NSA and SA series for a given variable and the second (e.g., Figure 1b) plots the difference between them, denoted the seasonal component. Figures 1–6 show m, p, $m-p$, y, $y+p$ (nominal TFE), and $m-p-y$ (inverse velocity) over 1963(1)–1989(2). Visually, all NSA and SA series appear I(1) at least; the augmented Dickey-Fuller (1979, 1981) [ADF] test statistics in Table 1 support this.[5] For m and p in particular, system analysis in Johansen (1992*c*) suggests that they are I(2), but the evidence is not conclusive. The hypothesis of a unit root in Δm (and in Δp) can not be rejected with the ADF statistic at standard significance levels, but the estimated root for Δm is only +0.31 (+0.87 for Δp). Thus, agnosticism on the order of integration for m and p seems appropriate, so Ericsson, Hendry, and Tran (1993, Appendix C) consider the implications that m and p being I(2) or I(1) has for this analysis.[6]

---

[5] Here and below, a maximum of five lags is chosen, thereby allowing for possible stochastic seasonality in the quarterly data while not being too profligate in parameters. For instance, the fifth-order VAR for System I below entails 29 coefficients in each of five equations estimated on 100 observations.

[6] The order of integratedness need not even be an inherent property of a time series; a series's order of integration could differ for different time periods.

Figure 1. Nominal Money Stock

(a) Logs of NSA (—) and SA (···) data $[m_t \text{ and } m_t^a]$ (b) The seasonal component $[m_t - m_t^a]$

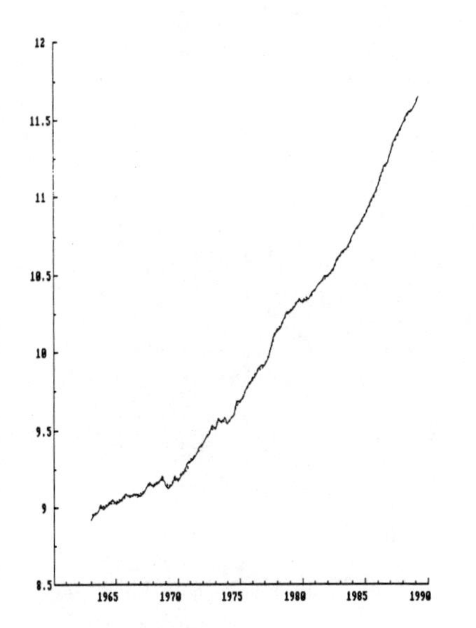

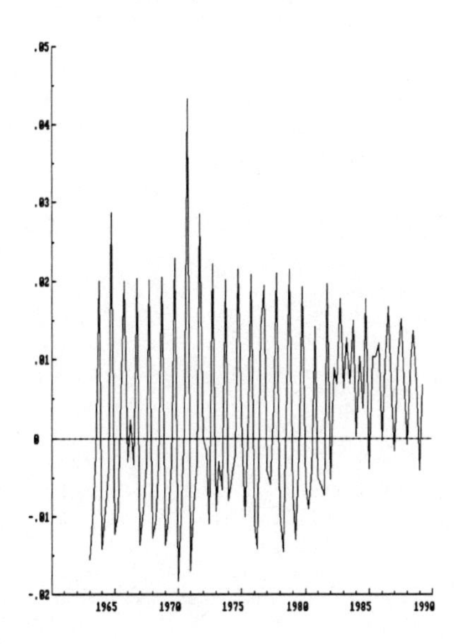

Figure 2. Prices

(a) Logs of NSA (—) and SA (···) data $[p_t \text{ and } p_t^a]$ (b) The seasonal component $[p_t - p_t^a]$

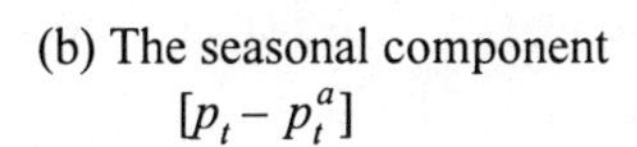

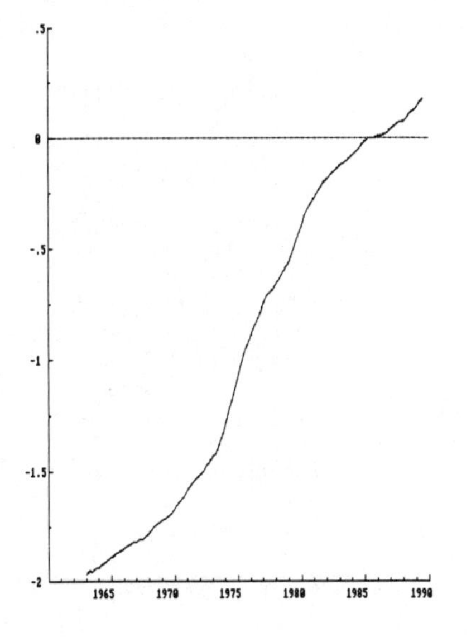

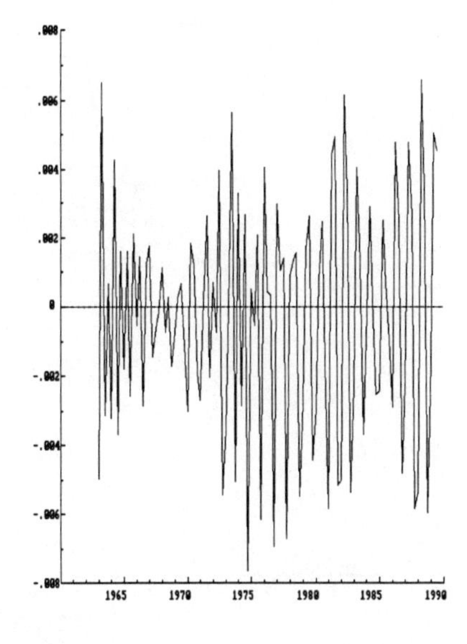

Figure 3. Real Money Stock

(a) Logs of NSA (—) and SA (···) data $[(m-p)_t$ and $(m^a - p^a)_t]$

(b) The seasonal component $[(m-p)_t - (m^a - p^a)_t]$

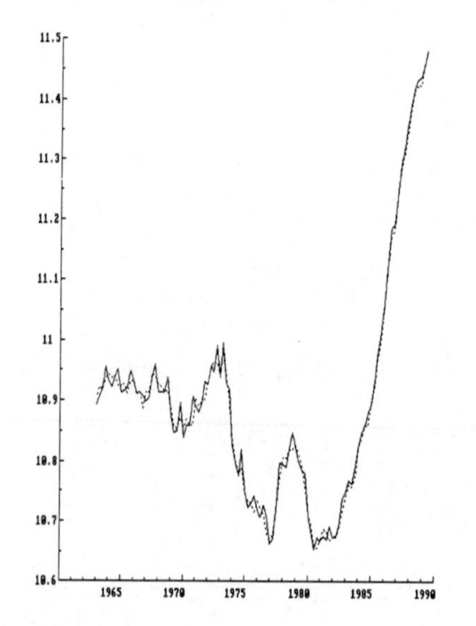

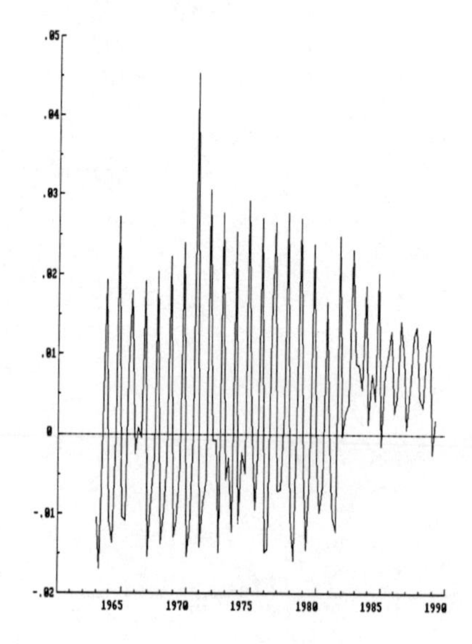

Figure 4. Real Income

(a) Logs of NSA (—) and SA (···) data $[y_t$ and $y_t^a]$

(b) The seasonal component $[y_t - y_t^a]$

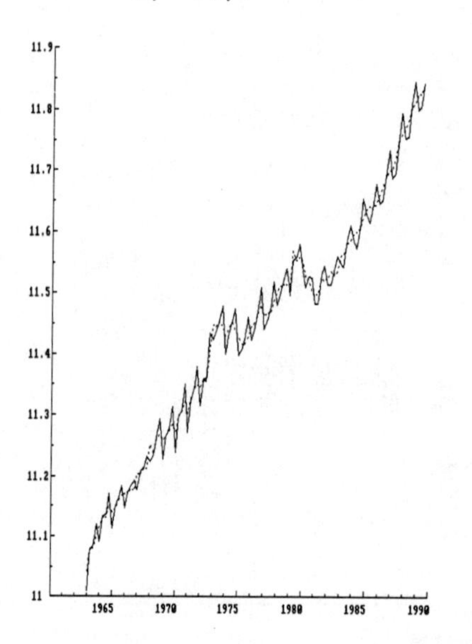

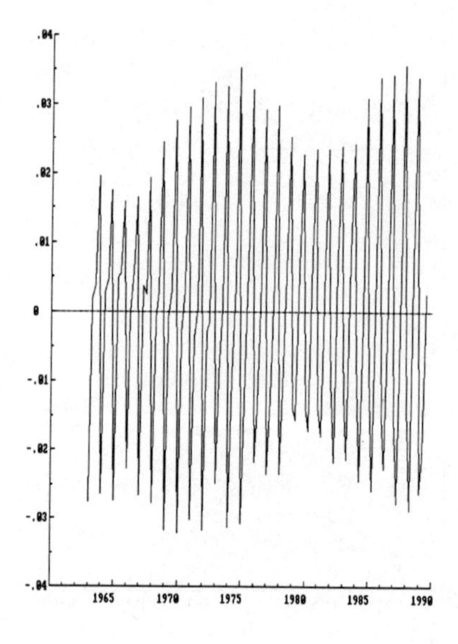

Figure 5. Nominal Income

(a) Logs of NSA (—) and SA (···) data $[(y+p)_t$ and $(y^a+p^a)_t]$

(b) The seasonal component $[(y+p)_t - (y^a+p^a)_t]$

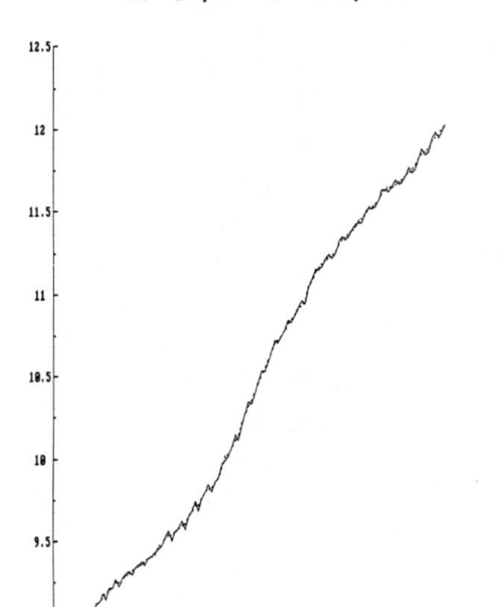

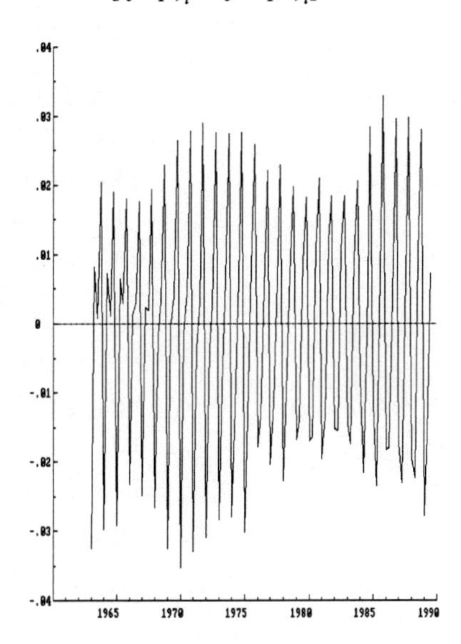

Figure 6. Inverse Velocity

(a) Logs of NSA (—) and SA (···) data $[(m-p-y)_t$ and $(m^a-p^a-y^a)_t]$

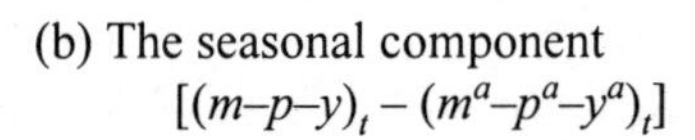
(b) The seasonal component $[(m-p-y)_t - (m^a-p^a-y^a)_t]$

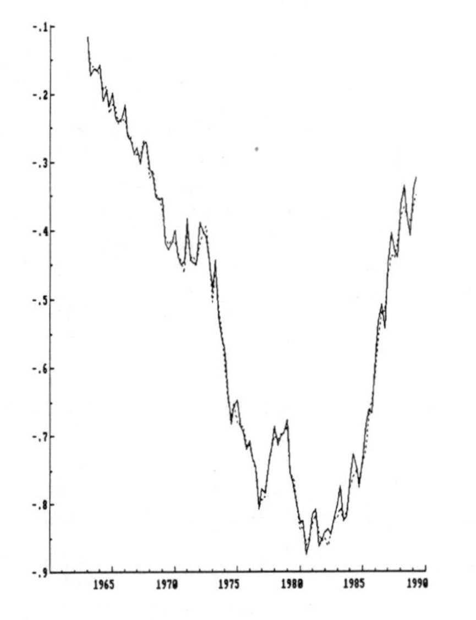

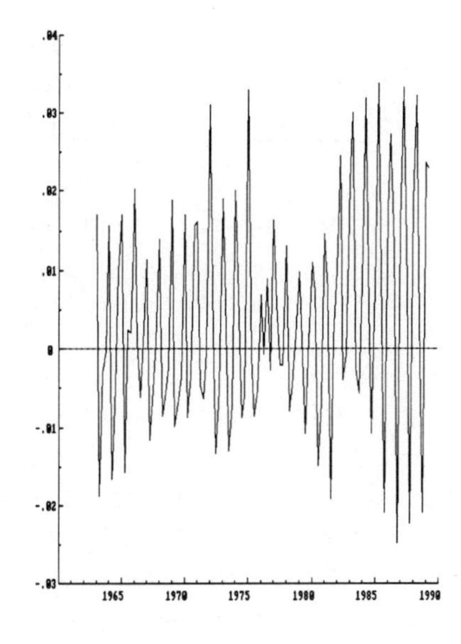

Table 1. Augmented Dickey-Fuller Test Statistics

Variable	Type of data	
	NSA	SA
m	−1.57 (−0.02)	−1.37 (−0.02)
p	−2.48 (−0.02)	−2.16 (−0.01)
$m-p$	−0.81 (−0.01)	−0.23 (−0.00)
y	−1.80 (−0.10)	−1.61 (−0.08)
$m-p-y$	−0.17 (−0.00)	+0.26 (+0.00)
Δm	−3.37 (−0.69)	−3.44 (−0.63)
Δp	−2.06 (−0.13)	−2.00 (−0.13)
$\Delta(m-p)$	−2.84 (−0.42)	−2.84 (−0.38)
Δy	−4.70 (−1.30)	−4.86 (−1.13)
$\Delta(m-p-y)$	−3.11 (−0.69)	−3.39 (−0.66)
$\Delta^2 m$	−5.77 (−2.84)	−6.63 (−2.90)
$\Delta^2 p$	−4.93 (−1.61)	−4.90 (−1.59)
$R3$	−3.45 (−0.22)	
Rra	−1.38 (−0.03)	
R^*	−2.07 (−0.08)	
$\Delta R3$	−5.19 (−1.15)	
ΔRra	−5.50 (−0.68)	
ΔR^*	−5.04 (−1.08)	

Notes: Each pair of entries includes the fourth-order ADF statistic with a constant term and trend and (in parentheses) the estimated coefficient associated with the ADF statistic. The ADF regressions for NSA data (except those for $R3$, Rra and R^*) also include seasonal dummies. The sample is 1964(3)–1989(2) [$T = 100$], except for $\Delta^2 m$, $\Delta^2 p$, $\Delta^2 m^a$ and $\Delta^2 p^a$, which use 1964(4)–1989(2) [$T = 99$]. The 90%, 95% and 99% critical values for $T = 100$ from MacKinnon (1991, Table 1) are −3.15, −3.45 and −4.05 respectively.

All six NSA series display strong seasonality, as evidenced by the seasonal components (of the form $x - x^a$). In Figure 1b, $m - m^a$ is often 1 – 3% in absolute value, compared to an average quarterly growth rate for m of about 3%. The seasonal component for prices (Figure 2b) is nearly an order of magnitude smaller, often 0.2 – 0.6%, contrasting with an average growth rate for p of about 3% per quarter. Thus, the seasonal component for real money (Figure 3b) is virtually identical to that for nominal money, noting that

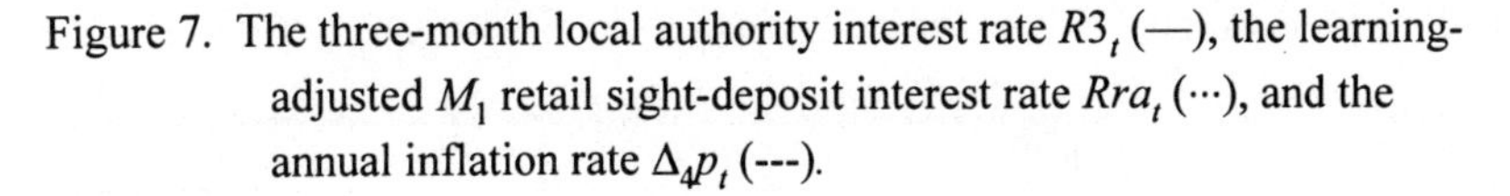

Figure 7. The three-month local authority interest rate $R3_t$ (—), the learning-adjusted M_1 retail sight-deposit interest rate Rra_t (···), and the annual inflation rate $\Delta_4 p_t$ (---).

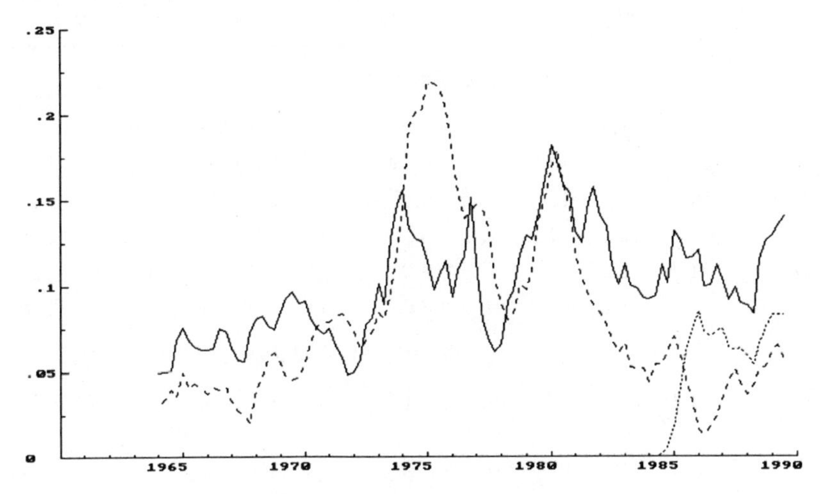

$(m-p)-(m^a-p^a)=(m-m^a)-(p-p^a)$. The seasonality in real and nominal TFE (Figures 4b and 5b) is the most regular of those observed, with a strong cyclical pattern superimposed on the seasonality. Real TFE also has the largest seasonal component of the series examined: often 2–3%, compared to an average growth rate of under 1% per quarter.

The two dominant seasonal components are from nominal money and real income, and are of the same magnitude but contrast in pattern. The seasonality of inverse velocity (Figure 6b) is essentially the difference of those two components. Seasonal patterns in the levels of $m-p$, y, $y+p$, and $m-p-y$ are clearly detectable in Figures 3a, 4a, 5a, and 6a. Seasonal patterns for m and p are not apparent in Figures 1a and 2a because the range of the data is so great.

Figure 7 graphs the local-authority interest rate $R3$, the learning-adjusted M_1 retail sight-deposit interest rate (or 'own rate') Rra, and the annual inflation rate $\Delta_4 p$. The two oil price increases are evident, and inflation declines to single-digit levels in the 1980s. Substantial differences between $\Delta_4 p$ and $R3$ persist over long episodes. Thus, inflation may play a role in the money demand function as the return on an illiquid, non-financial asset. The interest rate offered on M_1 (Rra) was introduced in 1984(3) and parallels $R3$ with a relatively constant spread. Importantly for money demand, the opportunity cost ($R3-Rra$) declines from (often) double-digit figures in the 1970s and early 1980s to only a couple percent in the late 1980s.

Table 2. Cointegration Analysis for Pairs of NSA and SA Series

Variable	Statistic					First row of:	
	λ_{max}	λ^{df}_{max}	λ_{trace}	λ^{df}_{trace}	$\chi^2(1)$	β'	α'
			Full sample: 1964(3)–1989(2)				
m	13.0	11.7	17.2	15.4	0.01	(1 –0.9998)	(+1.20 +1.66)
p	40.3	36.3	40.8	36.8	0.09	(1 –1.0000)	(–3.40 –0.68)
y	19.7	17.7	20.0	18.0	0.63	(1 –0.9994)	(–4.11 –2.07)
$m-p$	18.0	16.2	19.2	17.2	0.35	(1 –0.9974)	(+1.57 +2.18)
$y+p$	24.7	22.3	24.8	22.3	0.06	(1 –1.0000)	(–1.09 +1.31)
$m-p-y$	16.9	15.2	18.6	16.8	0.01	(1 –1.0000)	(+2.22 +2.77)
			Short sample: 1964(3)–1982(4)				
m	20.7	17.9	22.9	19.8	2.19	(1 –0.9985)	(+2.34 +3.44)
$m-p$	17.6	15.2	20.8	18.0	0.58	(1 –0.9957)	(+1.03 +2.23)
$m-p-y$	16.2	14.0	17.2	14.9	0.02	(1 –0.9997)	(+1.30 +2.58)
			Critical values				
90%	12.1	12.1	13.3	13.3	2.71		
95%	14.1	14.1	15.4	15.4	3.84		
99%	18.6	18.6	20.0	20.0	6.63		

4.1 Evidence on pairs of series

Table 2 presents Johansen's maximum likelihood cointegration test statistics, estimated cointegrating vectors (β'), and estimated weighting coefficients (α) for x and x^a, where x is any of m, p, y, $m-p$, $y+p$ (nominal TFE), or $m-p-y$. In each case, a fifth-order bivariate VAR is estimated for x and x^a with unconstrained seasonal dummies and a constant. The first four statistics in Table 2 are the maximal eigenvalue and trace eigenvalue statistics (λ_{max} and λ_{trace}) and those statistics adjusted for degrees of freedom lost in estimation (λ^{df}_{max} and λ^{df}_{trace}), with all four statistics being based on the largest eigenvalue.[7] Those statistics test the hypothesis that x and x^a are not cointegrated against the alternative that there is at least one cointegrating vector. Critical values for the cointegration statistics are from Osterwald-Lenum (1992, Table 1). The fifth statistic in the table, denoted $\chi^2(1)$, tests the hypothesis $\beta' = (1:-1)$ and is asymptotically distributed as $\chi^2(1)$

[7] The hypothesis that there is at most one cointegrating vector (i.e., at least one zero eigenvalue) was not rejected for any of the series except m, and then only marginally so.

under that null hypothesis. For variables involving m and m^a, the statistics are also calculated over a 'short sample', noting that the properties of the seasonal factor $m - m^a$ change after 1982; see Figure 1b.

Beginning with nominal money, the 'max' and 'trace' statistics reject at the 90% level; and the estimated cointegrating vector for $(m : m^a)'$ is $(1 : -0.9998)$, which is statistically and numerically insignificantly different from $(1 : -1)$. That is, $m - m^a$ appears I(0). Similar, even stronger evidence for cointegration appears for the other five variables and for nominal money over the short sample. In no case is the hypothesis $\beta' = (1 : -1)$ rejected.

For nominal money, the estimated weighting coefficients are +1.20 and +1.66, which are the coefficients on (approximately) $(m - m^a)_{t-1}$ in the equations for Δm_t and Δm_t^a respectively. Weighting coefficients in the other bivariate VARs are similar or larger in magnitude and of either sign. These coefficients for the bivariate VARs are numerically much larger than in typical empirical analyses of sets of variables; see Sections 4.2 and 4.3.

4.2 System cointegration analysis of the unadjusted data

This subsection tests for cointegration among the unadjusted series (m, p, y, $R3$, Rra). For both the NSA data set and the SA data set, inference could be affected by whether $R3$ and Rra enter separately or only via the opportunity cost R^*, and by whether m and p are I(1) or I(2). To assess the sensitivity of the cointegration tests to these factors, Ericsson, Hendry, and Tran (1993, Appendix C) examine four systems with the following variables:

System I: $m, p, y, R3, Rra$;

System II: m, p, y, R^*;

System III: $m - p, \Delta p, y, R3, Rra$; and

System IV: $m - p, \Delta p, y, R^*$.

The cointegrating vector and the weighting coefficients are little affected by the choice of system, so this subsection focuses on System IV. The system is a fifth-order vector autoregression with a constant term and seasonal dummies, but no trend. The estimation period is 1964(3)–1989(2).

Table 3 summarizes the cointegration results. It lists the eigenvalues related to $\hat{\pi}$ from largest to smallest, the max and trace statistics, the standardized estimated α and β', and statistics for testing restrictions on α and β'. The cointegration tests strongly reject the null of no cointegration ($r = 0$), but not the null of at most one cointegrating vector ($r \leq 1$), so there appears to be a single cointegrating vector for $(m - p, \Delta p, y, R^*)'$. The estimated cointegrating vector implies a long-run solution

Table 3. A Cointegration Analysis of NSA Data: $\{m-p, \Delta p, y, R^*\}$

Eigenvalues	0.345	0.121	0.049	0.015	
Hypotheses	$r=0$	$r\leq 1$	$r\leq 2$	$r\leq 3$	
λ_{max}	42.4	12.9	5.0	1.5	
λ^{df}_{max}	33.9	10.3	4.0	1.2	
95% critical value	27.1	21.0	14.1	3.8	
λ_{trace}	61.7	19.3	6.5	1.5	
λ^{df}_{trace}	49.4	15.5	5.2	1.2	
95% critical value	47.2	29.7	15.4	3.8	
	Standardized eigenvectors β'				
Variable	$m-p$	Δp	y	R^*	
	1	6.46	–0.99	6.76	
	–0.05	1	–0.04	–0.46	
	–0.89	16.33	1	–4.95	
	–1.53	–4.82	–0.18	1	
	Standardized adjustment coefficients α				
$m-p$	–0.18	–0.08	–0.00	0.00	
Δp	0.03	–0.09	–0.00	–0.00	
y	–0.00	0.34	–0.01	–0.00	
R^*	0.03	0.17	0.00	–0.01	
	Test statistics for restrictions on β'				
Variable	$m-p$	Δp	y	R^*	Joint
$\chi^2(\cdot)$	—	—	0.0	—	—
p-value			[0.917]		
	Test statistics for zero restrictions on α				
Variable	$m-p$	Δp	y	R^*	Joint
$\chi^2(\cdot)$	28.6	5.0	0.0	1.3	5.6
p-value	[0.000]	[0.025]	[0.928]	[0.260]	[0.130]

of:

$$m-p = 0.99y - 6.46\Delta p - 6.76R^*, \qquad (37)$$

with income, inflation, and interest rate elasticities in line with theory. A unit long-run homogeneity restriction on income cannot be rejected.

Table 4. A Cointegration Analysis of SA Data: $\{m^a - p^a, \Delta p^a, y^a, R^*\}$

Eigenvalues	0.386	0.128	0.050	0.009	
Hypotheses	$r = 0$	$r \leq 1$	$r \leq 2$	$r \leq 3$	
$\lambda_{\max}$	48.8	13.7	5.1	0.9	
$\lambda_{\max}^{df}$	39.0	11.0	4.1	0.8	
95% critical value	27.1	21.0	14.1	3.8	
λ_{trace}	68.6	19.8	6.1	0.9	
$\lambda_{\text{trace}}^{df}$	54.9	15.9	4.9	0.8	
95% critical value	47.2	29.7	15.4	3.8	
	Standardized eigenvectors β'				
Variable	$m^a - p^a$	Δp^a	y^a	R^*	
	1	7.22	–1.08	7.16	
	–0.08	1	–0.04	–0.79	
	–1.26	16.03	1	–7.00	
	1.33	6.58	–0.12	1	
	Standardized adjustment coefficients α				
$m^a - p^a$	–0.18	–0.03	0.00	–0.00	
Δp^a	0.02	–0.05	–0.00	0.00	
y^a	–0.00	0.23	–0.01	–0.00	
R^*	0.03	0.14	0.00	0.01	
	Test statistics for restrictions on β'				
Variable	$m^a - p^a$	Δp^a	y^a	R^*	Joint
$\chi^2(\cdot)$	—	—	0.8	—	—
p-value			[0.380]		
	Test statistics for zero restrictions on α				
Variable	$m^a - p^a$	Δp^a	y^a	R^*	Joint
$\chi^2(\cdot)$	34.6	3.4	0.0	2.3	4.5
p-value	[0.000]	[0.065]	[0.996]	[0.133]	[0.215]

The weighting coefficient on the cointegrating vector is –0.18 in the equation for money and is virtually zero in each of the other equations. While the coefficient in the inflation equation appears to be statistically significantly different from zero, the coefficients in the income, inflation, and the net interest rate equations jointly appear to be zero.[8] Those zeros are necessary for inflation, income, and the interest rate to be weakly exogenous for the parameters in the money equation; cf. Johansen (1992*a*). Section 5.1 assumes weak exogeneity and develops a model conditional on those variables.

4.3 System cointegration analysis of the adjusted data

Ericsson, Campos, and Tran (1990) test for cointegration in the SA data using Johansen's procedure. Table 4 replicates and adds to their System IV results, in which $m^a - p^a$, Δp^a, and y^a replace $m - p$, Δp, and y, and where a constant term (but no seasonal dummy) is included in the VAR. The eigenvalues, test statistics, and the estimated cointegrating vector in Table 4 are strikingly similar to those in Table 3, as implied by the theoretical analysis in Section 2.[9] Surprisingly, the estimated weighting matrices in Tables 3 and 4 are nearly identical. In general, they need not be, although they would be if, for all the SA filters, f_0 [in (2)] were approximately unity and the other f_i's were relatively small. Weights in the SA filters might have those values even for very seasonal data (such as m or y) if the seasonal component were well approximated by seasonal dummies and the SA filter included adjustment by dummies.

Figure 8 plots the estimated disequilibria $\beta' x_t$ and $\beta' x_t^a$, where β is numerically different for x_t and x_t^a. The choice of NSA or SA data matters little for the properties of the estimated disequilibrium. Shocks to the system and subsequent adjustment toward equilibrium are evident for higher inflation and interest rates (in 1973 and 1980), lower inflation and interest rates (in 1977 and 1982), and a lower opportunity cost (in 1985). Further, the disequilibria were substantial, sometimes exceeding ±40%.

5. SINGLE-EQUATION ANALYSIS

This section obtains a single-equation conditional model for money demand on NSA data (Section 5.1), summarizes Hendry and Ericsson's (1991*b*, equation (6)) model on the SA data (Section 5.2), and tests each against the other with the encompassing tests designed in Section 3 above (Section 5.3).

---

[8] The 'joint' test of zero restrictions on α does not include the zero restriction on α for the equation determining money. That latter restriction is resoundingly rejected, so money cannot be assumed weakly exogenous in an equation determining inflation, income, or the interest rate.

[9] Hendry and Mizon (1993) analyze this SA data over a sample for which *Rra* is zero. They obtain a cointegrating vector for money demand that is virtually identical to the one for System IV.

Figure 8. The NSA and SA disequilibrium measures $\beta' x_t$ (—) and $\beta' x_t^a$ (···).

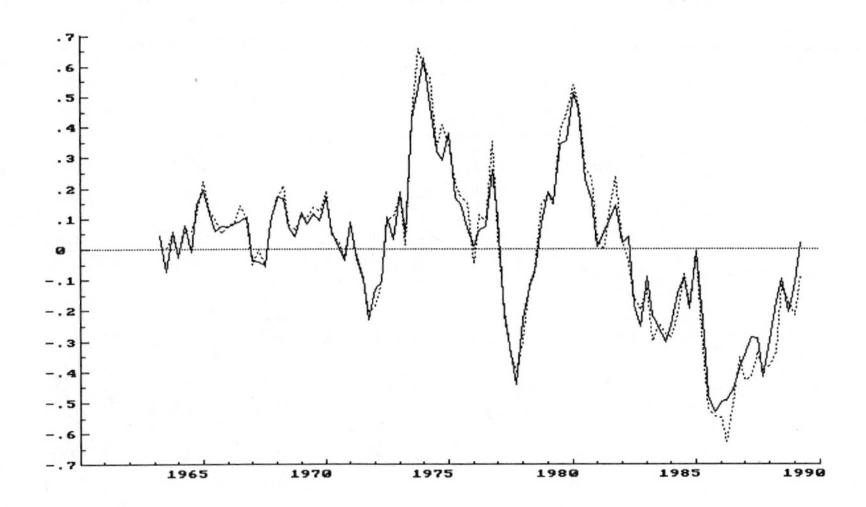

5.1 Single-equation analysis of the unadjusted data

As noted above, income, prices, and interest rates appear weakly exogenous, so single-equation modeling starts with an unrestricted autoregressive distributed lag (ADL) model for money. A parsimonious, economically interpretable, data-coherent simplification of that model is obtained via sequential reduction.

To match $\ell = 5$ in the VAR, the unrestricted single-equation model is a fifth-order ADL of m, conditional on p, y, $R3$, and Rra.[10] Table 5 lists coefficient estimates and their estimated standard errors (in parentheses). The long-run, static, non-stochastic solution to the model in Table 5 is:

$$\begin{aligned} m_t = \underset{(0.09)}{0.98}\, p_t + \underset{(0.33)}{1.05}\, y_t - \underset{(1.6)}{6.4}\, R3_t + \underset{(0.9)}{7.0}\, Rra_t \\ - \underset{(3.8)}{0.4} - \underset{(0.08)}{0.12}\, S_{1t} + \underset{(0.08)}{0.10}\, S_{2t} + \underset{(0.08)}{0.15}\, S_{3t}, \end{aligned} \tag{38}$$

where S_{it} is a seasonal dummy for the ith quarter, and estimated standard errors are in parentheses. Equation (38) corresponds to (36). The estimates in (38) closely match the system estimates of the first cointegrating vector when $\Delta p = 0$.

[10] This specification is the least restrictive of the conditional ADLs associated with the four systems in that it does not impose restrictions on the long-run coefficients for prices and interest rates. Those restrictions *are* imposed early on in the sequential reduction; and their imposition from the start of the sequential reduction does not affect the final outcome.

Table 5. A General Autoregressive Distributed Lag for Nominal Money, Conditional on Prices, Incomes, and Interest Rates (NSA Data)

	lag i (or summation over lags)						
Variable	0	1	2	3	4	5	$\Sigma_{i=0}^{5}$
m_{t-i}	−1 (—)	0.588 (0.126)	0.127 (0.142)	−0.193 (0.132)	0.367 (0.130)	−0.042 (0.114)	−0.152 (0.043)
p_{t-i}	0.430 (0.258)	0.024 (0.404)	−0.213 (0.380)	−0.031 (0.348)	−0.342 (0.341)	0.282 (0.214)	0.149 (0.039)
y_{t-i}	0.022 (0.108)	0.219 (0.117)	0.044 (0.116)	−0.372 (0.116)	0.107 (0.123)	0.138 (0.116)	0.159 (0.057)
$R3_{t-i}$	−0.433 (0.137)	−0.259 (0.204)	−0.342 (0.208)	0.081 (0.204)	0.187 (0.201)	−0.209 (0.150)	−0.974 (0.211)
Rra_{t-i}	0.223 (0.515)	−0.024 (0.995)	2.250 (1.111)	−2.249 (1.143)	0.491 (1.089)	0.370 (0.621)	1.060 (0.327)
constant	−0.055 (0.565)						
S_{it}		−0.018 (0.011)	0.016 (0.011)	0.023 (0.010)			

$T = 100$ [1964(3)–1989(2)]	$R^2 = 0.9998$	$\hat{\sigma} = 1.403\%$
$dw = 2.09$	AR: $F(4,63) = 1.37$	$Normality$: $\chi^2(2) = 2.48$
$ARCH$: $F(4,59) = 0.87$	$Hetero$: $F(61,5) = 0.12$	$RESET$: $F(1,66) = 1.70$

Table 5 (and also the regressions (39) and (41) below) includes diagnostic statistics for testing against various alternative hypotheses: residual autocorrelation (*dw* and *AR*), skewness and excess kurtosis (*Normality*), autoregressive conditional heteroscedasticity (*ARCH*), RESET (*RESET*), heteroscedasticity (*Hetero*), and heteroscedasticity quadratic in the regressors (alternatively, misspecification of functional form) (*Form*).[11] The null distribution is designated by $\chi^2(\cdot)$ or $F(\cdot\ ,\ \cdot)$, where the degrees of freedom fill the parentheses. For *AR* and *ARCH*, the first degree of freedom is the maximal lag. No statistic in Table 5 is significant at its 95% critical value.

The model in Table 5 has an equivalent error correction representation, which may be simplified to (39) below. Details of the simplification process appear in Appendix B.

[11] For references on the test statistics, see Durbin and Watson (1950, 1951), Box and Pierce (1970), Godfrey (1978), and Harvey (1981, p. 173); Jarque and Bera (1980); Engle (1982); Ramsey (1969); and White (1980, p. 825) and Nicholls and Pagan (1983) (the latter two on both *Hetero* and *Form*).

$$
\begin{aligned}
\Delta(m-p)_t = & -\underset{[0.13]}{0.95}\left[\Delta_3(m-p)_{t-1}/3\right] - \underset{[0.12]}{1.07}\left[(\Delta p_t + \Delta p_{t-4})/2\right] + \underset{[0.05]}{0.16}\,\Delta^2 y_{t-2} \\
& - \underset{[0.080]}{1.189}\left[(R_t^* + R_{t-1}^* + R_{t-2}^*)/3\right] - \underset{[0.011]}{0.174}\,(m-p-y)_{t-1} \\
& + \underset{[0.006]}{0.038} - \underset{[0.005]}{0.012}\,S_{1t} + \underset{[0.005]}{0.010}\,S_{2t} + \underset{[0.008]}{0.018}\,S_{3t} + \hat{u}_{1t}
\end{aligned} \tag{39}
$$

$T = 100$ [1964(3) – 1989(2)]	$R^2 = 0.84$ $\hat{\sigma} = 1.348\%$
AR: $F(4,87) = 1.07$	$dw = 1.95$ *ARCH*: $F(4,83) = 0.44$
Normality: $\chi^2(2) = 4.08$	*RESET*: $F(1,90) = 1.64$
Hetero: $F(13,77) = 1.12$	*Form*: $F(38,52) = 0.62$.

Jack-knife, heteroscedasticity-consistent, estimated standard errors appear in square brackets under coefficient estimates; see White (1980), Nicholls and Pagan (1983), and MacKinnon and White (1985).

The first three terms on the right-hand side of (39) capture the effects of the lagged dependent variable, inflation, and income growth. Noting the near (negative) unit coefficient on $(\Delta p_t + \Delta p_{t-4})/2$, (39) can be transformed to have nominal money growth Δm_t as the dependent variable and $+0.5\Delta_4\Delta p_t$ on the right-hand side. That is, inflation per se has no immediate effect on nominal money demand; only its acceleration does. In the long run, both inflation and its acceleration affect money demand via the error correction. From the interest rate and error correction coefficients, the long-run solution of (39) is virtually (38), with unit income and price elasticities imposed as part of the sequential reduction.

Statistically, (39) appears satisfactory. None of the diagnostic tests reject, and the F-statistic for testing (39) against Table 5 is $F(24,67) = 0.70$ [83%], where the tail probability is in square brackets.

Constancy is an additional, crucial statistical property, particularly in the context of money demand equations; see Judd and Scadding (1982) and Goldfeld and Sichel (1990). Recursive least squares and the associated sequences of test statistics provide incisive tools for investigating constancy; cf. Brown, Durbin, and Evans (1975) and Dufour (1982). Graphs efficiently summarize the large volume of output. Figure 9a records the one-step residuals and the corresponding calculated equation standard errors for (39), i.e., $\{y_t - \hat{\beta}_t' x_t\}$ and $\{0.0 \pm 2\hat{\sigma}_t\}$ in a common notation. The equation standard error $\hat{\sigma}$ varies little. Figure 9b plots the 'break-point' Chow (1960) statistics for the sequence {1969(3)–1989(2), 1969(4)–1989(2), 1970(1)–1989(2),...,1989(1)–1989(2), 1989(2)}, none of which are significant at even the 5% level. Figures 9c–9k show the numerical values of all the recursively estimated coefficients and plus-or-minus twice their recursively estimated standard errors, denoted $\hat{\beta}_t$ and $\hat{\beta}_t \pm 2ese(\hat{\beta}_t)$ respectively in the

Figure 9. Recursive least squares test statistics and estimates

(a) One-step residuals (—) and $0 \pm 2\hat{\sigma}_t$ (···)

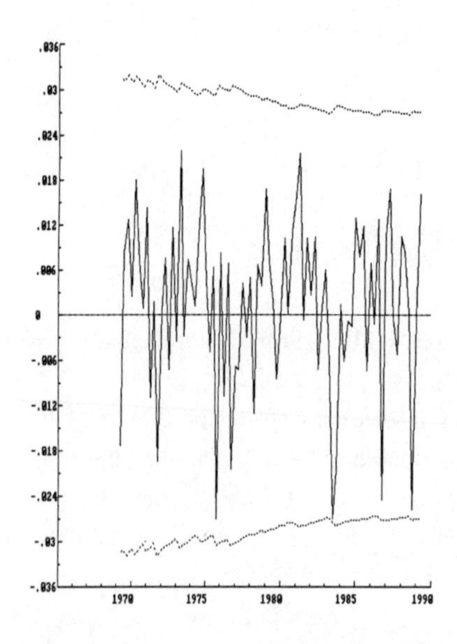

(b) Sequence of break-point Chow stats. (—), scaled by their one-off 5% critical values

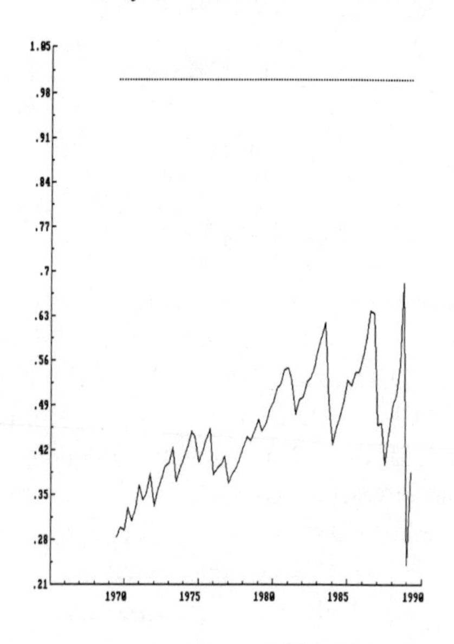

(c) Recursive estimates of the coefficient of $\Delta_3(m-p)_{t-1}/3$ (—), with $\hat{\beta}_t \pm 2ese(\hat{\beta}_t)$ (···)

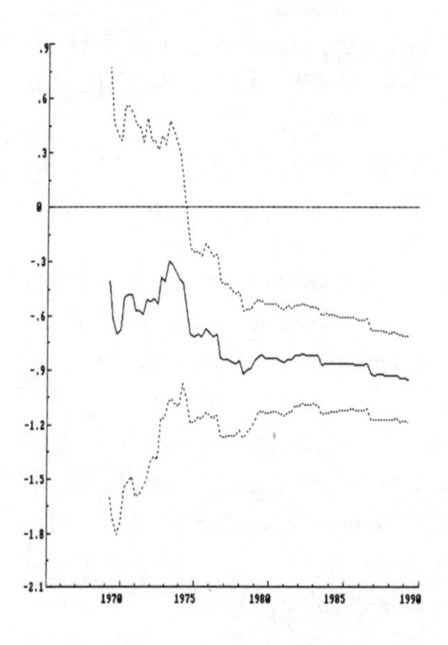

(d) Recursive estimates of the coefficient of $(\Delta p_t+\Delta p_{t-4})/2$ (—), with $\hat{\beta}_t \pm 2ese(\hat{\beta}_t)$ (···)

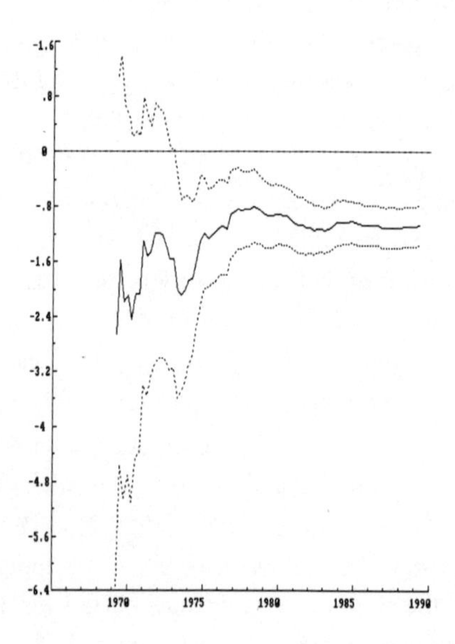

(e) Recursive estimates of the coefficient of $\Delta^2 y_{t-2}$ (—), with $\hat{\beta}_t \pm 2ese(\hat{\beta}_t)$ (···)

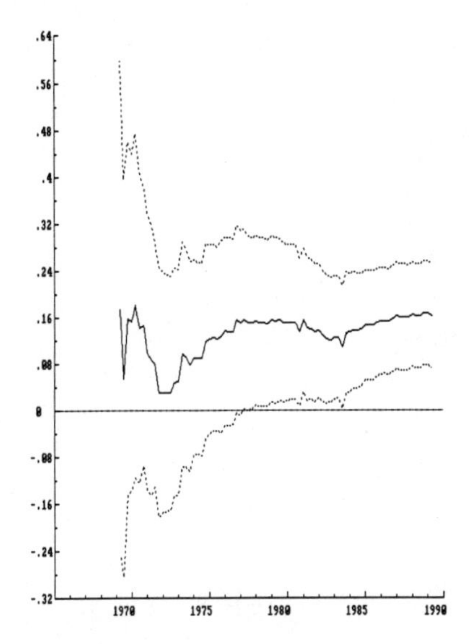

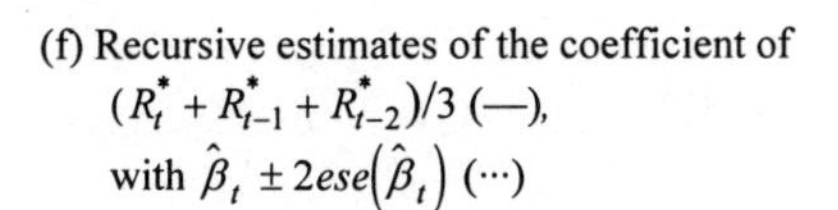

(f) Recursive estimates of the coefficient of $(R_t^* + R_{t-1}^* + R_{t-2}^*)/3$ (—), with $\hat{\beta}_t \pm 2ese(\hat{\beta}_t)$ (···)

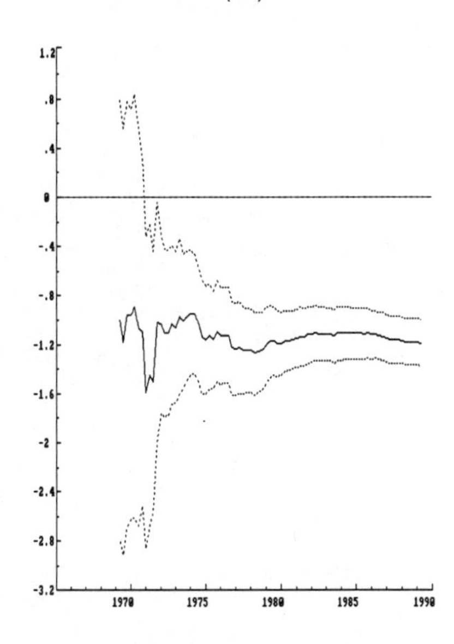

(g) Recursive estimates of the coefficient of $(m-p-y)_{t-1}$ (—), with $\hat{\beta}_t \pm 2ese(\hat{\beta}_t)$ (···)

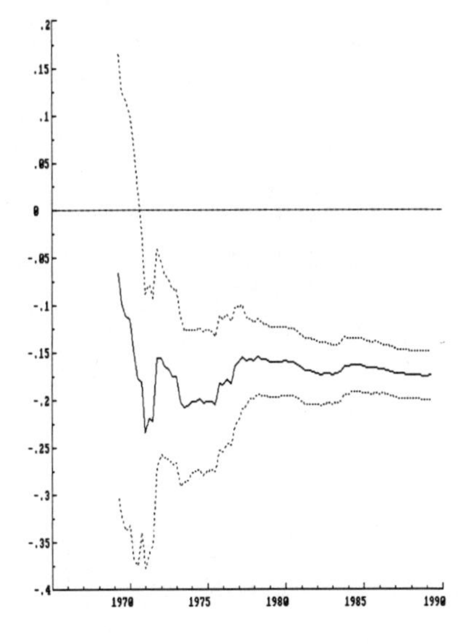

(h) Recursive estimates of the coefficient of the constant term (—), with $\hat{\beta}_t \pm 2ese(\hat{\beta}_t)$ (···)

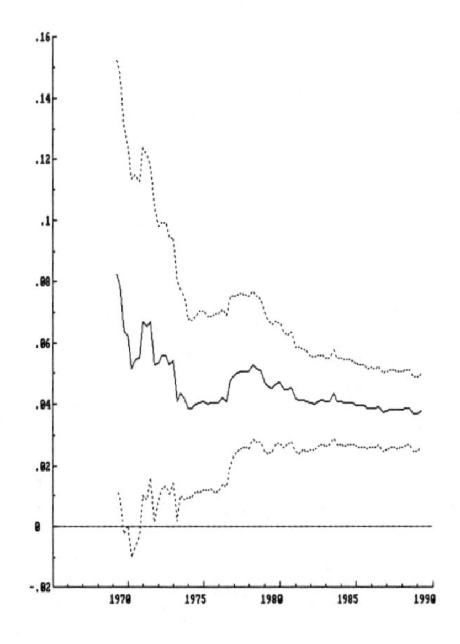

(i) Recursive estimates of the coefficient of S_{1t} (—), with $\hat{\beta}_t \pm 2ese(\hat{\beta}_t)$ (⋯)

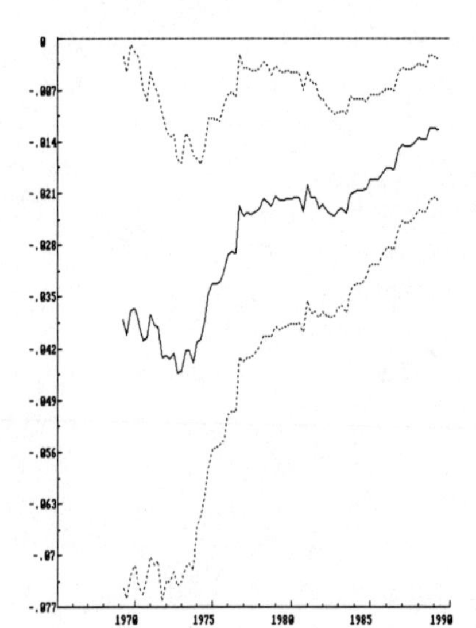

(j) Recursive estimates of the coefficient of S_{2t} (—), with $\hat{\beta}_t \pm 2ese(\hat{\beta}_t)$ (⋯)

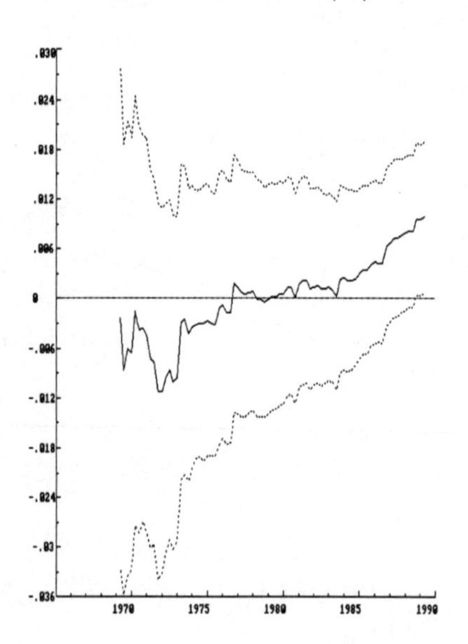

(k) Recursive estimates of the coefficient of S_{3t} (—), with $\hat{\beta}_t \pm 2ese(\hat{\beta}_t)$ (⋯)

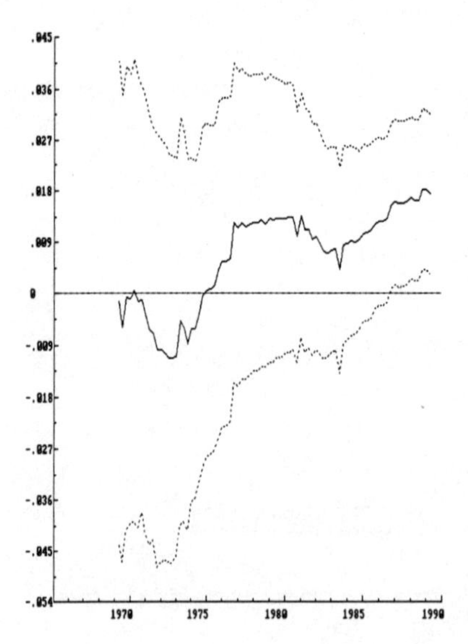

(ℓ) Actual (—) and fitted (⋯) values of $\Delta(m-p)_t$

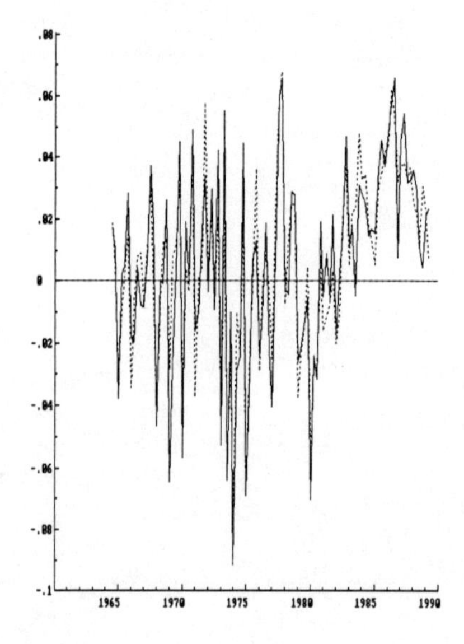

graphs. Coefficients on economic variables vary only slightly relative to their *ex ante* standard errors, and all those variables except $\Delta^2 y_{t-2}$ are highly significant by 1980. The quarterly dummies are statistically constant, but appear to drift numerically. Even with the full sample, their coefficients are only marginally significant. Figure 9ℓ plots the actual and fitted values for $\Delta(m-p)_t$ and shows how well (39) explains the data.

Hendry and Ericsson (1991*b*) document large changes in the data properties. Together with the constancy of (39), those changes imply the super exogeneity of expenditure, prices, and interest rates for the parameters in (39); cf. Engle, Hendry, and Richard (1983), Hendry (1988), and Engle and Hendry (1993). To summarize, (39) is a constant, economically interpretable, data-coherent model of NSA money demand in the United Kingdom.

5.2 Single-equation analysis of the adjusted data

The SA data have been thoroughly studied in a sequence of papers, starting with Hacche (1974), Coghlan (1978) and Hendry (1979). The latter develops a constant, parsimonious error correction model over 1964(1)–1977(4). Subsequent models by Trundle (1982), Hendry (1985), Davidson (1987), Cuthbertson (1988), Hendry (1988), Ericsson, Campos, and Tran (1990), Hall, Henry, and Wilcox (1990), Hendry and Ericsson (1991*b*), and Hendry and Mizon (1993) are similar in form and numerical parameter values, with the main differences arising from using different sample periods, and data series with different base years.

From a fifth-order ADL on the current SA data set, Ericsson, Campos, and Tran (1990) obtain the following long-run, static, non-stochastic solution:

$$m_t^a = \underset{(0.08)}{0.96}\, p_t^a + \underset{(0.30)}{1.17}\, y_t^a - \underset{(1.5)}{6.7}\, R3_t + \underset{(0.8)}{7.0}\, Rra_t - \underset{(3.4)}{1.7}\,, \tag{40}$$

$$T = 100\ [1964(3)\text{–}1989(2)].$$

These estimates closely match those in (38) on NSA data, and the system estimates of the first cointegrating vector on both SA and NSA data.

Hendry and Ericsson (1991*b*, equation (6)) obtain the following error correction model (ECM).

$$\begin{aligned}\Delta\left(m^a - p^a\right)_t = &- \underset{[0.14]}{0.69}\, \Delta p_t^a - \underset{[0.06]}{0.17}\, \Delta\left(m^a - p^a - y^a\right)_{t-1} - \underset{[0.053]}{0.630}\, R_t^* \\ &- \underset{[0.008]}{0.093}\left(m^a - p^a - y^a\right)_{t-1} + \underset{[0.004]}{0.023} + \hat{u}_{0t}\end{aligned} \tag{41}$$

$T = 100$ [1964(3) – 1989(2)] $R^2 = 0.76$ $\hat{\sigma} = 1.313\%$
AR: $F(4,91) = 1.94$ $dw = 2.18$ *ARCH*: $F(4,87) = 0.74$
Normality: $\chi^2(2) = 1.53$ *RESET*: $F(1,94) = 0.08$
Hetero: $F(8,86) = 1.36$ *Form*: $F(14,80) = 1.05$.

Ericsson, Campos, and Tran (1990) find that (41) is a valid sequential reduction from the ADL, with an F-statistic of $F(25,70)=0.97$ [51%]. From the coefficients on the interest rate and the error correction term, the long-run solution of (41) is little changed from (40). Extensive evaluation by Ericsson, Campos, and Tran (1990) and Hendry and Ericsson (1991*b*) shows that (41) is a constant, data-coherent, conditional model with sensible economic properties. Prices, expenditure, and interest rates appear super exogenous.

5.3 Encompassing tests in practice

From the studies to date, (41) appears a well-specified model of SA money demand. However, the encompassing tests from Section 3 will show that (41) does not encompass the NSA model (39). Other than mixing data sets, no obvious respecification of (41) results in an improved model of the SA data. Because of computational and modeling issues, the formally correct encompassing test in the reverse direction is not performed. However, evidence from the 'invalid' encompassing test suggests that the NSA model does encompass the SA model.

To test whether or not the SA model encompasses the NSA model, the seasonal component $\Delta(m-p)_t - \Delta(m^a - p^a)_t$ and the regressors from the NSA model are added to the SA model, and the significance of the added variables is tested. The resulting estimated equation is the following, which implements (34).

$$
\begin{aligned}
\Delta\left(m^a - p^a\right)_t = & -\underset{[0.20]}{0.51}\,\Delta p_t^a - \underset{[0.07]}{0.09}\,\Delta\left(m^a - p^a - y^a\right)_{t-1} - \underset{[0.164]}{0.241}\,R_t^* \\
& - \underset{[0.116]}{0.048}\left(m^a - p^a - y^a\right)_{t-1} + \underset{[0.007]}{0.026} \\
& - \underset{[0.19]}{0.29}\left[\Delta(m-p)_t - \Delta\left(m^a - p^a\right)_t\right] - \underset{[0.16]}{0.53}\left[\Delta_3(m-p)_{t-1}/3\right] \\
& - \underset{[0.18]}{0.39}\left[\left(\Delta p_t + \Delta p_{t-4}\right)/2\right] + \underset{[0.05]}{0.09}\,\Delta^2 y_{t-2} \\
& - \underset{[0.23]}{0.70}\left[\left(R_t^* + R_{t-1}^* + R_{t-2}^*\right)/3\right] - \underset{[0.116]}{0.088}\,(m-p-y)_{t-1} \\
& + \underset{[0.008]}{0.008}\,S_{1t} + \underset{[0.005]}{0.012}\,S_{2t} + \underset{[0.007]}{0.015}\,S_{3t} + \hat{e}_{0t}
\end{aligned}
\tag{42}
$$

$T = 100$ [1964(3)–1989(2)] $R^2 = 0.81$ $\hat{\sigma} = 1.225\%$

The F-statistic for the significance of the additional regressors is $F(9,86) = 2.57$ [1.1%]. If the seasonal component $\Delta(m-p)_t - \Delta(m^a - p^a)_t$ is not included, the F-statistic is still large: $F(8,87) = 2.50$ [1.7%].

As discussed in Section 3, reversing the roles of SA and NSA data is not valid because it generally entails invalid conditioning for the NSA model. Even so, the results from reversal may be of interest, to the extent that future values play a small role in the added regressors. Adding $\Delta(m-p)_t - \Delta(m^a - p^a)_t$ and the SA regressors in (41) to the NSA model (39) obtains (42), except that $\Delta(m-p)_t$ is the dependent variable and so the coefficient on $\Delta(m-p)_t - \Delta(m^a - p^a)_t$ is +0.71 (implied by Section 3.2). The F-statistic for the additional regressors is $F(5,86) = 4.83$ [0.06%]. However, rejection appears entirely explained by the presence of the seasonal component $\Delta(m-p)_t - \Delta(m^a - p^a)_t$. If that seasonal component is not included in the regression, the F-statistic is $F(4,87) = 1.54$ [19.9%].

In summary, the SA model does not encompass the NSA model, whereas the NSA model appears to encompass the SA model. The NSA model appears well specified otherwise and is sensible economically, so it represents the currently best available empirical model for UK narrow money demand. While the long-run solutions for the SA and NSA models are virtually identical empirically (as implied by the theoretical analysis), estimated short-run dynamics do differ. For instance, inflation and the net interest rate enter the SA model current-dated only, whereas they appear as time averages in the NSA model. Development of the NSA model also shows how new tests and new data play a central role in a progressive research program, whereby existing models are supplanted by new models that encompass the existing models and offer some 'value added'.

6. CONCLUDING REMARKS

Seasonality and seasonal adjustment procedures have stimulated a wealth of theoretical and empirical studies. With tools from the cointegration literature, this paper derives central relationships between pairs of SA and NSA data, and between sets of SA and NSA data. The encompassing framework provides a basis for comparing models developed on SA data with those developed on NSA data. This paper extends the analysis of Wallis and Sims to cointegrated series and develops methods for evaluating NSA and SA models directly against each other. Contrasting with results from Sims and Wallis in a stationary framework, there *are* invariants to seasonal adjustment of cointegrated data: the number of cointegrating vectors and the cointegrating vectors themselves. That said, inference may well be affected by the choice of data type.

A substantive model development of NSA narrow money demand in the United Kingdom illustrates the analytical results. Hendry and Ericsson's (1991*b*) SA model of narrow money demand appeared well-specified on existing tests, but was found deficient in the presence of the new NSA model. This result demonstrates the value of the new tests and of the NSA data. It also highlights the importance of statistical agencies providing NSA data even if they already provide SA data, since the latter need not be the appropriate data for empirical economic modeling.

APPENDIX A. THE DATA

Sources. The data sources are: *Bank of England Quarterly Bulletin*, various issues (*BEQB*); *Economic Trends Annual Supplement*, 1990 Edition, No. 15 (*ETAS*); *Financial Statistics*, various issues (*FS*); and *Monthly Digest of Statistics*, various issues (*MDS*). The first is a publication of the Bank of England, London; the other three are published by the Central Statistical Office (CSO), Her Majesty's Stationery Office, London. The four-character sequence is the CSO databank series number.

GDP, *GDP*85, *IMP*, and *IMP*85 are from *ETAS* (Table 3), with minor changes for data revisions from *MDS* (Table 1.2). *M* is the M_1 series in *FS* (January 1989, Supplementary Table S32, Columns 6 [NSA] and 7 [SA]) and *BEQB* (November 1989, Table 11.1, Columns 4 [NSA] and 14 [SA]). *R*3 is from various issues of the *BEQB* (e.g., May 1989, Table 9.2) and *FS* (e.g., February 1990, Table 13.14). *Rr* is zero prior to 1984(3), and as listed in Hendry and Ericsson (1991b, Table A.2) thereafter. We are grateful to Stephen Hall at the Bank of England for providing *Rr*.

All data are quarterly and span 1963(1)–1989(2), unless otherwise noted.

Adjustments. Topping and Bishop (1989) document numerous breaks in the series for M_1. We account for the four primary breaks in M_1, proportionately rescaling data before the break to match the post-break value of M_1 for the quarter in which the break occurred. Adjusting the data for these breaks is critical, statistically as well as economically. The breaks range from –1.5% to +6.3%, but $\hat{\sigma}$ is only 1.3% in (39) and (41). See also Healey *et al.* (1990).

Topping and Bishop's breaks are for NSA data, and are reported in Table A2 below. We use the same breaks for SA data, as suggested by Topping and Bishop (1989, p. 11).

Table A1. Data Definitions and Sources

Variable	Definition	Source
GDP	Gross domestic product (expenditure-based) at market prices [£ million, current prices]	DJAF (NSA) DJBB (SA)
*GDP*85	Gross domestic product (expenditure-based) at market prices [£ million, 1985 prices]	DJCX (NSA) DJDI (SA)
IMP	Imports of goods and services at market prices [£ million, current prices]	DJAG (NSA) DJBC (SA)
*IMP*85	Imports of goods and services at market prices [£ million, 1985 prices]	DJCY (NSA) DJDJ (SA)
M	Monetary aggregate M_1: notes and coin in circulation with the public plus UK private sector sterling sight bank deposits, both non-interest-bearing and interest-bearing (financial year constrained when seasonally adjusted) [£ million, current prices]	AGAF (NSA) AGBA (SA)
P	Implicit deflator for total final expenditure [$= (GDP+IMP)/(GDP85+IMP85)$] [1985 = 1.00]	—
*R*3	Interest rate on deposits with local authorities, for a minimum of three months and thereafter at seven days' notice (quarterly average of the rate on the last Friday of each month) [fraction]	AJOI
R^*	Learning-adjusted net interest rate ($= R3 - Rra$) [fraction]	—
Rr	Interest rate on (M_1) sterling retail sight deposits at banks [fraction]	Hendry and Ericsson ('91b)
Rra	Learning-adjusted interest rate on retail sight deposits at banks ($= w_t \cdot Rr_t$) [fraction]	—
w_t	Weighting function representing agents' learning about interest-bearing retail sight deposits [$= \{1 + \exp[\kappa_0 - \kappa_1 (t - t_0 + 1)]\}^{-1}$ for $t \geq t_0$, zero otherwise; $t_0 = 1984(3)$, $\kappa_0 = 5.0$, and $\kappa_1 = 1.2$]	Hendry and Ericsson (1991b)
Y	Total final expenditure at market prices ($= GDP85 + IMP85$) [£ million, 1985 prices]	—

Table A2. The Four Primary Breaks in M_1

Date	Break $£\times10^6$	Break per cent	M_1 after break		Explanation
1971(4)	+403	+3.8	11088	(NSA)	A break occurs '... due to the incorporation of new information collected from the London clearing banks ... on the sector split of current and deposit accounts ...'. (p. 25)
		+3.9	10765	(SA)	
1975(2)	+618	+4.1	15791	(NSA)	'New, more comprehensive, statistical returns introduced in May 1975 further reduced the estimation necessary to calculate M_1 ...'. (p. 26)
		+4.0	15929	(SA)	
1976(1)	−266	−1.5	17421	(NSA)	'This is due to the incorporation of data on public corporations' holdings of notes and coin ...', i.e., which are *not* included in M_1. (pp. 26–27)
		−1.5	17588	(SA)	
1981(4)	+208	+6.1	35956	(NSA)	'... the 'monetary sector' was introduced in place of the 'banking sector'; amongst others, this brought the [Trustee Savings Banks] into the monetary sector.' (pp. 12, 28)
	1	+6.3	35257	(SA)	

Source for quotes and breaks: Topping and Bishop (1989); see their Table 2(a) for breaks.
Units: £ million, unless otherwise noted.

APPENDIX B. SEQUENTIAL REDUCTION ANALYSIS

This appendix describes a sequential reduction from the general ADL model in Table 5 to the parsimonious ECM in (39). Many other 'routes' for the reduction are possible. However, the values of the F-statistics for this sequence are small, implying that the F-statistics for other routes are unlikely to be statistically significant.

First, the model in Table 5 is rewritten as an equivalent ECM representation. Two types of transformations are used. Levels and lagged levels are written as differences and a current or lagged level; and levels on two variables are written as a differential between the two variables and the level of one of the variables. See Ericsson, Campos, and Tran (1990) for a motivation and further discussion. The specific transformations are:

1. nominal money m and prices p are transformed to real money $m-p$ and prices;
2. the interest rates $R3$ and Rra are transformed to the spread R^* and Rra;

3. each of the variables $m-p$, p, y, and *Rra* is transformed to a single log-level (or level) and a set of current and lagged differences, with the log-levels $m-p$, p, and y at the first lag and the level *Rra* current; and
4. the variables $(m-p)_{t-1}$ and y_{t-1} are transformed to $(m-p-y)_{t-1}$ and y_{t-1}, where $(m-p-y)_{t-1}$ is the potential error correction term.

The resulting coefficient estimates and estimated standard errors appear in Table B1, and provide the starting point for the sequential reduction.

To aid in the sequential reduction of the model in Table B1, we list several variables in Table B1 with highly statistically significant coefficients and which are economically reasonable to retain, as well as several variables whose coefficients appear numerically and statistically insignificant. The following are highly significant. The error correction term $(m-p-y)_{t-1}$ enters with a coefficient of –0.152, close to the first term in the α matrix for the system analysis (Table 3). The current net interest rate R_t^* and the current inflation rate Δp_t each enter with large negative coefficients, interpretable as reflecting costs to holding money when other assets (or goods) yield a return. The first and third lags of the dependent variable $\Delta(m-p)_t$ are statistically significant.

The following do not appear either numerically or statistically significant:

(i) the variables p_{t-1} and y_{t-1};

(ii) the variable Rra_t; and

(iii) all current and lagged values of ΔRra.

Four additional reductions are considered:

(iv) the coefficients on R_t^*, R_{t-1}^*, and R_{t-2}^* are equal, and those on R_{t-3}^*, R_{t-4}^*, and R_{t-5}^* are zero;

(v) the coefficients on $\Delta(m-p)_{t-1}$, $\Delta(m-p)_{t-2}$, and $\Delta(m-p)_{t-3}$ are equal, and that on $\Delta(m-p)_{t-4}$ is zero;

(vi) the coefficients on Δy_t, Δy_{t-1}, and Δy_{t-4} are zero, and those on Δy_{t-2} and Δy_{t-3} are equal and opposite; and

(vii) the coefficients on Δp_t and Δp_{t-4} have equal coefficients, and those on Δp_{t-1}, Δp_{t-2}, and Δp_{t-3} are zero.

Eight models arise from treating these seven restrictions sequentially:

Model 1. The unrestricted ECM in Table B1 (equivalently, in Table 5);

Model 2. Model 1, excluding p_{t-1} and y_{t-1} (long-run unit homogeneity of prices and income imposed);

Model 3. Model 2, excluding Rra_t (long-run restriction of 'opposite sign, equal magnitude' coefficients on *R*3 and *Rra* imposed);

Table B1. The Unrestricted Error Correction Model for NSA Data

	lag i					
Variable	0	1	2	3	4	5
$\Delta(m-p)_{t-i}$	–1 (—)	–0.260 (0.111)	–0.133 (0.119)	–0.325 (0.113)	0.042 (0.114)	
Δp_{t-i}	–0.570 (0.258)	0.045 (0.273)	–0.041 (0.272)	–0.265 (0.247)	–0.240 (0.247)	
Δy_{t-i}	0.022 (0.108)	0.083 (0.137)	0.126 (0.125)	–0.246 (0.131)	–0.138 (0.116)	
R^*_{t-i}	–0.432 (0.137)	–0.259 (0.204)	–0.342 (0.208)	0.082 (0.204)	0.187 (0.201)	–0.209 (0.150)
ΔRra_{t-i}	–0.297 (0.485)	–0.580 (0.619)	1.328 (0.642)	–0.840 (0.642)	–0.161 (0.617)	
$(m-p-y)_{t-i}$		–0.152 (0.043)				
p_{t-i}		–0.003 (0.014)				
y_{t-i}		0.007 (0.049)				
Rra_{t-i}	0.088 (0.276)					
constant	–0.055 (0.565)					
S_{it}		–0.018 (0.011)	0.016 (0.011)	0.023 (0.010)		

T = 100 [1964(3)–1989(2)] $R^2 = 0.87$ $\hat{\sigma}$ = 1.403%
Note: All residual-based statistics are identical to those in Table 5.

Model 4. Model 3, excluding current and lagged values of ΔRra (short-run restriction of 'opposite sign, equal magnitude' coefficients on $R3$ and Rra imposed);

Model 5. Model 4, excluding R^*_{t-1}, R^*_{t-2}, R^*_{t-3}, R^*_{t-4}, and R^*_{t-5} [once R^*_t, R^*_{t-1}, and R^*_{t-2} are transformed to $\sum_{i=0}^{2} R^*_{t-i}/3$, R^*_{t-1}, and R^*_{t-2}];

Model 6. Model 5, excluding $\Delta(m-p)_{t-2}$, $\Delta(m-p)_{t-3}$, and $\Delta(m-p)_{t-4}$ [once $\Delta(m-p)_{t-1}$, $\Delta(m-p)_{t-2}$, and $\Delta(m-p)_{t-3}$ are transformed to $\Delta_3(m-p)_{t-1}/3$, $\Delta(m-p)_{t-2}$, and $\Delta(m-p)_{t-3}$];

Model 7. Model 6, excluding Δy_t, Δy_{t-1}, Δy_{t-3}, and Δy_{t-4} [once Δy_{t-2} and Δy_{t-3} are transformed to $\Delta^2 y_{t-2}$ and Δy_{t-3}]; and

Model 8. Model 7, excluding Δp_{t-1}, Δp_{t-2}, Δp_{t-3}, and Δp_{t-4} [once Δp_t and Δp_{t-4} are transformed to $(\Delta p_t + \Delta p_{t-4})/2$ and Δp_{t-4}].

Table B2. *F*- and Related Statistics for the Sequential Reduction from the Fifth-order ADL Model in Table 5 (NSA Data)

Null Hypothesis				Maintained Hypothesis (Model Number)						
Model	k	$\hat{\sigma}$	*SC*	1	2	3	4	5	6	7
1 ↓ (i)	33	1.403%	−7.41	— — —						
2 ↓ (ii)	31	1.383%	−7.50	0.02 [0.98] (2,67)						
3 ↓ (iii)	30	1.378%	−7.54	0.15 [0.93] (3,67)	0.43 [0.51] (1,69)					
4 ↓ (iv)	25	1.392%	−7.69	0.85 [0.56] (8,67)	1.16 [0.34] (6,69)	1.31 [0.27] (5,70)				
5 ↓ (v)	20	1.388%	−7.86	0.86 [0.59] (13,67)	1.04 [0.42] (11,69)	1.12 [0.36] (10,70)	0.90 [0.49] (5,75)			
6 ↓ (vi)	17	1.382%	−7.97	0.84 [0.64] (16,67)	0.99 [0.48] (14,69)	1.04 [0.43] (13,70)	0.85 [0.56] (8,75)	0.77 [0.51] (3,80)		
7 ↓ (vii)	13	1.360%	−8.14	0.73 [0.78] (20,67)	0.84 [0.65] (18,69)	0.87 [0.61] (17,70)	0.67 [0.78] (12,75)	0.51 [0.83] (7,80)	0.31 [0.87] (4,83)	
8	9	1.348%	−8.29	0.70 [0.83] (24,67)	0.79 [0.73] (22,69)	0.81 [0.70] (21,70)	0.64 [0.84] (16,75)	0.53 [0.88] (11,80)	0.44 [0.89] (8,83)	0.59 [0.67] (4,87)

Note: The first four columns report the model number (with reduction), and for that model: the number of unrestricted parameters *k*, the estimated equation standard error $\hat{\sigma}$, and the Schwarz criterion *SC*, defined as $\ln(RSS_T/T) + k \cdot (\ln T)/T$. The text of Appendix B defines the models. The three entries within a given block of numbers are: the *F*-statistic for testing the null hypothesis (indicated by the model number to the left of the entry) against the maintained hypothesis (indicated by the model number above the entry), the tail probability associated with that value of the *F*-statistic (in square brackets), and the degrees of freedom for the *F*-statistic (in parentheses).

Table B2 reports $\hat{\sigma}$ and the Schwarz criterion (*SC*) for each model, the *F*-statistics for the reductions between all model pairs, and the associated tail probability values. Throughout the reduction sequence, $\hat{\sigma}$ remains relatively constant, the Schwarz criterion is always declining, and no reductions are statistically significant at the 5% level, whether considered individually or as sub-sequences. The complete reduction appears valid, with $F(24,67) = 0.70$ [0.83].

REFERENCES

BABA, Y., D. F. HENDRY, and R. M. STARR (1992), 'The Demand for M1 in the U.S.A., 1960–1988', *Review of Economic Studies*, 59, 1, pp. 25–61.

BANERJEE, A., J. J. DOLADO, J. W. GALBRAITH, and D. F. HENDRY (1993), *Co-integration, Error Correction, and the Econometric Analysis of Non-stationary Data*, Oxford, Oxford University Press.

BARTELSMAN, E. J. and W. P. CLEVELAND (1993), 'Joint Seasonal Adjustment of Economic Time Series', mimeo, Board of Governors of the Federal Reserve System, Washington, D.C.

BAUMOL, W. J. (1952), 'The Transactions Demand for Cash: An Inventory Theoretic Approach', *Quarterly Journal of Economics*, 66, 4, pp. 545–56.

BEAULIEU, J. J. and J. A. MIRON (1993), 'Seasonal Unit Roots in Aggregate U.S. Data', *Journal of Econometrics*, 55, 1/2, pp. 305–28.

BELL, W. R. (1992), 'On Some Properties of X–11 Symmetric Linear Filters', mimeo, Statistical Research Division, U.S. Bureau of the Census,Washington, D.C.

BELL, W. R. and D. WILCOX (1990), 'A Note on Seasonal Models, Seasonal Adjustment, and Nonseasonal Unit Roots', mimeo, Board of Governors of the Federal Reserve System, Washington, D.C.

BIRCHENHALL, C. R., R. C. BLADEN-HOVELL, A. P. L. CHUI, D. R. OSBORN, and J. P. SMITH (1989), 'A Seasonal Model of Consumption', *Economic Journal*, 99, 397, pp. 837–43.

BOX, G. E. P. and D. A. PIERCE (1970), 'Distribution of Residual Autocorrelations in Autoregressive-integrated Moving Average Time Series Models', *Journal of the American Statistical Association*, 65, 332, pp. 1509–26.

BROWN, R. L., J. DURBIN, and J. M. EVANS (1975), 'Techniques for Testing the Constancy of Regression Relationships over Time', *Journal of the Royal Statistical Society, Series B*, 37, 2, pp. 149–92 (with discussion).

BURRIDGE, P. and K. F. WALLIS (1984), 'Unobserved-components Models for Seasonal Adjustment Filters', *Journal of Business and Economic Statistics*, 2, 4, pp. 350–9.

CAMPOS, J., N. R. ERICSSON, and D. F. HENDRY (1990), 'An Analogue Model of Phase-averaging Procedures', *Journal of Econometrics*, 43, 3, pp. 275–92.

CHONG, Y. Y. and D. F. HENDRY (1986), 'Econometric Evaluation of Linear Macro-economic Models', *Review of Economic Studies*, 53, 4, pp. 671–90.

CHOW, G. C. (1960), 'Tests of Equality between Sets of Coefficients in Two Linear Regressions', *Econometrica*, 28, 3, pp. 591–605.

CLEVELAND, W. P. and G. C. TIAO (1976), 'Decomposition of Seasonal Time Series: A Model for the Census X–11 Program', *Journal of the American Statistical Association*, 71, 355, pp. 581–7.

COGHLAN, R. T. (1978), 'A Transactions Demand for Money', *Bank of England Quarterly Bulletin*, 18, 1, pp. 48–60.

COX, D. R. (1961), 'Tests of Separate Families of Hypotheses' in J. Neyman (ed.), *Proceedings of the Fourth Berkeley Symposium on Mathematical Statistics and Probability*, Berkeley, University of California Press, Volume 1, pp. 105–23.

—— (1962), 'Further Results on Tests of Separate Families of Hypotheses', *Journal of the Royal Statistical Society, Series B*, 24, 2, pp. 406–24.

CUTHBERTSON, K. (1988), 'The Demand for M1: A Forward Looking Buffer Stock Model', *Oxford Economic Papers*, 40, 1, pp. 110–31.

DAVIDSON, J. E. H. (1987), 'Disequilibrium Money: Some Further Results with a Monetary Model of the UK', Chapter 6 in C. A. E. Goodhart, D. Currie, and D. T. Llewellyn (eds.), *The Operation and Regulation of Financial Markets*, London, Macmillan Press, pp. 125–49.

DAVIDSON, J. E. H., D. F. HENDRY, F. SRBA, and S. YEO (1978), 'Econometric Modelling of the Aggregate Time-series Relationship between Consumers' Expenditure and Income in the United Kingdom', *Economic Journal*, 88, 352, pp. 661–92.

DICKEY, D. A. and W. A. FULLER (1979), 'Distribution of the Estimators for Autoregressive Time Series with a Unit Root', *Journal of the American Statistical Association*, 74, 366, pp. 427–31.

—— (1981), 'Likelihood Ratio Statistics for Autoregressive Time Series with a Unit Root', *Econometrica*, 49, 4, pp. 1057–72.

DICKEY, D. A., D. P. HASZA, and W. A. FULLER (1984), 'Testing for Unit Roots in Seasonal Time Series', *Journal of the American Statistical Association*, 79, 386, pp. 355–67.

DOORNIK, J. A. and D. F. HENDRY (1992), *PcGive Version 7: An Interactive Econometric Modelling System* (*Version 8.00*), Oxford, Institute of Economics and Statistics, University of Oxford.

—— (1994), *PcFiml Version 7: An Interactive Econometric Modelling System*, Oxford, Institute of Economics and Statistics, University of Oxford, in press.

DUFOUR, J.-M. (1982), 'Recursive Stability Analysis of Linear Regression Relationships: An Exploratory Methodology', *Journal of Econometrics*, 19, 1, pp. 31–76.

DURBIN, J. and G. S. WATSON (1950), 'Testing for Serial Correlation in Least Squares Regression. I', *Biometrika*, 37, 3 and 4, pp. 409–28.

—— (1951), 'Testing for Serial Correlation in Least Squares Regression. II', *Biometrika*, 38, 1 and 2, pp. 159–78.

ENGLE, R. F. (1982), 'Autoregressive Conditional Heteroscedasticity with Estimates of the Variance of United Kingdom Inflation', *Econometrica*, 50, 4, pp. 987–1007.

ENGLE, R. F. and C. W. J. GRANGER (1987), 'Co-integration and Error Correction: Representation, Estimation, and Testing', *Econometrica*, 55, 2, pp. 251–76.

ENGLE, R. F. and D. F. HENDRY (1993), 'Testing Super Exogeneity and Invariance in Regression Models', *Journal of Econometrics*, 56, 1/2, pp. 119–39.

ENGLE, R. F., D. F. HENDRY, and J.-F. RICHARD (1983), 'Exogeneity', *Econometrica*, 51, 2, pp. 277–304.

ERICSSON, N. R. (1983), 'Asymptotic Properties of Instrumental Variables Statistics for Testing Non-nested Hypotheses', *Review of Economic Studies*, 50, 2, pp. 287–304.

—— (ed.), (1992*a*), *Cointegration, Exogeneity, and Policy Analysis*, Special Issue, *Journal of Policy Modeling*, 14, 3 (*Part I: Expenditure and Money Demand*) and 4 (*Part II: Inflation and Parameter Constancy*).

—— (1992*b*), 'Parameter Constancy, Mean Square Forecast Errors, and Measuring Forecast Performance: An Exposition, Extensions, and Illustration', *Journal of Policy Modeling*, 14, 4, pp. 465–95.

ERICSSON, N. R., J. CAMPOS, and H.-A. TRAN (1990), 'PC-GIVE and David Hendry's Econometric Methodology', *Revista de Econometria*, 10, 1, pp. 7–117.

ERICSSON, N. R., D. F. HENDRY, and H.-A. TRAN (1993), 'Cointegration, Seasonality, Encompassing, and the Demand for Money in the United Kingdom', International Finance Discussion Paper No. 457, Board of Governors of the Federal Reserve System, Washington, D.C., October.

ERMINI, L. and C. W. J. GRANGER (1993), 'Some Generalizations on the Algebra of I(1) Processes', *Journal of Econometrics*, 58, 3, pp. 369–84.

FAVERO, C. and D. F. HENDRY (1992), 'Testing the Lucas Critique: A Review', *Econometric Reviews*, 11, 3, pp. 265–306.

FRANSES, P. H. and T. KLOEK (1991), 'A Periodic Cointegration Model of Quarterly Consumption in Austria and Japan', Report 9172/A, Econometric Institute, Erasmus University Rotterdam, Rotterdam, The Netherlands.

FRIEDMAN, M. (ed.), (1956), *Studies in the Quantity Theory of Money*, Chicago, University of Chicago Press.

FRIEDMAN, M. and A. J. SCHWARTZ (1982), *Monetary Trends in the United States and the United Kingdom: Their Relation to Income, Prices, and Interest Rates, 1867–1975*, Chicago, University of Chicago Press.

GHYSELS, E., H. S. LEE, and J. NOH (1991), 'Testing for Unit Roots in Seasonal Time Series: Some Theoretical Extensions and a Monte Carlo Investigation', mimeo, Department of Economics, University of Montreal, Montreal, Canada.

GHYSELS, E. and P. PERRON (1993), 'The Effect of Seasonal Adjustment Filters on Tests for a Unit Root', *Journal of Econometrics*, 55, 1/2, pp. 57–98.

GODFREY, L. G. (1978), 'Testing Against General Autoregressive and Moving Average Error Models when the Regressors Include Lagged Dependent Variables', *Econometrica*, 46, 6, pp. 1293–301.

GOLDFELD, S. M. and D. E. SICHEL (1990), 'The Demand for Money', Chapter 8 in B. M. Friedman and F. H. Hahn (eds.), *Handbook of Monetary Economics*, Amsterdam, North-Holland, Volume 1, pp. 299–356.

GOODHART, C. A. E. (1984), *Monetary Theory and Practice: The UK Experience*, London, Macmillan Press.

—— (1989), 'The Conduct of Monetary Policy', *Economic Journal*, 99, 396, pp. 293–346.

GRANGER, C. W. J. (1978), 'Seasonality: Causation, Interpretation, and Implications' in A. Zellner (ed.), *Seasonal Analysis of Economic Time Series*, Economic Research Report ER–1, Washington, D.C., Bureau of the Census, pp. 33–45.

GRANGER, C. W. J. and J. HALLMAN (1991), 'Nonlinear Transformations of Integrated Time Series', *Journal of Time Series Analysis*, 12, 3, pp. 207–24.

GRETHER, D. M. and M. NERLOVE (1970), 'Some Properties of *Optimal* Seasonal Adjustment', *Econometrica*, 38, 5, pp. 682–703.

HACCHE, G. (1974), 'The Demand for Money in the United Kingdom: Experience Since 1971', *Bank of England Quarterly Bulletin*, 14, 3, pp. 284–305.

HALL, S. G., S. G. B. HENRY, and J. B. WILCOX (1990), 'The Long-run Determination of the UK Monetary Aggregates', Chapter 5 in S. G. B. Henry and K. D. Patterson (eds.), *Economic Modelling at the Bank of England*, London, Chapman and Hall, pp. 127–66.

HARVEY, A. C. (1981), *The Econometric Analysis of Time Series*, Oxford, Philip Allan.

HEALEY, J., C. MANN, R. CLEWS, and G. HOGGARTH (1990), 'Monetary Aggregates in a Changing Environment: A Statistical Discussion Paper', Bank of England Discussion Paper No. 47, London, England.

HENDRY, D. F. (1979), 'Predictive Failure and Econometric Modelling in Macroeconomics: The Transactions Demand for Money', Chapter 9 in P. Ormerod (ed.), *Economic Modelling*, London, Heinemann Education Books, pp. 217–42.

—— (1983), 'Comment', *Econometric Reviews*, 2, 1, pp. 111–14.

—— (1985), 'Monetary Economic Myth and Econometric Reality', *Oxford Review of Economic Policy*, 1, 1, pp. 72–84.

—— (ed.), (1986), *Economic Modelling with Cointegrated Variables*, *Oxford Bulletin of Economics and Statistics*, *Special Issue*, 48, 3.

—— (1988), 'The Encompassing Implications of Feedback versus Feedforward Mechanisms in Econometrics', *Oxford Economic Papers*, 40, 1, pp. 132–49.

HENDRY, D. F. and N. R. ERICSSON (1991*a*), 'An Econometric Analysis of U.K. Money Demand in *Monetary Trends in the United States and the United Kingdom* by Milton Friedman and Anna J. Schwartz', *American Economic Review*, 81, 1, pp. 8–38.

—— (1991*b*), 'Modeling the Demand for Narrow Money in the United Kingdom and the United States', *European Economic Review*, 35, 4, pp. 833–81.

HENDRY, D. F. and G. E. MIZON (1978), 'Serial Correlation as a Convenient Simplification, Not a Nuisance: A Comment on a Study of the Demand for Money by the Bank of England', *Economic Journal*, 88, 351, pp. 549–63.

—— (1993), 'Evaluating Dynamic Econometric Models by Encompassing the VAR', in P. C. B. Phillips (ed.), *Models, Methods, and Applications of Econometrics*, Oxford, Basil Blackwell, pp. 272–300.

HENDRY, D. F. and J.-F. RICHARD (1989), 'Recent Developments in the Theory of Encompassing', Chapter 12 in B. Cornet and H. Tulkens (eds.), *Contributions to Operations Research and Economics: The Twentieth Anniversary of CORE*, Cambridge, Massachusetts, MIT Press, pp. 393–440.

HYLLEBERG, S. (1986), *Seasonality in Regression*, Orlando, Florida, Academic Press.

HYLLEBERG, S., R. F. ENGLE, C. W. J. GRANGER, and B. S. YOO (1990), 'Seasonal Integration and Cointegration', *Journal of Econometrics*, 44, 1/2, pp. 215–238.

HYLLEBERG, S., C. JØRGENSEN, and N. K. SØRENSEN (1991), 'Seasonality in Macroeconomic Time Series', Working Paper 1991–9, Økonomisk Institut, Aarhus University, Aarhus, Denmark.

JARQUE, C. M. and A. K. BERA (1980), 'Efficient Tests for Normality, Homoscedasticity and Serial Independence of Regression Residuals', *Economics Letters*, 6, 3, pp. 255–9.

JOHANSEN, S. (1988), 'Statistical Analysis of Cointegration Vectors', *Journal of Economic Dynamics and Control*, 12, 2/3, pp. 231–54.

—— (1991), 'Estimation and Hypothesis Testing of Cointegration Vectors in Gaussian Vector Autoregressive Models', *Econometrica*, 59, 6, pp. 1551–80.

—— (1992*a*), 'Cointegration in Partial Systems and the Efficiency of Single-equation Analysis', *Journal of Econometrics*, 52, 3, pp. 389–402.

—— (1992*b*), 'A Representation of Vector Autoregressive Processes Integrated of Order 2', *Econometric Theory*, 8, 2, pp. 188–202.

—— (1992*c*), 'Testing Weak Exogeneity and the Order of Cointegration in UK Money Demand Data', *Journal of Policy Modeling*, 14, 3, pp. 313–34.

JOHANSEN, S. and K. JUSELIUS (1990), 'Maximum Likelihood Estimation and Inference on Cointegration — With Applications to the Demand for Money', *Oxford Bulletin of Economics and Statistics*, 52, 2, pp. 169–210.

JUDD, J. P. and J. L. SCADDING (1982), 'The Search for a Stable Money Demand Function: A Survey of the Post-1973 Literature', *Journal of Economic Literature*, 20, 3, pp. 993–1023.

KENNY, P. B. and J. DURBIN (1982), 'Local Trend Estimation and Seasonal Adjustment of Economic and Social Time Series', *Journal of the Royal Statistical Society, Series A*, 145, 1, 1–41 (with discussion).

LAIDLER, D. E. W. (1985), *The Demand for Money: Theories, Evidence, and Problems*, New York, Harper and Row, Third Edition.

LEE, H. S. (1992), 'Maximum Likelihood Inference on Cointegration and Seasonal Cointegration', *Journal of Econometrics*, 54, 1–3, pp. 1–47.

LEE, H. S. and P. L. SIKLOS (1991), 'Seasonality in Time Series: Money-Income Causality in U.S. Data Revisited', mimeo, Department of Economics, Tulane University, New Orleans, Louisiana.

—— (1993), 'The Influence of Seasonal Adjustment on the Canadian Consumption Function, 1947–1991', *Canadian Journal of Economics*, 26, 3, pp. 575–89.

LOVELL, M. C. (1963), 'Seasonal Adjustment of Economic Time Series and Multiple Regression Analysis', *Journal of the American Statistical Association*, 58, 304, pp. 993–1010.

MACKINNON, J. G. (1991), 'Critical Values for Cointegration Tests', Chapter 13 in R. F. Engle and C. W. J. Granger (eds.), *Long-run Economic Relationships: Readings in Cointegration*, Oxford, Oxford University Press, pp. 267–76.

MACKINNON, J. G. and H. WHITE (1985), 'Some Heteroskedasticity-consistent Covariance Matrix Estimators with Improved Finite Sample Properties', *Journal of Econometrics*, 29, 3, pp. 305–25.

MARAVALL, A. (1993), 'Stochastic Linear Trends: Models and Estimators', *Journal of Econometrics*, 56, 1/2, 5–37.

MARAVALL, A. and D. A. PIERCE (1987), 'A Prototypical Seasonal Adjustment Model', *Journal of Time Series Analysis*, 8, 2, pp. 177–93.

MILBOURNE, R. (1983), 'Optimal Money Holding Under Uncertainty', *International Economic Review*, 24, 3, pp. 685–98.

MILLER, M. H. and D. ORR (1966), 'A Model of the Demand for Money by Firms', *Quarterly Journal of Economics*, 80, 3, pp. 413–35.

MIZON, G. E. and J.-F. RICHARD (1986), 'The Encompassing Principle and Its Application to Testing Non-nested Hypotheses', *Econometrica*, 54, 3, pp. 657–78.

NERLOVE, M. (1964), 'Spectral Analysis of Seasonal Adjustment Procedures', *Econometrica*, 32, 3, pp. 241–86.

NICHOLLS, D. F. and A. R. PAGAN (1983), 'Heteroscedasticity in Models with Lagged Dependent Variables', *Econometrica*, 51, 4, pp. 1233–42.

OSBORN, D. R. (1988), 'Seasonality and Habit Persistence in a Life Cycle Model of Consumption', *Journal of Applied Econometrics*, 3, 4, pp. 255–66.

—— (1991), 'The Implications of Periodically Varying Coefficients for Seasonal Time-series Processes', *Journal of Econometrics*, 48, 3, pp. 373–84.

—— (1993), 'Moving Average Detrending of Integrated Processes', mimeo, Department of Econometrics and Social Statistics, University of Manchester, Manchester, England.

OSTERWALD-LENUM, M. (1992), 'A Note with Quantiles of the Asymptotic Distribution of the Maximum Likelihood Cointegration Rank Test Statistics', *Oxford Bulletin of Economics and Statistics*, 54, 3, pp. 461–72.

PHILLIPS, P. C. B. (1991), 'Optimal Inference in Cointegrated Systems', *Econometrica*, 59, 2, pp. 283–306.

RAMSEY, J. B. (1969), 'Tests for Specification Errors in Classical Linear Least-squares Regression Analysis', *Journal of the Royal Statistical Society, Series B*, 31, 2, pp. 350–71.

SIMS, C. A. (1974), 'Seasonality in Regression', *Journal of the American Statistical Association*, 69, 347, pp. 618–26.

SMITH, G. W. (1986), 'A Dynamic Baumol-Tobin Model of Money Demand', *Review of Economic Studies*, 53, 3, pp. 465–9.

TOBIN, J. (1956), 'The Interest Elasticity of Transactions Demand for Cash', *Review of Economics and Statistics*, 38, 3, pp. 241–7.

TOPPING, S. L. with S. L. BISHOP (1989), 'Breaks in Monetary Series', Bank of England Discussion Paper No. 23, Technical Series, London, England.

TRUNDLE, J. M. (1982), 'The Demand for M1 in the UK', mimeo, Bank of England, London, England.

WALLIS, K. F. (1974), 'Seasonal Adjustment and Relations Between Variables', *Journal of the American Statistical Association*, 69, 345, pp. 18–31.

—— (1978), 'Seasonal Adjustment and Multiple Time Series Analysis' in A. Zellner (ed.), *Seasonal Analysis of Economic Time Series*, Economic Research Report ER–1, Washington, D.C., Bureau of the Census, pp. 347–57.

—— (1982), 'Seasonal Adjustment and Revision of Current Data: Linear Filters for the X–11 Method', *Journal of the Royal Statistical Society, Series A*, 145, 1, pp. 74–85.

—— (1983), 'Models for X–11 and X–11 Forecast Procedures for Preliminary and Revised Seasonal Adjustments' in A. Zellner (ed.), *Applied Time Series Analysis of Economic Data*, Economic Research Report ER–5, Washington, D.C, Bureau of the Census, pp. 3–11.

—— (1993), 'Henderson Detrending, Symmetric or Asymmetric, Reduces I(4) Series to Stationarity', mimeo, Department of Economics, University of Warwick, Warwick, England.

WHITE, H. (1980), 'A Heteroskedasticity-consistent Covariance Matrix Estimator and a Direct Test for Heteroskedasticity', *Econometrica*, 48, 4, pp. 817–38.

8

Evaluating a real business cycle model

Fabio Canova, Mary Finn and Adrian R. Pagan[*]

Most real business cycle models have been assessed by studying the correspondence of their predictions to a set of stylised facts. This paper argues that such tests are not extensive enough and proposes to evaluate the models using standard econometric procedures. Specifically it is argued that these models should be studied by eliciting the restricted VAR representation underlying them and comparing it with the VAR estimated in an unrestricted way from the underlying data. Allowance is made for cases where the driving forces are integrated and when they are stationary. When forces such as technology shocks are integrated these models produce a specific set of predictions about the cointegrating vectors as well as a set of restrictions upon the dynamics. The approach is illustrated using a real business cycle model estimated by Burnside, Eichenbaum and Rebelo. This model has been subjected to some formal testing based upon stylised facts, and it therefore seems an appropriate one upon which to utilize the formal econometric procedures.

1. INTRODUCTION

In the last decade, real business cycles (RBC) models have gone from the preliminary explorations of Long and Plosser (1983) and Kydland and Prescott (1982) to well developed and tested models such as Burnside, Eichenbaum and Rebelo (1993) and McGrattan (1991). Early models could be regarded as 'idealised', in the sense adopted in the philosophical literature summarised in

[*] Financial support from the U.K. Economic and Social Research Council under grant R000233447 is gratefully acknowledged by both authors. We are indebted to Neil Ericsson for helpful comments.

Hoover (1991*a*), in that they were 'simplifications that were designed to isolate an *essential* core' — in this instance, attempts to capture the characteristics of fluctuations within industrial economies. Given such an objective it was appropriate that the method employed to determine whether the 'essence' of an economy had been captured or not was the method of 'stylised facts'. In this procedure a certain number of key 'facts' are identified and subsequently used to gauge the performance of the model. Thus Long and Plosser concentrated upon the idea that business cycles generated co-movements between de-trended variables, and they asked whether it was possible to obtain such a feature with the very simple RBC model that they had constructed. Others have been somewhat more precise, asking if the variances and covariances between variables such as output, consumption and real wages observed in the US economy agreed with the predictions of their model. In extensions of this early work, e.g. King Plosser and Rebelo (1988), a similar strategy was adopted to that of Prescott and Kydland, but with a more extensive range of stylised facts to be explained.

Stylised facts are obviously a good way of evaluating idealised models. By their very nature the latter models are not meant to provide a complete description of any time series such as consumption or output, but rather attempt to emulate a few of the major characteristics of those variables. Nevertheless, even with such a limited objective, there still remains an important practical problem of determining just how well the models are emulating reality, and this necessitates the development of some 'metric' for that task. Because RBC models are explicitly stochastic a number of measures have been proposed that involve computing standard errors for the model predictions, either by analytic means or by computer based simulation, e.g. Gregory and Smith (1993), Canova (1990).

Early comparisons of model projections with stylised facts revealed that the models did not adequately account for the latter. Perhaps the most striking failures were the correlation of productivity with hours worked and of government consumption with the Solow residual. Stimulated by this fact, researchers in the area began to develop the models in a number of different directions, with the aim of getting a better match with the stylised facts. As discussed by Hoover (1991*a*) this development can be thought of as 'concretising ' the idealised models so as to make for a better correspondence with the 'real world'. 'Concretisation' has now been performed in many different directions and there has been substantial success in clearing up some of the striking failures of the early models.

The developments described above are reminiscent of early work with macroeconometric models. Initially, the desire was to explain some very broad characteristics of the data. As ambitions rose and simple models were replaced by large scale ones in an attempt to capture real world complexities, it was necessary to devise tests of the latter that were much more demanding, so as to try to isolate where the deficiencies of the models lay. It seems appropriate therefore that the attempts at evaluation via stylised facts, which have characterized most RBC studies to date, should also be replaced by more demanding and comprehensive tests, particularly since these models are progressively 'concretised' in order to account for specific 'stylised facts'. What makes this task different to the older econometric literature is that RBC models are models with a great deal of internal coherence, and it is very hard to evaluate the components separately; one is inevitably faced with the need to work with the whole model. Consequently, many of the 'single equation' tests that have been used so effectively when evaluating large scale macroeconometric models are difficult to apply, since one could not make a modification to a 'part' of the RBC model without affecting it somewhere else. Complete model evaluation methods are the logical way to proceed.

This paper is an attempt to do the requisite analysis along these lines. It is well known that RBC models involve a VAR in the variables — see Long and Plosser (1983) for example. Furthermore, as we observe in Section 2, it is a highly restricted VAR. Thus, just as for the rational expectations models considered by Sargent (1978), it seems as if a sensible way to evaluate the models is to test the restrictions on the implied VAR. Although the idea is straightforward, one has to be somewhat more cautious. Frequently, the driving forces in these models are integrated, and the VAR is actually a vector ECM, due to there being a smaller number of driving forces than variables being explained. If the driving forces are integrated, analysis suggests that there are two types of restrictions that might be tested. First, there are the cointegrating restrictions stemming from the fact that there are generally more variables to be modelled than there are independent integrated forcing processes. Second, there are restrictions upon the dynamics which apply to the system written after factoring out the cointegrating relations. Section 2 develops these ideas.

Section 3 of the paper takes a particular RBC model, that due to Burnside *et al.* and applies the ideas developed earlier to it. This model was chosen because there have been a number of concretising steps taken to make it emulate the real world, although there remains some doubt over whether it actually agrees with a comprehensive range of stylised facts. Our claim is that consideration of the two types of restrictions described above, and a determination of whether they are acceptable, can be a very useful input into a modelling exercise. In particular, such information can highlight deficiencies in the models and may suggest suitable re-specifications. In our example, we find that the BER model is strongly rejected by the data and we enquire into what changes might be made to the model to produce a VAR that more closely approximates what is seen in the data.

In Section 4, we ask the question of how well the RBC model functions relative to a simple model such as a multiplier-accelerator mechanism and discuss whether the latter is any more successful in reproducing the VAR than the RBC model is. The viewpoint of this section is that, ultimately, the relevant question pertains to the relative rather than absolute quality of a model. Such comparisons are also likely to yield better information about potential respecifications. Finally, Section 5 concludes with some suggestions about how the RBC model might be modified to produce a better fit to the data.

2. TESTING THE RESTRICTIONS OF AN RBC MODEL

Define y_t as the $(q \times 1)$ vector of variables of interest, z_t as the $(n \times 1)$ vector of controlled and uncontrolled state variables, and x_t as the $(p \times 1)$ vector of exogenous or forcing variables (the uncontrolled states). Most RBC models can be regarded as conforming to a linear structure of the form

$$y_t = \mathrm{A}z_t \tag{1}$$

$$z_t = \mathrm{F}z_{t-1} + \mathrm{G}e_t, \tag{2}$$

where e_t are the innovations into the forcing variables and G is a matrix showing how these innovations impinge upon the state variables. Generally G is a matrix that only has rank p, i.e. there are more state variables than there are stochastic elements in (2). The linearity of the system stems from the fact that these systems are frequently solved by either linearising the Euler equations around the steady state, as in King, Plosser, and Rebelo (1988), or solving the Riccati equation associated with the linear/quadratic control problem, a method employed by McGrattan (1991). It is possible to argue that (1) and (2) are more general than they might appear to be in that some types of non-linearities might be accommodated, e.g. z_t might be functions of state variables. Some of the solution methods, such as Marcet (1989) or Chow (1992*a*), can allow this interpretation. Higher order dynamics can also be incorporated, but, since the application given later has first order dynamics, the discussion will focus upon the special case.

An especially important characteristic of many RBC models is that F and A are functions of a smaller number of parameters such as utility and production function parameters, and the latter are typically selected by some 'calibration' strategy. It is hard to be precise about exactly what the latter is as it ranges from selecting parameter values gleaned from micro or macro studies — estimated either by sample averages or by methods such as GMM and FIML — to 'guesstimates'. We will simply assume that A and F have precise numercial values assigned to them, so that an RBC model is both a set of relationships as in (1) and (2) and a specific set of values for the parameters A and F. Of course, this is true of any macroeconomic model. Nevertheless, one might argue over whether the parameter values should be taken as capturing the 'essence' of an economy or are simply concretising assumptions, i.e. perhaps what should be tested is the general format in (1) and (2) rather than the particular structure coming from specific values of A and F. As an example of the difference, suppose y_t was consumption and z_t was output. Then (1) can be interpreted as either saying that the average propensity to consume is *exactly* A or that the average propensity to consume is simply some unknown *constant* A. Although there are some testable implications of the latter viewpoint, they are obviously very weak, and it is likely that many models would yield such a prediction, e.g. there were many early consumption relations that were not intertemporal but which would imply constancy of the consumption ratio. Hence, as a way of distinguishing between different theories, it seems necessary to maintain that the numbers assigned to A and F are parts of the model. One could plausibly argue against this strategy if A and F were estimated directly from data, but since they are functions of a much smaller number of 'deep parameters', the power of the RBC model presumably derives from just this fact. Indeed that seems to have been an essential ingredient in the original arguments put forth for such models in Long and Plosser (1983) and by Prescott (1991) who imposes parameter values as a consequence of steady state relations.

It is necesssary to distinguish between two scenarios for (1) and (2) depending upon the nature of the forcing variables, x_t. In many applications of RBC models x_t are made I(1) processes, generally independent of each other, i.e.

$$x_t = x_{t-1} + e_t. \tag{3}$$

Under this specification, the structure of F is $F = \begin{bmatrix} \gamma & \delta \\ 0 & I_p \end{bmatrix}$ so that p of the eigen-values of F are unity while the remaining $(n-p)$ are the eigenvalues of γ. In RBC models the latter are less than unity, implying that there must be $(n-p)$ cointegrating vectors among the z_t. Defining the elements of z_t which exclude x_t as z_{1t}, if the z_{1t} are I(1) then it follows immediately that the cointegrating vectors are $[(I-\gamma) \quad -\delta]$; alternatively, if any of the z_{1t} is I(0), the corresponding row, $(\gamma_1 \quad \delta_1)$, must be a cointegrating vector. Identifying z_t with some observed data this would be a *first prediction* of the RBC model. It is also apparent that there are some Granger Causality predictions which stem from (2.3).

Equation (1) predicts that an exact relation should hold between y_t and z_t. Such an exact relation is unlikely to be observed with any set of data and it is important to weaken (1) so as to allow it to be non-exact. The most appropriate extension would seem to be to assume that $y_t - Az_t$ is an I(0) process. There are two arguments one might make in favour of this stance. The first is that the RBC model aims at capturing the essential mechanisms at work in the economy, and, *prima facie*, this suggests that what is left out should be distinguishable as something of less importance than what is retained. When z_t is integrated it is natural therefore to think that what has been ignored should be non-integrated. Second, if one thought of observed data as being different from the model constructs due to measurement error, it is natural to make the measurement error an I(0) process when the variable being incorrectly measured is I(1). Therefore, in terms of either argument, $(I_q \quad -A)$ should be a set of cointegrating vectors, and this is a *second testable implication* of an RBC model. Note that what we have is not only the requirement that y_t and z_t be cointegrated but that they be cointegrated with the numerical values assigned to A.

A second set of restrictions implied by RBC models involves the dynamic structure, or what will be termed the 'non-cointegrating restrictions'. To derive these write (1) and (2) as

$$\Delta y_t = A\Delta z_t, \tag{4}$$

where

$$\Delta z_t = (F - I)z_{t-1} + Ge_t \tag{5}$$

$$= \Pi z_{t-1} + Ge_t \tag{6}$$

$$= \alpha\beta' z_{t-1} + Ge_t \tag{7}$$

and β are the cointegrating vectors existing among the z_t. Substituting (6) into (5) yields

$$\Delta y_t = A\alpha\beta' z_{t-1} + AGe_t \tag{8}$$

and, forming the cointegrating error $v_t = \beta' z_{t-1}$, we have

$$\Delta y_t = A\alpha v_{t-1} + AGe_t \tag{9}$$

which is a relation solely between I(0) variables. Defining $w_t' = (\Delta y_t' \; v_t')$ the VAR in w_t implied by an RBC model therefore has two characteristics. First, unless y_t is a state variable, Δy_{t-1} is excluded from it. Second, the coefficients of v_{t-1} are given by $A\alpha$. These are the *third set of testable predictions*, and they concern the non-cointegrating restrictions. Notice that the restrictions stem from the dynamic nature of the model, provided we have previously accepted that the cointegrating restrictions are valid ones.

When the forcing variables are not I(1) the distinction between the two types of restrictions ceases to be valid. In these cases, although (1) would still be a restriction, it would be very hard to test it, as any variables left out are of the same order of integration, zero, and one would be faced with the prospect of doing a regression in the presence of specification error. Hence, in these cases, it is logical to combine the two directly, substituting (1) into (2) to get

$$y_t = \mathrm{AF}z_{t-1} + \mathrm{AG}e_t. \tag{10}$$

Viewed as a VAR, now in the I(0) variables $\overline{w}_t' = (y_t' \; z_t')$, one finds a similar set of restrictions to the non-cointegrating set found above. Specifically, y_{t-1} does not appear in the VAR and the coefficients of z_{t-1} should be AF.

Basically, the argument for testing the restrictions upon the VAR advanced above is that it may be possible to identify suitable re-specifications of the RBC model in the event that rejections of the restrictions are encountered. For example, if the prediction that Δy_{t-1} is excluded from the VAR is false, attention is immediately directed to how the RBC model might be modified so as to induce such a variable into the implied VAR. The VAR is therefore being used as a 'reduced form' and, indeed, the evaluation strategy being followed here is the modern equivalent of the classical precepts laid down by the Cowles Commission researchers when testing the structural equation restrictions upon the reduced form — see Byron (1974). All that has changed is the substitution of the reduced form by its time series construct, the VAR. This idea has been mentioned or exploited by a number of authors, e.g. Spanos (1986), Monfort and Rabemananjara (1990) and Hendry and Mizon (1993), the latter being the most complete treatment in that it allows for variables to be either I(1) or I(0).

Although the systems approach to testing set out above is an attractive one, there may be advantages to focusing upon more restricted implications of the RBC model. One of these is the nature of the final equations for y_t, i.e. if y_t is a scalar, finding the ARMA process

$$\mathrm{C}(L)y_t = \mathrm{D}(L)\varepsilon_t \tag{11}$$

implied by (1) and (2). Comparing this derived equation to the ARMA models estimated from the data may be used to indicate how good a representation the RBC model is. Tinbergen (1939) was an early user of the final equations for summarising the properties of a system, and the idea was subsequently formalised and utilised in Zellner and Palm (1974), Wallis (1977) and Evans (1989). Cogley and Nason (1992) apply the idea to a variety of RBC models, showing that, with the exception of the Burnside *et al.* model, such models do not reproduce the higher order autocorrelation features of GDP data for the US. Obviously, such a comparison may be extremely valuable in revealing how well the system mimics the data on selected variables. Its principal disadvantage is that the information gleaned from such a comparison may be extremely difficult to use in re-specifying any RBC model, simply because C(L) and D(L) will inevitably be complex functions of all parts of the original model.

A related procedure, after making y_t a vector, is to determine the VAR in y_t alone, i.e. to reduce the VAR in y_t, z_t to one in terms of y_t alone. Such a construct may be of interest because of our familiarity with many bivariate and trivariate relations. For example, if y_t is composed of net investment and output, the accelerator mechanism is a well known bivariate relation linking those two variables, and it might therefore be profitable to enquire into whether there is an accelerator mechanism at work within RBC models. To perform this task requires a number of steps. First, after computing the autocovariances of y_t from (1) and (2), an approximating VAR can be fitted by solving the multivariate version of the Yule-Walker equations. This would lead to

$$\mathrm{C}(L)y_t = \varepsilon_t. \tag{12}$$

Second, suppose that y_t was bivariate with elements y_{1t} and y_{2t}. To investigate relations like the accelerator necessitates relating y_{1t} to y_{2t} as well as their past histories. The error term in ε_t has to be decomposed to isolate the contemporaneous effect. To this end let the VAR in (12) be re-expressed as

$$\begin{aligned} y_{1t} &= \mathrm{C}_1(L)y_{t-1} + \varepsilon_{1t} \\ y_{2t} &= \mathrm{C}_2(L)y_{t-1} + \varepsilon_{2t}. \end{aligned} \tag{13}$$

Owing to the linear structure ε_{1t} can be written as $\varepsilon_{1t} = \rho\varepsilon_{2t} + \eta_{1t}$, where $\rho = \sigma_{22}^{-1}\sigma_{12}$, $\sigma_{ij} = \mathrm{E}(\varepsilon_{it}\varepsilon_{jt})$, and η_{1t} is an innovation with respect to $\{y_{t-j}, y_{2t}\}_{j=1}^{\infty}$. Consequently,

$$y_{1t} = [\mathrm{C}_1(L) - \rho\mathrm{C}_2(L)]y_{t-1} + \rho y_{2t} + \eta_{1t} \tag{14}$$

gives the desired relationship. In (13) the polynomials $\mathrm{C}_j(L)$ and the correlation coefficient ρ are by-products from fitting the VAR (12) to the autocovariances of y_t coming from (1) and (2). Operationally, one simply has to decide upon the order of the approximating VAR. Of course, the relation under study might also be a trivariate one, e.g. if y_t contains real money, interest rates and output, a 'money demand function' could be elicited. Perhaps the main use of this device is when comparisons are made between RBC and alternative business cycle models such as multiplier-accelerator, as conversion of the RBC model to resemble the alternative model allows an easier assessment of the relative performance of the two contenders. Another use is if one wants to compute quantities such as the Kullback-Liebler Information Criterion (KLIC) in order to compare models. Because there are fewer shocks than variables in most RBC models, the density for z_t would be singular, and hence the KLIC is not defined. However, by restricting attention to a VAR system whose order equals the number of shocks one can define the KLIC for such a system.

Although what should be tested when evaluating RBC models seems to be fairly clear, exactly how it is to be done is much more controversial. The source of the controversy resides in the fact that the variables y_t and z_t in the model may not be accurately measured by data, i.e. there are errors in variables. When testing the cointegrating restrictions such a difficulty can be ignored, provided that the errors are I(0), but the same cannot be said for tests of the non-cointegrating restrictions. Here what is being tested is whether the coefficients of $v_{t-1} = \beta' z_{t-1}$ in (9) have the values predicted by the RBC model. But if the errors in z_t and y_t are linearly related to z_{t-1}, the observed value could validly deviate from that predicted by the RBC model. Without some statement about the mapping of the errors into z_{t-1}, it would therefore be impossible to follow the testing strategy outlined above. Within the literature on calibrated models, this point appears to be regarded as the critical one that prohibits formal econometric testing — Kydland and Prescott (1991) and Watson (1993). There is little that can be said about this objection. It could be applied to any model and, taken to its extreme, would result in nothing being testable. If it is adopted the only consistent attitude would seem to be one in which all quantitative modelling was eschewed. However, such consistency is a rare phenomenon; it is not uncommon to find proponents of RBC models rejecting competitive scenarios as incompatible with the data but failing to apply the same test to their preferred approach on the grounds that the models are too idealised. For example, Kydland and Prescott (1991) regurgitate the Lucas-Sargent criticism that large scale Keynesian models of the 1970's were inadequate due to a failure to correctly predict the observed unemployment-inflation correlations of that decade, but immediately exempt RBC models from a similar test by stating that 'the issue of how confident we are in the econometric answer is a subtle one which cannot be resolved by computing some measure of how well the model economy mimics historical data. The degree of confidence in the answer depends on the confidence that is placed in the economic theory being used'. (1991, p.171).

The only way out of this morass is to place some constraint upon the relationship of any errors in variables to z_{t-1}. Traditionally, this has been to insist upon the errors being white noise. Such errors in y_t would result in a white noise disturbance for (9), whereas a similar assumption for errors in z_t would create an MA(1) disturbance. In the first instance, estimation and testing would proceed in the normal way; in the second, some form of instrumental variables estimation would need to be performed to allow for the correlation between z_{t-1} and the MA(1) disturbance. Of course, the disturbance in (9) could be uncorrelated with z_{t-1} under weaker conditions than white noise in the errors in variables. The situation is reminiscent of rational expectations modelling where forward looking behaviour creates disturbance terms that are MA's but which are still orthogonal to any regressors that appear in agents' information sets. If this extension is envisaged allowance needs to be made for the effects of such serial correlation upon inferences by adopting robust measures of the variances of estimators.

If the errors in variables are to be allowed to be functions of z_{t-1}, it may still be possible to find some measures of fit of the model to data, even though inference is highly unlikely. This is Watson's (1993) approach. He takes the deviation between model output and data to be an 'error', u_t, and then finds an expression for it when the objective is to reproduce the autocovariance function (a.c.f.) of the data. Thus, distinguishing data by means of an asterisk, $y_t^* = y_t + u_t$ $= \text{AF}z_{t-1} + \text{AG}e_t + u_t$, and the task is to determine u_t. For convenience in exposition it will be assumed that z_t is perfectly measured and that y_t is a scalar. Approximating the observed a.c.f. of y_t^* with a VAR in y_t^* and z_t gives $y_t^* = \text{C}_1(L)y_{t-1}^* + \text{C}_2(L)z_{t-1} + \varepsilon_t$. By equating the two expressions for y_t^*, u_t is found to be $u_t = \text{C}_1(L)y_{t-1}^* + (\text{C}_2(L) - \text{AF})z_{t-1} - \text{AG}e_t + \varepsilon_t$, and this choice of u_t means that the augmented model output reproduces the a.c.f. of the data (at least up to the chosen order of VAR). Watson's proposal is then to compute an 'R^2', equal to $1-(\text{var}(u_t)/\text{var}(y_t^*))$, as a measure of fit of the model. As it stands this latter measure is indeterminate as the var(u_t) depends upon an unknown, the covariance of ε_t with u_{t-j}. Because this is a free parameter, Watson proposes to choose it such that var(u_t) is minimised. To see how this is done take $\text{C}_1(L) = c_1 > 0$. Then the smallest value of σ_u^2 occurs when $\text{cov}(\varepsilon_t u_{t-1})$ attains its largest negative value $-\sigma_\varepsilon\sigma_u$ (this corresponding to a correlation between the two variables y_t^* and y_t of -1). A low 'R^2' would presumably be taken as indicating that there is much left unexplained by the RBC model. In practice there are significant complications coming from the fact that y_t will generally be a vector, as the variance of u_t will become a matrix and there is no longer a unique measure of fit.

Watson's idea is certainly ingenious and, given the concern expressed about the idealised nature of these models, has to be useful information for anyone wishing to assess them. However, one cannot escape the feeling that the criterion has to be augmented with supplementary information. One problem that arises is the decision to take the minimum value of σ_u^2 as the basis of the 'R^2'. This is arbitrary, as many values of $\text{cov}(\varepsilon_t u_{t-1})$ would reproduce the a.c.f. of y_t^*, and it is unclear why the one minimising σ_u^2 is to be preferred. Obviously a model with a low R^2 would not be satisfactory, but it is conceivable that a high R^2 could be produced solely due to the particular selection made for $\text{cov}(\varepsilon_t u_{t-1})$, while other choices of this parameter may produce low R^2. Since the parameter, $\text{cov}(\varepsilon_t u_{t-1})$, has nothing to do with the model, and is essentially 'unidentified', it would seem misleading to conclude from the evidence of a high R^2 that the RBC model was satisfactory. At the very least it would seem important that the R^2 be provided for the values of $\text{cov}(\varepsilon_t u_{t-1})$ that both maximise and minimise R^2. If this range is narrow, and the minimum R^2 is a high one, it might be appropriate to conclude that, *prima facie*, the RBC model provides a satisfactory description of the data.

A second problem with the measure is that it does not provide information that may be useful in re-specifying the model. The variance of u_t may be large for a variety of reasons; a high σ_ε^2, a large gap between $C_2(L)$ and A, a large value for $C_1(L)$ etc., but this information is lost in the aggregative measure. However, our attitude towards the model is likely to be significantly affected by which one of these is the principal contributor. If it was due to a high value of $C_1(L)$, we would be led to enquire into whether the RBC model might be re-specified so as to induce the variable y_{t-1} into the VAR. In contrast, if it was a consequence of a large value for σ_ε^2, we are less likely to feel that there is something inadequate in the idealised model, as this parameter represents the extent to which variables exogeneous to the model are unpredictable, and all models would have a similar deficiency e.g. a Keynesian model also has to make some assumption about how government expenditure is to evolve over time.

3. EVALUATING AN RBC MODEL

The model chosen for the evaluation exercise is due to Burnside, Eichenbaum and Rebelo (1990) (BER). It represents a modification of that described in Christiano and Eichenbaum (1992). Appendix 1 presents the principal equations underlying it. The controlled state variables are the capital stock and employment and the uncontrolled states are the technology and government expenditure shocks. When measured as deviations from a steady state growth path these variables are designated as k_t, n_t, a_t and g_t respectively. Other variables explained by the model, also as deviations from steady state, are output (y_t), private consumption (c_t), and investment (i_t). An assumption of the model is that the forcing factors are AR(1) processes. Parameter values for the model were estimated by BER from data over 1955(3) to 1984(1) using various moment conditions.

To evaluate the model BER compared the numerical values of selected variable correlations predicted by the model with the estimated values from the data. The vector of discrepancies can be formally compared with zero using the J-test of over-identifying restrictions. The principal comparison BER made involved the cross correlation of productivity and hours worked at L leads and lags. When all the sample was used there was strong rejection if $L = 2$ (the p value of the test being 0.001). This outcome encouraged them to split the sample at 1969(1) and to perform validation of the model on two different samples. They then concluded that the model seemed satisfactory for the first period (p-value = 0.278) but not for the second period (p-value = 0.001). Because of this diversity of outcomes the discussion below concentrates upon the two sub-samples separately. We also avoid the emphasis upon the relation between productivity and hours that characterizes BER's evaluation work, as an important ingredient of the way in which their model manages to emulate the data is by making the assumption that the employment data is subject to errors of measurement. That modification seems to be very important to their success along the productivity hours correlation dimension, even though it is hard to think of it as part of a 'model'.

3.1. Sample period 1955(3) to 1969(4)

As reviewed in the preceding section any RBC model makes a number of predictions, either about the cointegrating vectors expected to hold between variables or the dynamic behaviour of the variables. Our strategy will be to determine if the predictions made by the RBC model are consistent with the data.

A first item to check is whether the assumption made pertaining to the evolution of the uncontrolled states is valid. BER's point estimates for the AR(1) coefficients of a_t and g_t are 0.87 and 0.94 respectively. Although these are different from unity the ADF tests recorded in Table 1 point to the fact that the hypothesis of the series being integrated is accepted fairly easily. Furthermore, the correlation between the residuals from the AR(1)'s fitted to g_t and a_t is only 0.12, which suggests the processes are uncorrelated, as specified in BER's model. Based on this outcome, and the evidence of integration for k_t and n_t in Table 1, it is anticipated that the state vector comprising k_t, n_t, g_t and a_t should have two cointegrating vectors, as there are two common trends driving the RBC model (see

Table 1. Tests for Integration in Data using ADF(4)

Variable	Without trend	With trend
k_t	–1.43	–1.80
n_t	–1.25	–2.97
a_t	–2.44	–2.42
g_t	–2.13	–2.12
c_t	–2.19	–2.18
y_t	–2.21	–2.31
i_t	–2.37	–2.88
crit. val.	–2.92	–3.50

the brief description of the main features of the model in the Appendix).

Using the parameter values provided by BER it is possible to compute $F = \begin{bmatrix} \gamma & \delta \\ 0 & I_p \end{bmatrix}$ in (2) and hence to derive the predicted cointegrating vectors among vectors among the four states viz. $[(I - \gamma) \quad -\delta]$. Logically, there are two distinct questions here. One is whether there are two cointegrating vectors or not. Using a VAR(4), Johansen's likelihood ratio test (LR) for the hypothesis of r cointegrating vectors easily indicates that there are two (the test of $r = 1$ versus $r = 2$ gives LR = 25.2 while $r = 2$ versus $r = 3$ has LR = 7.75, where the critical values corresponding to the 5% significance levels are 21.0 and 14.0 respectively). Exactly the same conclusion is reached with Johansen's trace test. Thus the number of cointegrating vectors agrees with the model prediction. A more demanding test is to assess whether the predicted numeric values (0.0435 –0.0295 –0.1434 0.0062) and (0.5627 1.0174 –1.5008 –0.1974) are compatible with the data. For this query, a likelihood ratio test of the restrictions gives a value of 45.53 which, when referred to a $\chi^2(4)$, soundly rejects the constraint. Consequently, a basic property of the model is rejected. Figure 1 plots $0.5627k_t + 1.0174n_t - 1.5008a_t - 0.1974g_t$, the projected second cointegrating error, and the lack of cointegration shows quite clearly (actually ADF tests applied to each cointegrating error separately shows that neither series is I(0)).

In addition to the state variables being I(1), Table 1 shows that three 'output' variables — consumption, output and investment — also possess this property. Therefore, RBC models conjecture that there are further cointegrating restrictions, now between the 'outputs' and the states — see (1). King *et al.* (1991) and Neusser

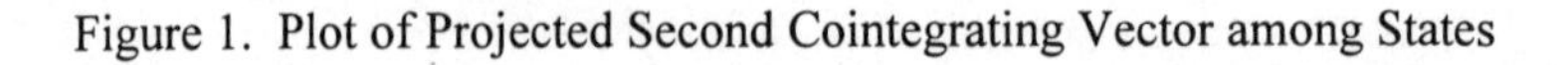

Figure 1. Plot of Projected Second Cointegrating Vector among States

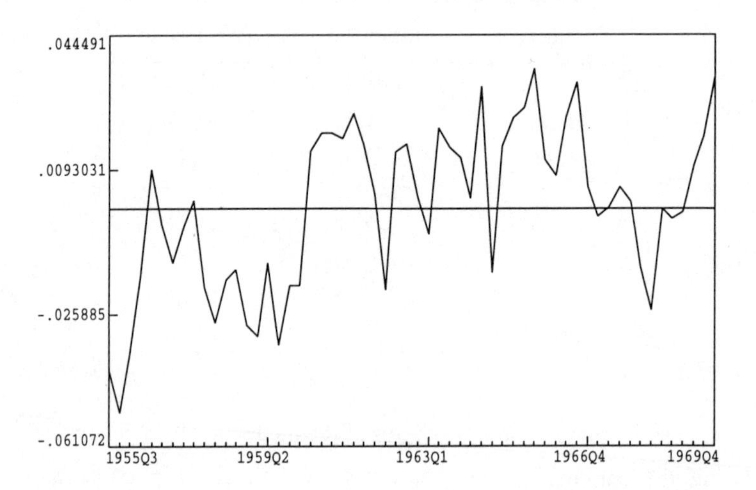

(1991) considered the long-run implications of neoclassical growth models for relations *between* the three 'output' variables above . In particular they argued that consumption and income and investment and income should be cointegrating pairs with cointegrating vectors (1 –1). In this model it is consumption, income and government expenditure which should be cointegrated, as well as investment and output. Johansen's tests indicate that the first of these relations is satisfied, but the likelihood and trace tests are in conflict over whether investment and output are cointegrated.[1] Moreover, for two reasons, tests of an RBC model performed in this way are rather weak. First, information is being discarded. The RBC model makes a direct prediction about the cointegrating relations between states and 'outputs' but only an indirect one about the connection between 'outputs'. Thus we might have $c_t - z_t'\alpha_1$ and $y_t - z_t'\alpha_1$ both being I(0), so that $c_t - y_t$ is I(0), but the α_1 may not coincide with that indicated by the RBC model. Second, many models have the property that $c_t - y_t$ and $i_t - y_t$ are cointegrated, e.g. the multiplier accelerator model can be designed to produce this effect by an appropriate choice of ECM format, and therefore cointegration between 'output' variables cannot be taken as validating the RBC viewpoint. In summary, what should be tested are the *direct* implications of the RBC model and not the *indirect* ones.

[1] Robert King has suggested that this failure may well be a consequence of the way in which data is constructed by BER.

Choosing BER's parameter values, the specific cointegrating relations from equation (1) are

$$c_t = 0.55k_t + 0.017n_t + 0.348a_t - 0.07g_t + \eta_{ct} \quad (15a)$$

$$i_t = -0.65k_t + 1.12n_t + 5.45a_t - 0.24g_t + \eta_{it} \quad (15b)$$

$$y_t = 0.13k_t + 0.31n_t + 1.64a_t + 0.07g_t + \eta_{yt}, \quad (15c)$$

and our objective is to test if the $\eta_{\cdot t}$ are I(1). This could be done in one of two ways.[2] A first possibility is to apply an ADF test to the errors from (15*a*–*c*); since no parameters are being estimated, many of the problems of using this test for cointegration are absent. An alternative is to use the fact that cointegration between variables means the existence of an ECM relationship — Engle and Granger (1985). Therefore, assuming (say) c_t and z_t are I(1) and cointegrated with vector $(1 \quad -\alpha)$, an ECM of the form

$$\Delta c_t = a\Delta z_t'\alpha_1 + b(c_{t-1} - z_{t-1}'\alpha_1) \quad (16)$$

would connect c_t and z_t. If c_t and z_t are not cointegrated, $b = 0$ making the *t*-ratio for $H_0: b = 0$ a suitable test of no cointegration. This test is proposed in Banerjee *et al.* (1986) and has been dubbed the 'ECM test' by Kremers *et al.* (1992). The latter have argued that it has much better power than the ADF test whenever the latter imposes an invalid common factor restriction. Unfortunately, the distribution of the ECM test varies between the Dickey-Fuller density and the standard normal as $\text{var}[(a-1)\Delta z_t]/\text{var}(\Delta z_t)$ tends from zero to infinity. Because Δz_t in our situation is a vector it is difficult to determine exactly what the critical values are. One plan of action would be to be conservative and to adopt the DF critical values. Note that there are no tests of (15*c*). The reason is that the unobserved variable a_t is effectively computed from data on y_t, k_t etc. by inverting (15*c*), and therefore η_{yt} is identically zero. Unless a separate estimate of a_t can be made it is therefore impossible to test this cointegrating restriction in an RBC model.[3]

The evidence from Table 2 is that the cointegrating restrictions are most likely invalid. The problematic outcome is for consumption. Referred to an N(0,1) random variable one would opt for cointegration, but this would not be true if the comparison was made with a 5% critical value from the DF density (–2.91).

[2] A third method would be to employ Johansen's test, but the fact that the states do not have the co-integrating relations implied by the RBC model makes it more convenient to perform 'single equation' tests.

[3] Provided a unit root is specified for the a_t process it would be possible to generate data on a_t using a random number generator and thereupon one could test (15*c*). Smith (1990) advocates this approach when there is a latent variable.

Table 2. Tests of Cointegrating Relations in (15a) and (15b)

Variable	ADF(4) with trend	ECM Test
c_t	–3.12	–2.69
i_t	–2.66	–2.26
crit. val.	–3.50	

Nevertheless, Table 2 does hint at specification difficulties with the BER model. To see why the restrictions are being rejected it is useful to fit relations such as (15*a–c*) using the data to give

$$c_t = 0.89k_t + 0.12n_t + 0.56a_t - 0.03g_t \tag{17a}$$

$$i_t = -0.18k_t + 0.69n_t + 3.61a_t - 0.31g_t. \tag{17b}$$

Comparing (15*a*) and (15*b*) with (17*a*) and (17*b*), it seems as if the weight given to k_t in the model is too low for both variables, whereas the influence of a_t is too low for consumption but far too high for investment. As the R^2 from the regressions in (17*a*) and (17*b*) are 0.94 and 0.93 respectively, provided the series are I(1) there is likely to be only small bias in the estimated cointegrating vectors.[4]

Although it seems unlikely, let us suppose that the cointegrating restrictions are satisfied. Then the third set of restrictions imposed by an RBC model are those relating to dynamics — equation (8). These involve testing if the coefficients of the cointegrating errors v_{t-1} are $A\alpha$ in the regression of Δy_t on v_{t-1}. A simple way to compute the statistic for such a test is to regress Δy_t on z_{t-1} and test if the coefficients are equal to Π (equation (6)). One has to be careful to refer the resulting test statistic to a $\chi^2(2)$ since the distribution of $\hat{\Pi}$ is singular owing to the cointegration, i.e. as v_{t-1} is a 2×1 vector, only two coefficients are really being tested. With Δy_t set to Δc_t and Δi_t the test statistics are 3.5 and 98.3 respectively, showing that, although the dynamics of consumption seem to be accounted for, the investment dynamics are missed badly (there is some serial correlation in the regression for Δi_t but there is only a minor change in the value of the test statistic when computed robustly). Unlike the situation for cointegration tests, it is also possible to check the output dynamics, and the test statistic there is 28.58, again showing some problems with the model. Equations (18*a–c*) list the

---

[4] A more serious problem is that the parameters being estimated may not be identified. If there are only two stochastic trends then it is impossible to estimate the four parameters here as the number of identified parameters can be no larger than the number of trends.

predicted dynamic relations along with the estimated relations (in brackets) for each of the series

$$\Delta c_t = \underset{(-0.100)}{-0.033}\, k_{t-1} \underset{(-0.038)}{-0.001}\, n_{t-1} \underset{(0.086)}{+0.059}\, a_{t-1} \underset{(-0.017)}{+0.005}\, g_{t-1} \qquad (18a)$$

$$\Delta i_t = \underset{(-0.597)}{-0.603}\, k_{t-1} \underset{(-0.24)}{-1.16}\, n_{t-1} \underset{(-0.120)}{+0.877}\, a_{t-1} \underset{(+0.03)}{+0.24}\, g_{t-1} \qquad (18b)$$

$$\Delta y_t = \underset{(-0.19)}{-0.18}\, k_{t-1} \underset{(-0.086)}{-0.309}\, n_{t-1} \underset{(-0.075)}{-0.266}\, a_{t-1} \underset{(-0.016)}{+0.055}\, g_{t-1}, \qquad (18c)$$

As revealed by $(18a-c)$, the major problem with the RBC model in its forecasts of dynamics is that it ascribes far too much weight to the productivity shock and lagged employment.

It is now appropriate to consider some objections that might be made to the above analysis. One of these is that the restrictions being tested are found by using the parameter values in BER and these are $\rho_a = 0.8691$ and $\rho_g = 0.938$ rather than the values of unity needed if we are to argue that the series are I(1). For this reason it is logically more correct to re-compute what the implied restrictions would be if unit roots are imposed upon the two forcing processes and to then test if the resulting restrictions are compatible with the data. This means that $(15a-c)$ become

$$c_t = 0.55k_t + 0.017n_t + 0.888a_t - 0.14g_t + \eta_{ct} \qquad (19a)$$

$$i_t = -0.65k_t + 1.12\,n_t + 2.31\,a_t + 0.13g_t + \eta_{it} \qquad (19b)$$

$$y_t = 0.13k_t + 0.31n_t + 1.11\,a_t + 0.13g_t + \eta_{yt}\,. \qquad (19c)$$

Doing so does not change any of the conclusions reached previously however. For example, the ADF tests for cointegration among the states now become –2.89 and –3.32 (with a $\chi^2(4) = 44.79$ when testing using Johansen's estimator), while ADF test values of –1.19 (c_t) and –2.64 (i_t) are found when directly testing the restrictions in (19a) and (19b). Tests of the dynamic restrictions yield $\chi^2(2)$ test statistics of 3.0 (c_t), 118.4 (i_t) and 35.2 (y_t)

Another objection to the analysis could be that the series are not integrated and that the power of the ADF test is low. There is some merit to this argument. If $\rho_a = 0.8691$, simulation of the ADF(4) test (with trend) for 58 observations shows that 55% of the time one gets an ADF test larger than –2.44 (the value of the ADF tests using the data on a_t). Hence one would falsely conclude that the series is I(1) 55% of the time. In the same vein, with $\rho_g = 0.93$, one would invalidly conclude there was a unit root 45% of the time (using the ADF value of –2.13 found from the data). Hence it may be more reasonable to conduct tests that assume the processes are I(0) rather than I(1). In this case we will test the restrictions from (9), i.e. that the coefficients of z_{t-1} are AF. Equations (20*a*–*c*) set out the theoretical coefficients for the 'reduced' VAR.

$$c_t = 0.518k_{t-1} + 0.016n_{t-1} + 0.407a_{t-1} - 0.069g_{t-1} \quad (20a)$$
$$i_t = -1.255k_{t-1} - 0.039n_{t-1} + 6.322a_{t-1} + 0.003g_{t-1} \quad (20b)$$
$$y_t = -0.045k_{t-1} - 0.001n_{t-1} + 1.907a_{t-1} + 0.128g_{t-1}\,. \quad (20c)$$

Corresponding empirical estimates (where we have added in missing terms from the VAR in (20*a*–*c*) if the *t* ratio was greater than 2) are

$$c_t = \underset{(2.500)}{0.403}\,k_{t-1} + \underset{(0.570)}{0.026}\,n_{t-1} + \underset{(3.490)}{0.411}\,a_{t-1} \underset{(-1.800)}{-0.033}\,g_{t-1} + \underset{(2.6)}{0.433}\,c_{t-1} \quad (21a)$$

$$i_t = \underset{(-2.870)}{-0.673}\,k_{t-1} + \underset{(0.260)}{0.047}\,n_{t-1} + \underset{(2.200)}{1.401}\,a_{t-1} \underset{(-1.320)}{-0.106}\,g_{t-1} + \underset{(3.57)}{0.579}\,i_{t-1} \quad (21b)$$

$$y_t = \underset{(-0.430)}{-0.057}\,k_{t-1} + \underset{(2.670)}{0.222}\,n_{t-1} + \underset{(11.11)}{1.566}\,a_{t-1} + \underset{(1.570)}{0.056}\,g_{t-1} \quad (21c)$$

The results in (21*a*–*c*) constitute a strong rejection of the restrictions implied by the RBC model. Testing that the parameters in (21*a*–*c*) equal those in (20*a*–*c*) gives χ^2 statistics of $\chi^2(5) = 104.1(c_t)$, $\chi^2(5) = 625.0(i_t)$, and $\chi^2(4) = 28.59(y_t)$. A comparison of the two sets of equations shows there are some variables missing from the former, c_{t-1} in the c_t equation, and i_{t-1} in the i_t — and that the model accords productivity too great an influence in determining investment and output. Others have remarked upon such a 'missing variable' feature, specifically for consumption (Chow, 1992*b*), but a casual comparison of the equations emphasises that there are many factors responsible for the failure of the model to explain output and investment variations.

The outcomes observed above bring to the fore a question raised in the introduction; is the rejection being caused by the model or by the parameter values being supplied to it? That is, does there exist an RBC model of this form that would be compatible with the data but which had a different set of parameter values? It might be argued that the essence of the model is the type of functional forms fed in and not the values of the parameters chosen to calibrate it. Earlier we remarked why we feel that this view should be rejected, but it is worth exploring what would happen if we adopted it. One can say immediately that the non-zero coefficients seen for c_{t-1} and i_{t-1} in (21*a*) and (21*b*) cannot be matched by calibration changes, as the model design automatically assigns a zero coefficient to these variables. Only re-specification of the RBC model would change this fact. Some of the other parameters in (21*a–c*) can be modified by changing the calibration settings. By studying the sensitivity of (21*a–c*) to variations in the parameters of BER's model, it was found that we could improve the approximation by increasing ρ_a and reducing α. However, it was necessary to make ρ_a almost unity if the weight on the productivity variable was to be reduced to the required magnitude. This would mean that we are dealing with processes that are very close to being integrated and so it would be appropriate to test the cointegrating restrictions. As mentioned earlier however, these are rejected when we impose I(1) behaviour upon the forcing variables. Hence, it does not seem as if the essentials of the economy are captured by the BER formulation. It is worth emphasising here that the rejections of the RBC model using the techniques above are far stronger than those encountered by BER, where what evidence there was against their model in this period was very mild. This fact emphasises that different types of information are being gathered by the different methods of evaluation.

3.2 Sample period 1970(1) to 1984(1)

In the second period there is evidence that the evolutionary pattern for the variables identified in the first period has changed. Looking first at the forcing processes, there is some doubt that they are now I(1). The ADF tests (with trend) are –3.40 and –2.85 for g_t and a_t respectively, while the ADF (without trend) for a_t of –2.86 is very close to the 5% critical value of –2.92. Examination of the estimates of the autoregressive parameters upon which the ADF test is based reveal them to be 0.47 (g_t) and 0.76 (a_t), below the values of 0.87 and 0.81 found in the first period. It seems very likely therefore that the processes are I(0); certainly one would only be comfortable with a single common trend, due to a_t, as the autoregressive parameter for g_t is far too low. Turning to the other series, here the evidence of I(1) behaviour is stronger, but even then the autoregressive parameter is (at best) just above 0.8.

What is to be done about these features? One possibility is to proceed with the tests outlined in the previous sub-section, maintaining that there is a single common trend. When this is done one encounters rejections of all the cointegrating restrictions. In the interests of economising on space, and recognizing the doubt raised over the integration properties of the data, our preference has been to only report results derived under the assumption that the series are all I(0). This means that we perform and report the tests of the dynamic restrictions appearing in (9).

Equations (22*a–c*) provide the estimated equations, along with the predicted values of the coefficients in brackets (variables not entering the model VAR have been deleted if their t ratio is less than 2, while estimated intercept terms have also been suppressed).

$$\underset{}{c_t} = \underset{(-0.535)}{-0.112}\, k_{t-1} \underset{(0.014)}{-0.071}\, n_{t-1} \underset{(0.396)}{-0.109}\, a_{t-1} \underset{(-0.021)}{-0.001}\, g_{t-1} + \underset{(0)}{1.153}\, c_{t-1} \tag{22a}$$

$$i_t = \underset{(-1.246)}{-1.371}\, k_{t-1} \underset{(-0.034)}{-0.537}\, n_{t-1} \underset{(6.363)}{-1.340}\, a_{t-1} \underset{(-0.294)}{-0.342}\, g_{t-1} + \underset{(0)}{1.970}\, c_{t-1} + \underset{(0)}{1.033}\, i_{t-1} \tag{22b}$$

$$y_t = \underset{(-0.034)}{-0.685}\, k_{t-1} \underset{(-0.001)}{-0.153}\, n_{t-1} \underset{(1.913)}{-0.113}\, a_{t-1} \underset{(0.027)}{-0.151}\, g_{t-1} + \underset{(0)}{1.053}\, c_{t-1} + \underset{(0)}{0.297}\, i_{t-1} \tag{22c}$$

The task is to determine whether the predicted and estimated parameters are significantly different from each other, and the resulting test statistics are $\chi^2(5) = 225.2$, $\chi^2(6) = 638.6$ and $\chi^2(6) = 91.7$ for c_t, i_t and y_t respectively. If only the coefficients of k_{t-1}, n_{t-1}, a_{t-1} and g_{t-1} are tested for having their predicted values, the corresponding $\chi^2(4)$ statistics would be 110.3, 151.1 and 37.18. As before, this constitutes a very strong rejection of the model, although an important difference from the previous period is that the prediction of zero coefficients for c_{t-1} and i_{t-1} in the equations is now wildly at variance with the data, indicating that the dynamic structure of the model seems to have undergone some major shifts in the period. Looking at the estimates in (22*a*–*c*), the most striking feature is the fact that the technology shock a_t is estimated to have a *negative* impact on all variables in this period, which is in sharp contrast to the positive effect predicted by the model.

4. COMPARING MODELS

As mentioned in the introduction it is perhaps more reasonable to evaluate a model by its performance relative to others than to impose an absolute standard. For this reason it was decided to effect a comparison of the RBC model with a stylized version of the type of macro model that was popular in the 1960s. This generally featured a consumption relation dynamically connecting consumption and output, as well as an accelerator mechanism for investment. Although money featured in such models as well, here it is excluded in the interest of retaining comparability with the RBC model; the idea being to work with the same variables as BER did, but to provide a 'demand' rather than 'supply' side account of developments in the US economy.

Most of the models of the 1960s worked with levels of the variables and we therefore chose to do the same thing here. To make comparisons with the RBC model, the predictions of the latter had to be converted from deviations around steady state values back to levels. Levels of variables are distinguished by capital

Table 3. Estimates for Multiplier-Accelerator Model

Parameter	First Period	Second Period
α_1	0.695	0.848
β_1	0.135	0.128
β_2	–0.040	–0.106
c_1	352.3	362.7
ϕ_1	2.32	1.279
α_2	0.873	0.942
γ_1	0.267	0.430
γ_2	0.180	0.065
c_2	18.65	25.86
ϕ_2	0.523	–0.214
δ	0.020	0.022
ρ_g	0.934	0.560
c_3	62.32	583.1
ϕ_3	0.613	–0.469

letters. The multiplier-accelerator model (MPA) that was fitted is given in the equations below. No experimentation with lag lengths etc. was undertaken; the idea was just to take a simple model and to see how well it performs on the same data set. Some of the regressors in the equations were insignificant but were nevertheless retained.

$$C_t = \alpha_1 C_{t-1} + \beta_1 Y_t + \beta_2 Y_{t-1} + c_1 + \phi_1 t \tag{23a}$$

$$NI_t = \alpha_2 NI_{t-1} + \gamma_1 \Delta Y_t + \gamma_2 \Delta Y_{t-1} + c_2 + \phi_2 t \tag{23b}$$

$$NI_t = I_t - \delta K_{t-1} \tag{23c}$$

$$K_t = (1-\delta)K_{t-1} + I_t \tag{23d}$$

$$G_t = \rho_g G_{t-1} + c_3 + \phi_3 t \tag{23e}$$

$$Y_t = C_t + I_t + G_t + D_t \tag{23f}$$

The variable D_t is needed to make the series on output satisfy the national income identity. It is always a small fraction of output Y_t and rarely reaches 1% of that variable, so that its introduction would not seem to produce any distorting factors. Table 3 gives the parameter estimates of the unknown parameters of the multiplier-accelerator model for each of the two periods. Estimation was done by OLS, as that was also the most common way of doing 'calibration' at that time.

It is interesting to first ask whether the MPA model makes correct predictions of the VAR coefficients. As can be seen, the MPA model implies that the data should be a VAR(2), and the coefficients of each lag can be worked out by solving (23*a*–*f*). The $\chi^2(10)$ statistics testing the adequacy of the model during the first period were 21.4, 23.4 and 42.3, for C_t, I_t and Y_t respectively. The corresponding test statistics in the second period were 73.0, 112.0 and 162.7. Although the fact that we are working with levels, and hence potentially integrated data, makes the actual distribution of these 'χ^2' statistics unlikely to be exactly that, their magnitude has to make one seriously question the MPA model as a good representation of the data. This conclusion is especially true of the second period, a feature that is consistent with the notion that 'Keynesian' models broke down in the 1970s. If one takes the size of the χ^2 statistics as an index of how good the model is, then both the RBC and MPA models have noticeably worse performance in the second period.

Figures 2(a–c) and 3(a–c) provide plots of the one step predictions of C_t, I_t and Y_t, from both models for each of the time periods. The overall impression is that the MPA model is more successful than the RBC model in tracking all series. Because the models are non-nested, imposing different restrictions upon the same VAR, one way to check the above impression is to enquire into whether the explanation of variables of interest given by the MPA model can be improved upon by using information from the RBC model. To this end we regress data on the variable being studied against the predictions of it made by both the MPA and RBC models; if the RBC model is correct then the coefficient on the predictions from the MPA model should be zero, and conversely. This test is in the spirit of Davidson and MacKinnon's (1981) J-test for non-nested models. Selecting C_t, I_t and Y_t as the variables of interest, the results are given below in (24*a*–*c*) for the first period and in (25*a*–*c*) for the second period, with *t* ratios in brackets.

First Period

$$C_t = \underset{(2.15)}{0.21}\ \hat{C}_{RBC,t} + \underset{(8.08)}{0.79}\ \hat{C}_{MPA,t} \tag{24a}$$

$$I_t = \underset{(1.74)}{0.09}\ \hat{I}_{RBC,t} + \underset{(15.94)}{0.90}\ \hat{I}_{MPA,t} \tag{24b}$$

$$Y_t = \underset{(2.41)}{0.18}\ \hat{Y}_{RBC,t} + \underset{(11.94)}{0.83}\ \hat{Y}_{MPA,t} \tag{24c}$$

Second Period

$$C_t = \underset{(1.63)}{0.14}\ \hat{C}_{RBC,t} + \underset{(11.57)}{0.88}\ \hat{C}_{MPA,t} \tag{25a}$$

$$I_t = \underset{(3.45)}{0.15}\ \hat{I}_{RBC,t} + \underset{(14.47)}{0.82}\ \hat{I}_{MPA,t} \tag{25b}$$

$$Y_t = \underset{(7.57)}{0.59}\ \hat{Y}_{RBC,t} + \underset{(8.34)}{0.52}\ \hat{Y}_{MPA,t} \tag{25c}$$

The evidence in the above equations is that the RBC model rarely adds a great deal to the explanatory power of the MPA model. Perhaps the most striking exception to this statement is in (25*c*); it would seem that output in the second period cannot be satisfactorily explained by a pure demand side model like MPA. Given the oil-price shocks of the 1970s, part of which would be reflected in the *ex-post* measurements of productivity, such a conclusion may not be too surprising.

One way to understand the difference in the two models is to ask what the RBC model would look like if turned into an MPA type model. To do this we use the ideas in Section 2 for reducing the VAR implied by the RBC model into bivariate VAR's between the pairs (consumption, output) and (investment, output). Equations (26*a*–*b*) and (27*a*–*b*) give the implied bivariate VARs for the two periods, with the estimated parameters in brackets.

First Period

$$c_t = \underset{(0.79)}{1.08}\ c_{t-1} - \underset{(0.13)}{0.09}\ c_{t-2} + \underset{(0.24)}{0.18}\ y_t - \underset{(-0.12)}{0.19}\ y_{t-1} + \underset{(-0.08)}{0.07}\ y_{t-2} \tag{26a}$$

$$i_t = \underset{(0.94)}{0.87}\ i_{t-1} + \underset{(-0.10)}{0.04}\ i_{t-2} + \underset{(0.98)}{3.17}\ y_t - \underset{(-0.20)}{2.73}\ y_{t-1} - \underset{(-0.63)}{0.317}\ y_{t-2} \tag{26b}$$

Second Period

$$c_t = \underset{(1.12)}{0.99}\ c_{t-1} - \underset{(-0.16)}{0.05}\ c_{t-2} + \underset{(0.20)}{0.21}\ y_t - \underset{(-0.25)}{0.21}\ y_{t-1} + \underset{(0.05)}{0.07}\ y_{t-2} \tag{27a}$$

$$i_t = \underset{(0.41)}{0.73}\ i_{t-1} + \underset{(0.12)}{0.03}\ i_{t-2} + \underset{(1.81)}{3.31}\ y_t - \underset{(-0.55)}{2.39}\ y_{t-1} - \underset{(-0.23)}{0.24}\ y_{t-2} \tag{27b}$$

These equations encapsulate most of the information in Figures 2 and 3 regarding the behaviour of consumption and investment. A succinct summary of the latter is that investment is much too volatile whilst consumption is too smooth. In terms of (26*a*–*b*) and (27*a*–*b*) the impact of output upon investment is seen to be too large, while the lag distribution of consumption response to income changes is longer for the RBC model. Another interesting feature of (26*a*–*b*) and (27*a*–*b*) is that the accelerator mechanism is very clear in the data of the first period but is not in evidence in the second (in the sense that the coefficients of y_t, y_{t-1} and y_{t-2} do not sum to zero). This provides one explanation for the MPA model's deterioration in performance during the second period.

5. CONCLUSION

This paper has set out a strategy for evaluating small linear models via the restrictions they impose upon the VAR in the variables they are meant to explain. Three types of restrictions were elicited. First, there are cointegrating restrictions implied among the state variables. Second, there are the cointegrating constraints existing between the state and 'output' variables. Finally, there are restrictions upon the dynamics of the model when all variables are transformed to be I(0). It was recommended that evaluation should proceed by examining the constraints sequentially. The technology was then applied to an RBC model that had performed reasonably well when assessed relative to a set of 'stylised facts'; failure on all three counts was evident, pointing to the need for some re-specification of the model.

A failure of the cointegrating restrictions is generally the hardest feature to rectify as some variables need to be added to the system. Candidates could be the effects of taxation upon capital accumulation, the impact of the external sector via terms of trade movements, or monetary factors. Although a complete study of this phenomenon is beyond the paper, understanding the source of the cointegration failures seems critical to determining what course of action should be followed. However, even if the cointegrating restrictions were made acceptable there also appears to be some difficulties with the 'short run' responses within the model. Results presented in Sections 3 and 4 make a strong case for introducing adjustment costs into investment in order to reduce the magnitude of its short run reponse to fluctuations in output. The opposite is true of consumption, where the impact of current income needs to be strengthened.

Figure 2. First Period Values (—) and Predictions of them from the MPA (- - -) and RBC (·····) Models

(a) Consumption (C)

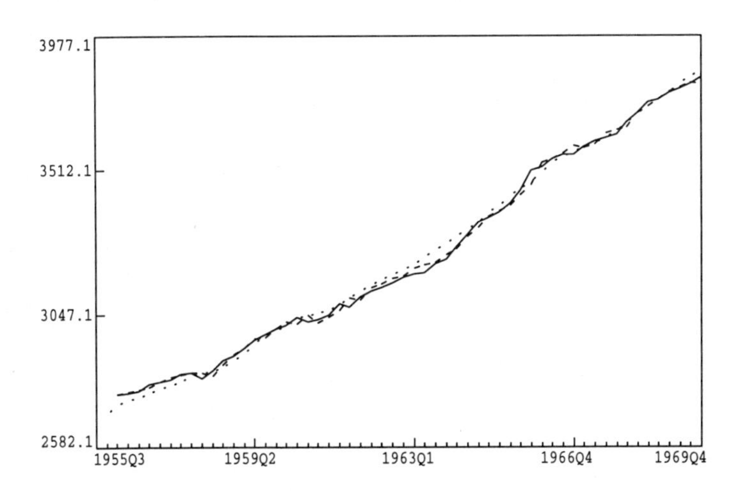

(b) Investment (I)

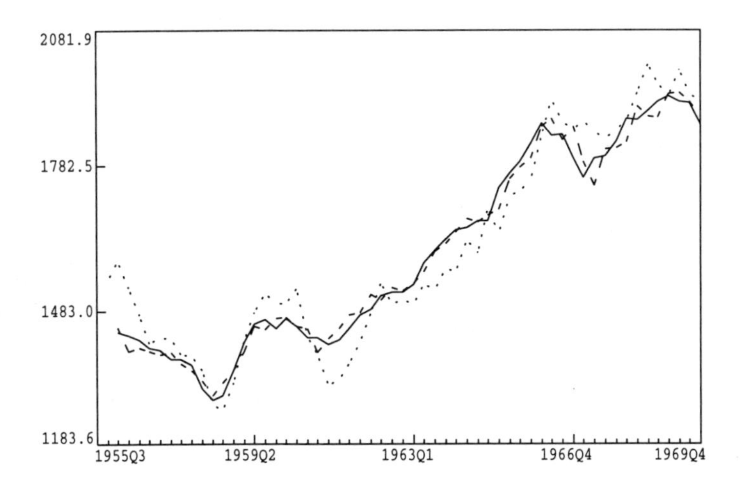

Figure 2. (continued)

(c) Income (Y)

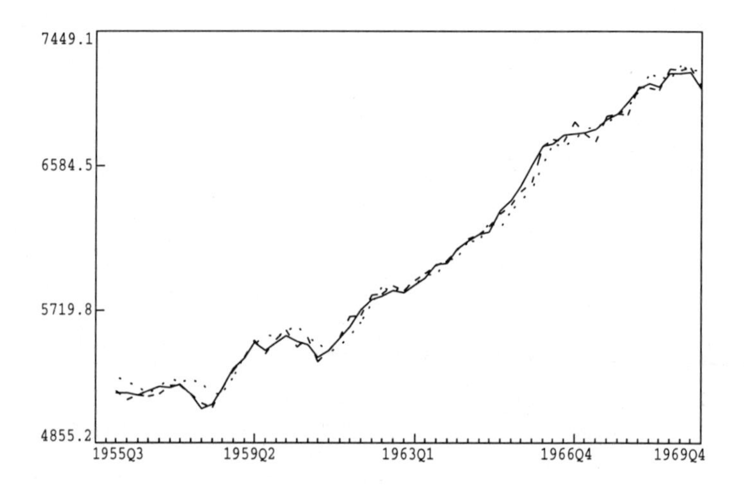

Figure 3. Second Period Values (—) and Predictions of them from the MPA (- - -) and RBC (·····) Models

(a) Consumption (C)

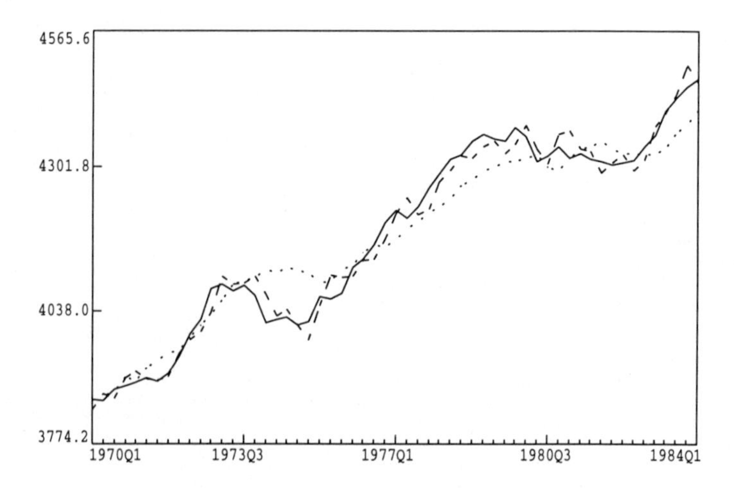

Figure 3. (continued)

(b) Investment (I)

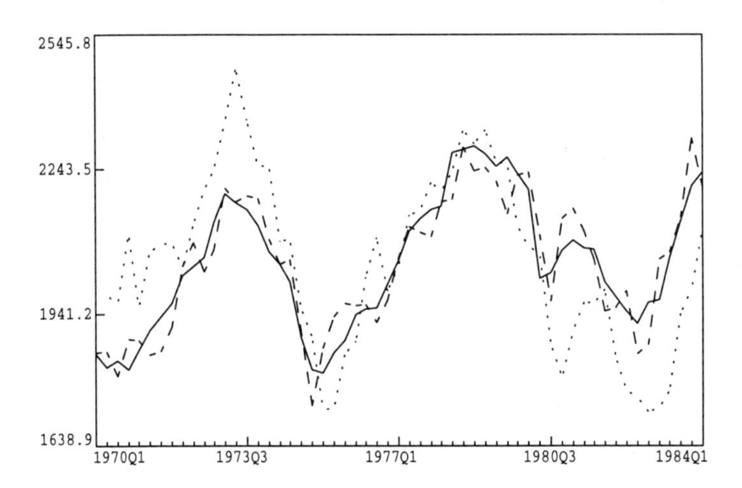

(c) Income (Y)

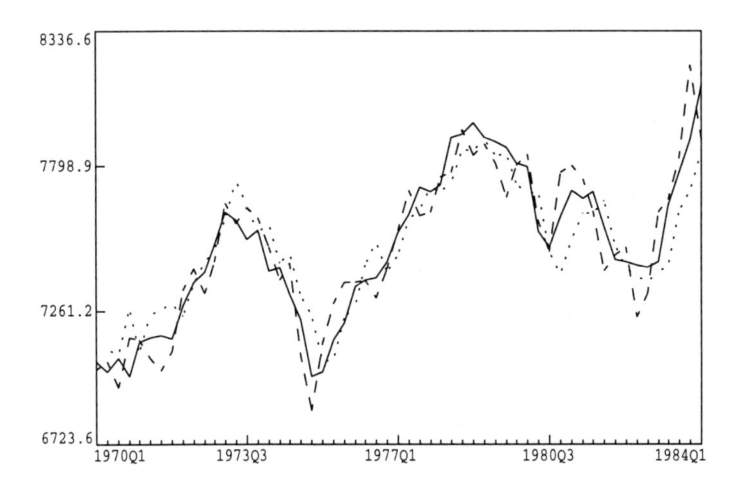

APPENDIX

The Burnside/Eichenbaum/Rebelo Model

Worker Utility

$$\ln(C_t^p) + \theta \ln(T - \xi - W_t f)$$

T = time endowment, C_t^p = private consumption, W_t = effort, f = hours worked per shift, ξ = fixed cost of work (in terms of hours of foregone leisure).

Non-Worker Utility

$$\ln(C_t^p) + \theta \ln(T)$$

Cobb-Douglas Production Function

$$Y_t = Z_t K_t^{1-\alpha}(N_t W_t f)^{\alpha}$$

Z_t = Technology, N_t = fraction of agents who are workers (the number of agents is normalized to unity), K_t = beginning of period capital stock.

Technology Change

$$Z_t = \gamma^{\alpha t} A_t$$

Productivity Shock

$$\ln(A_t) = (1-\rho_a)\ln(A) + \rho_a \ln(A_{t-1}) + \varepsilon_t$$

Aggregate Resource Constraint

$$C_t^p + K_{t+1} - (1-\delta)K_t + X_t \leq Y_t$$

X_t = government consumption

Fiscal Rule

$$X_t = \gamma_g^t G_t$$

$$\ln(G_t) = (1-\rho_g)\ln(G) + \rho_g \ln(G_{t-1}) + \mu_t$$

Table A. Parameter Values for the BER Model

Parameter	Period 1	Period 2
δ	0.0196	0.0221
θ	0.6593	0.6504
ρ_a	0.8691	0.8815
σ_ε	0.0042	0.0067
$\ln(A)$	8.4914	8.8733
$\ln(\gamma_y)$	0.0069	0.0015
$\ln(G)$	6.8090	7.1618
$\ln(\gamma_g)$	0.0073	–0.0013
ρ_g	0.938	0.6618
σ_μ	0.0143	0.0115

It is assumed that a social planner maximises

$$\mathrm{E}_0 \sum_{t=0}^{\infty} \beta^t \left\{ \ln\left(C_t^p\right) + \theta N_t \ln\left(T - \xi - W_t f\right) + \theta(1 - N_t)\ln(T) \right\}$$

subject to the constraints above and K_0 by choice of contingency plans for $\{C_t^p, K_{t+1}, N_t, W_t : t \geq 0\}$. $\mathrm{E}_0$is the time 0 conditional expectations operator, β is the subjective discount rate, $0 < \beta < 1$.

Certain transformations are made to the problem before it is solved. These are to express the variables as deviations from deterministic steady state growth paths. Thus $\overline{C}_t^p = C_t^p / \gamma^t$, $\overline{Y}_t = Y_t / \gamma^t$, $\overline{K}_t = K_t / \gamma^t$, $\overline{X}_t = X_t / \gamma^t$ which means that the constraints on the optimisation can be reduced to

$$\gamma \overline{K}_{t+1} = A \overline{K}_t^{1-\alpha} (N_t W_t f)^{\alpha} - \overline{C}_t^p + (1 - \delta)\overline{K}_t - \overline{X}_t$$

while the optimand becomes

$$\sum_{t=0}^{\infty} \beta^t \log(\gamma^t) + \mathrm{E}_0 \sum_{t=0}^{\infty} \beta^t \left\{ \ln(\overline{C}_t^p) + N_t \theta \ln(T - \xi - W_t f) + \theta(1 - N_t)\ln(T) \right\}$$

Finally small letters indicate deviations of variables from steady states. Thus $a_t = \log(A_t / A)$.

The solutions to this problem after linearization of the Euler equations are laws of motion for the state variables k_t, n_t, a_t and g_t as well as linear relations connecting other variables such as y_t, c_t^p to these states. With the parameter values in Table A it is possible to compute numerical values for these relations and they are presented in the text. The parameter σ_υ in the table arises from the assumption that there are measurement errors in hours worked.

REFERENCES

BANERJEE, A., J. J. DOLADO, D. F. HENDRY and G. W. SMITH (1986), 'Exploring Equilibrium Relationships Through Static Models: Some Monte Carlo Evidence', *Oxford Bulletin of Economics and Statistics*, 48, 3, pp. 253–77.

BURNSIDE, C., M. EICHENBAUM and S. REBELO (1993), 'Labor Hoarding and the Business Cycle', *Journal of Political Economy*, 101, pp. 245–73.

BYRON, R. P. (1974), 'Testing Structural Specification Using the Unrestricted Reduced Form', *Econometrica*, 42, pp. 869–83.

CANOVA, F. (1990), 'Simulating General Equilibrium Dynamic Models Using Bayesian Techniques', Mimeo, University of Rochester.

CHOW, G. C. (1992*a*), 'Dynamic Optimization without Dynamic Programming', *Economic Modelling*, 9, pp. 3–9.

—— (1992*b*), 'Statistical Estimation and Testing of a Real Business Cycle Model', Econometric Research Program Research Memorandum No. 365, Princeton University.

CHRISTIANO, L. J. and M. EICHENBAUM (1992), 'Current Real-Business-Cycle Theories and Aggregate Labor Market Fluctuations', *American Economic Review* , 82, pp. 430–50.

COGLEY, T. and J. M. NASON (1992), 'Do Real Business Cycle Models Pass the Nelson-Plosser Test?', Mimeo, University of British Columbia.

DAVIDSON, R. and J. G. MACKINNON (1981), 'Several Tests for Model Specification in the Presence of Alternative Hypotheses', *Econometrica*, 49, pp. 781–93.

EVANS, G. W. (1989), 'Output and Unemployment Dynamics in the United States: 1950–1985', *Journal of Applied Econometrics*, 4, pp. 213–37.

GREGORY, A. W. and G. W. SMITH (1993), 'Statistical Aspects of Calibration in Macroeconomics', in G. S. Maddala, C. R. Rao and H. D. Vinod (eds.), *Handbook of Statistics*, 11, pp. 703–19.

HENDRY, D. F. and G. M. MIZON (1993), 'Evaluating Dynamic Models by Encompassing the VAR', in P. C. B. Phillips (ed.), *Models, Methods and Applications of Econometrics*, Oxford: Basil Blackwell, pp. 272–300.

HOOVER, K. D. (1991*a*), 'Six Queries About Idealization in an Empirical Context', *Ponzan Studies in the Philosophy of Science and the Humanities* (forthcoming).

—— (1991*b*), 'Calibration and the Econometrics of the Macroeconomy', Mimeo, University of California at Davis.

KING, R. G., C. PLOSSER and S. REBELO (1988), 'Production, Growth and Business Cycles I: The Basic Neoclassical Model', *Journal of Monetary Economics*, 21, pp. 195–232.

KING, R. G., C. PLOSSER, J. STOCK and M. WATSON (1991), 'Stochastic Trends and Economic Fluctuations', *American Economic Review*, 81, pp. 819–46.

KREMERS, J. J. M., N. R. ERICSSON and J. J. DOLADO (1992), 'The Power of Cointegration Tests', International Finance Discussion Paper No. 431, Board of Governors of the Federal Reserve System.

KYDLAND, F. E. and E. C. PRESCOTT (1982), 'Time to Build and Aggregate Fluctuations', *Econometrica*, 50, pp. 1345–70.

—— (1991), 'The Econometrics of the General Equilibrium Approach to Business Cycles', *Scandinavian Journal of Economics*, 93, pp. 161–78.

LAIDLER, D. and B. BENTLEY (1983), 'A Small Macro-model of the Post-War United States', *Manchester School*, 51, pp. 317–40.

LONG, J. B. and C. I. PLOSSER (1983), 'Real Business Cycles', *Journal of Political Economy*, 91, pp. 39–69.

MARCET, A. (1989), 'Solving Non-linear Stochastic Models by Parameterizing Expectations', Mimeo, Carnegie-Mellon University.

MCGRATTAN, E. B. (1991), 'The Macroeconomic Effects of Distortionary Taxation', Discussion Paper No. 37, Institute for Empirical Macroeconomics.

MONFORT, A. and R. RABEMANANJARA (1990), 'From a VAR to a Structural Model, with an Application to the Wage Price Spiral', *Journal of Applied Econometrics*, 5, pp. 203–27.

NEUSSER, K. (1991), 'Testing the Long-Run Implications of the Neo-classical Growth Model', *Journal of Monetary Economics*, 27, pp. 3–37.

PRESCOTT, E. C. (1991), 'Real Business Cycle Theory: What Have We Learned?', Unpublished lecture, Latin American Meeting of the Econometric Society, Punta del Este, Uruguay.

SARGENT, T. J. (1978), 'Estimation of Dynamic Labour Demand Schedules Under Rational Expectations', *Journal of Applied Econometrics*, 4, pp. 213–37.

SPANOS, A. (1986), *Statistical Foundations of Econometric Modelling*, Cambridge University Press, Cambridge.

TINBERGEN, J. (1939), *Statistical Testing of Business Cycle Theories, Vol I: A Method and Its Application to Investment Activity*, League of Nations, Geneva.

WALLIS, K. F. (1977), 'Multiple Time Series Analysis and the Final Form of Econometric Models', *Econometrica*, 45, pp. 1481–97.

WATSON, M. W. (1993), 'Measures of Fit for Calibrated Models', *Journal of Political Economy*, 101, pp. 1011–41.

ZELLNER, A. and F. PALM (1974), 'Time Series Analysis and Simultaneous Equation Econometric Models', *Journal of Econometrics*, 2, pp. 17–54.

9

Misspecification versus bubbles in the Cagan hyperinflation model

Steven N. Durlauf and Mark A. Hooker[*]

Most test for rational bubbles using Cagan's hyperinflation model under rational expectations assume that the model is correctly specified. Rejections of the model are consequently treated as evidence of bubbles. This paper shows that it is possible to test model specification separately from the presence of bubbles. We provide a general framework for identifying all testable implications of the null hypothesis that the model is correctly specified as well as all implications of the null hypothesis that a given price sequence is a bubble-free realisation of the model. We find that in the German hyperinflation, empirical deviations from the fundamental price solution are primarily due to misspecification rather than bubbles, conditional on the same identification restrictions used in the parametric bubble-testing literature. Our analysis potentially reconciles some contradictory results in the literature.

1. INTRODUCTION

The Cagan hyperinflation model has provided a rich ground for the analysis of expectations-based models in macroeconomics. The reasons for interest in hyperinflations in general and the Cagan model in particular are twofold. First, hyperinflations represent periods where the role of expectations seems particularly crucial to the formation of endogenous variables. The explosive behaviour of the money supply in these episodes makes it almost certain that other observable

[*] Financial support from the U.K. Economic and Social Research Council under grant R000233447 is gratefully acknowledged by both authors. We are indebted to Neil Ericsson for helpful comments.

variables which usually appear in money demand functions, like income and interest rates, are of secondary importance. Second, the specific structure of the Cagan model admits a host of bubble equilibria in addition to the fundamental solution generated by expectations of future money supply growth. As a result, the model offers a clear opportunity to understand the dimensionality of restrictions which rational expectations places upon the sample paths of endogenous variables. Recent years have seen considerable interest in the use of the Cagan model to examine the empirical relevance of rational bubbles.

Perhaps the most salient feature of the empirical literature on the Cagan model is the contradictory nature of the conclusions, in both its major applications. In the first application, estimating the elasticity of money demand with respect to inflation, the parameter value obtained and the precision of the estimates have varied widely even when using the same basic specification. For instance, Salemi and Sargent (1979) obtained unstable parameter estimates with large standard errors while Casella's (1989) estimates are fairly precise and inconsistent with many of the Salemi-Sargent results using conventional confidence intervals. In the second main application, testing for rational bubbles, Burmeisterand Wall (1982, 1987) find strong evidence of bubbles in both the money exogenous case, where there is no feedback of prices into money, as well as the money endogenous case, where there is feedback, while Casella (1989) finds evidence of bubbles only in the money exogenous case. These disparate results were all obtained using essentially the same data set, that for Germany in the early 1920s. It appears that either the opportunities presented by hyperinflations for powerful inferences concerning monetary phenomena are illusory or they have not been fully exploited.

One difficulty with the extant literature is that tests of model specification are generally confounded with tests for the presence of bubbles. The general Cagan model under rational expectations requires additional identifying assumptions in order to estimate the elasticity of money demand and to construct bubble tests. These assumptions typically restrict the behaviour of the portfolio balance disturbance term. The identifying assumptions which have been most often used in the literature imply testable restrictions on comovements of money growth and inflation which we shall show are stronger than the model will bear empirically. A potential explanation for the disparate results mentioned above lies in the fact that this misspecification will in general invalidate conventional regression diagnostics (e.g. high R^2, small standard errors on coefficients) as measures of model fit and may combine with particular auxiliary equations used, such as those for the money supply, to produce idiosyncratic results. Moreover, many tests for bubbles take correct specification as given and interpret deviations from the fundamental solution as evidence of bubbles — an inference which is typically unjustified, since only the joint hypothesis of no bubbles and correct specification has been rejected.

The purpose of this paper is to design and implement a testing methodology which separates the validity of the specification of the Cagan model from the implications of a bubble component to prices, and which does so without needing to specify the rest of the model (such as the data generating process for the money supply). This is accomplished using sequential sets of tests. First, we perform tests to see whether price data conform to the fundamental solution to the Cagan model under rational expectations. Second, we conduct further tests to see whether the failure of price data to obey the fundamental solution is attributable to the presence of a rational bubble. We present evidence that the time series for prices in the German hyperinflation contains a misspecification term which does not have the time series properties of a rational bubble.

This approach has a second advantage over the various parametric bubbles tests. Those papers all employ and condition on a particular money supply specification in addition to the money demand (Cagan) equation. Relatively little is known about money supply processes in hyperinflation — often the specification is based on atheoretical Box-Jenkins methods — yet the estimates and tests employ cross-equation restrictions which make the money supply specification part of the joint null hypothesis. Furthermore, many of these tests assume an exogenous money supply, despite econometric and anecdotal evidence that money was actually endogenous. Both Sargent and Wallace (1973) and İmrohoroğlu (1989) provide strong statistical evidence of money supply endogeneity, and Webb (1985) confirms this conclusion through a study of the behaviour of the Reichsbank during the hyperinflation.

A nonparametric approach that is not susceptible to some of our criticisms is contained in Diba and Grossman (1988) and Hamilton and Whiteman (1985). These authors observe that when prices fulfil the fundamental solution of the Cagan model, then prices and money will possess the same order of integration. In order to identify the presence of bubbles, they calculate the order of differencing necessary to obtain a stationary money series (the observable fundamental in the model), and apply it to prices as well. Since a bubble grows faster than an integrated series, differenced prices will still be nonstationary if a bubble exists. They then judge whether both series appear stationary.

One virtue of the nonparametric approach, which our methods do not possess, lies in the mild restrictions they impose on the properties of the unobservable portfolio balance disturbance term. For such procedures, that disturbance must be stationary under a weakly lower order of differencing than the money series whereas for our tests we adopt the relatively restrictive identifying assumptions of the parametric bubbles testing literature.

Our approach, however, has two distinct advantages over the nonparametric procedure. First, the nonparametric procedure requires a nonexplosive money supply. However, if a bubble exists and the money supply is endogenous, i.e. inflation feeds into money growth, then the money supply may be explosive, which means that the procedure cannot be implemented.[1] In contrast, our tests remain consistent for any specification of the money supply under which the fundamental price solution to the Cagan model is well defined. Given evidence developed by Casella (1989) and others that the money supply process during the German hyperinflation was endogenous, any inferences which require an exogenous money supply process are clearly suspect. A second advantage of our approach is that it yields additional information about the model's specification. This may be useful for interpreting the equation as a money demand function and for interpreting bubble results. On the latter, since a bubble is a self-fulfilling nonfundamental equilibrium by definition, explosive noise relative to a misspecified equation is not a rational bubble, although the nonparametric approach will interpret it as such. Consequently, we believe that our tests provide a useful complement to previous work in this area.

The paper is organised as follows. Section 2 develops the unobserved components decomposition framework for the rational expectations Cagan model. In Sections 3 and 4, we use identifying restrictions to derive tests for the fundamental and general solutions. Section 5 relates our work to previous work on estimation and bubble testing. Section 6 presents empirical results using the German hyperinflation data. Section 7 contains summary and conclusions. A Technical Appendix contains some proofs.

2. TIMES SERIES RESTRICTIONS IN THE CAGAN MODEL

In this section, we consider the restrictions that the Cagan hyperinflation model places on data. There are two features of the model which give rise to testable restrictions; both of them rely crucially on the rational expectations assumption.

---

[1] The nonparametric procedure will be unable to detect bubbles when money is endogenous in two senses. First, the test cannot be consistent when bubbles are present since no order of differencing can actually render the money series stationary. Second, any order of differencing which makes the money series appear stationary in finite sample is very likely to make prices appear stationary as well since both prices and money will contain a common explosive component.

One is the martingale difference property of such models — that a particular transformation of equilibrium prices and fundamentals represents a forecast error which is orthogonal to any conditioning information set. The second is the explosive path that a rational bubble must obey. Our analysis identifies all testable implications of the model in the sense that we identify how the fundamental and fundamental plus bubbles solutions to prices constrain the autocorrelation function of the data. If none of the second moments of the data violate these constraints, then there exists a formulation of the null model which could have generated the data. The analysis employs the variables —

M_t = log of nominal money supply at t.
P_t = log of price level at t.
$\Im_t$ = information set consisting of all variables measurable at t.
$L_t(x)$ = econometrician's information set at t, defined as all linear combinations of variables x_t which are measurable at t. $L_t(x) \subseteq \Im_t$.

Cagan's hyperinflation model is a structural equation for money demand which depends upon the expected inflation rate. The linear form of the model is

$$M_t - P_t = \beta + \alpha \mathrm{E}\left((P_{t+1} - P_t)\middle|\Im_t\right) + \varepsilon_t \tag{1}$$

where β is a constant term, α is a constant less than zero, and ε_t is a random money demand disturbance. Rearranging this equation in terms of prices gives

$$P_t = \frac{\beta}{\alpha-1} + \frac{\alpha}{\alpha-1}\mathrm{E}\left(P_{t+1}\middle|\Im_t\right) - \frac{1}{\alpha-1}M_t + \frac{1}{\alpha-1}\varepsilon_t. \tag{2}$$

Recursively substituting forward for $\mathrm{E}(P_{t+i+1}|\Im_{t+1})$ and using the law of iterated expectations yields

$$\begin{aligned} P_t = -\beta + \left(\frac{1}{1-\alpha}\right)\sum_{j=0}^{\infty}\left(\frac{\alpha}{\alpha-1}\right)^j\left(\mathrm{E}\left(M_{t+j}\middle|\Im_t\right) - \mathrm{E}\left(\varepsilon_{t+j}\middle|\Im_t\right)\right) \\ + \lim_{i\to\infty}\left(\frac{\alpha}{\alpha-1}\right)^i \mathrm{E}\left(P_{t+i+1}\middle|\Im_t\right). \end{aligned} \tag{3}$$

Imposing the transversality condition, $\lim_{i\to\infty}(\alpha/\alpha-1)^i\,\mathrm{E}(P_{t+i+1}|\Im_t) = 0$, gives the fundamental price P_t^f

$$P_t^f = -\beta + \left(\frac{1}{1-\alpha}\right)\sum_{j=0}^{\infty}\left(\frac{\alpha}{\alpha-1}\right)^j\left(\mathrm{E}\left(M_{t+j}\middle|\Im_t\right) - \mathrm{E}\left(\varepsilon_{t+j}\middle|\Im_t\right)\right) \tag{4}$$

which depends upon the expected sequence of current and future money supply levels and money demand disturbances.

The general solution to (2) does not impose the transversality condition. Any process

$$B_t = \frac{\alpha - 1}{\alpha} B_{t-1} + \xi_t, \tag{5}$$

where

$$\mathrm{E}\left(\xi_t | \Im_{t-1}\right) = 0, \tag{6}$$

when added to a given solution of (2) will still solve that equation. The set of such admissible solutions, P_t^g, may be parameterised by

$$P_t^g = P_t^f + B_t. \tag{7}$$

B_t is called a rational bubble.

In analysing model misspecification, we wish to distinguish between the failure of the Cagan model and the failure of its fundamental solution to adequately characterise the data. In doing so, it is useful to decompose the actual price series into components which correspond to fundamentals, bubbles and model misspecification. We work with the identity

$$P_t = P_t^f + B_t + S_t. \tag{8}$$

In (8), $P_t^g = P_t^f + B_t$ characterises a set of solutions for P_t which are consistent with the logic of the Cagan model — P_t^f is unique, and the many B_t processes satisfying (5) and (6) may be thought of as indexing the solutions. The misspecification term S_t represents that component of price behaviour which cannot be attributed to the model. Equation (8) is an unobserved components representation of observed prices. All deviations of prices from the fundamental solution occur when model noise N_t defined as

$$N_t = B_t + S_t \tag{9}$$

is nonzero. The properties of the fundamental and bubbles components allow one to test which noise component, S_t or B_t, is responsible for nonzero noise.

In order to decompose the price series into its components in (8), we introduce some new variables. The first is a perfect foresight fundamental price,

$$P_t^* = -\beta + \left(\frac{1}{1-\alpha}\right) \sum_{j=0}^{\infty} \left(\frac{\alpha}{\alpha - 1}\right)^j \left(M_{t+j} - \varepsilon_{t+j}\right). \tag{10}$$

From (4) it may be seen that $\mathrm{E}(P_t^* | \Im_t) = P_t^f$ which means that

$$P_t^* = P_t^f + v_t \tag{11}$$

where v_t is a forecast error under the fundamental null hypothesis $\mathrm{H}_0^f : P_t = P_t^f$.

Observe that under H_1^f, the set of all models in which prices do not obey the fundamental solution,

$$P_t - P_t^* = B_t + S_t - v_t. \tag{12}$$

The projection of $P_t - P_t^*$ onto variables observable at t produces a time series which is made up of components related to both the bubble and misspecification terms, since v_t as a forecast error is by definition orthogonal to lagged observables.

The second variable we introduce is a forward quasi-difference[2] of (12) which eliminates the predictable part of the bubble term:

$$\begin{aligned} r_{t+1} &= -\left(1-\left(\frac{\alpha}{\alpha-1}\right)L^{-1}\right)(P_t - P_t^*) \\ &= -\left(1-\left(\frac{\alpha}{\alpha-1}\right)L^{-1}\right)(S_t - v_t) + \frac{\alpha}{\alpha-1}\xi_{t+1}. \end{aligned} \tag{13}$$

Projection of r_{t+1} onto $L_t(x)$ will capture only the misspecification component since v_t (and its forward quasi-difference) and ξ_{t+1} are both orthogonal to $\Im_t$. In other words, r_{t+1} is orthogonal to $L_t(x)$ under the general null H_0^g: P_t is consistent with (2). Durlauf and Hall (1989*a*) refer to analysis of $P_t - P_t^*$ as a stock approach and analysis of r_{t+1} as a flow approach to uncovering model noise.

The interactions of the time series $P_t - P_t^*$ and r_{t+1} with $L_t(x)$ identify all the testable restrictions of the rational expectations Cagan model. If the $P_t - P_t^*$ and r_{t+1} projections are zero, then prices obey the fundamental solution. If the $P_t - P_t^*$ projection is nonzero and the r_{t+1} projection is zero, then bubbles are present. If the $P_t - P_t^*$ and r_{t+1} projections are both nonzero, then misspecification is present and further analysis is required to determine whether a bubble component exists as well. These two types of projections therefore permit discrimination between different types of violations of the fundamental price solution.

These projections, however, cannot be directly implemented because $P_t - P_t^*$ and r_{t+1} are not necessarily observable. In the next two sections we construct variables which are observable yet retain the orthogonality properties of $P_t - P_t^*$ and r_{t+1} under two important and common identifying assumptions about ε_t.

3. SPECIFICATION RESTRICTIONS ON THE FUNDAMENTAL SOLUTION

We first consider the empirical implications of the null hypothesis that prices obey the fundamental solution to the Cagan model. In testing the fundamental solution

---

[2] r_{t+1} has a nonzero mean which we subtract out; h_{t+1} below is similarly demeaned since we want to project onto information sets which include a constant term.

to the Cagan model, the variable of interest, $P_t - P_t^*$, is not generally observable since the disturbances ε_t appear in

$$P_t - P_t^* = P_t + \beta - \left(\frac{1}{1-\alpha}\right)\sum_{j=0}^{\infty}\left(\frac{\alpha}{\alpha-1}\right)^j (M_{t+j} - \varepsilon_{t+j}). \tag{14}$$

Define the observable analog to P_t^* as P_t',

$$P_t' = -\beta + \left(\frac{1}{1-\alpha}\right)\sum_{j=0}^{\infty}\left(\frac{\alpha}{\alpha-1}\right)^j M_{t+j}; \tag{15}$$

an observable analog to $P_t - P_t^*$, $P_t - P_t'$, may be constructed,

$$P_t - P_t' = P_t - P_t^* - \left(\frac{1}{1-\alpha}\right)\sum_{j=0}^{\infty}\left(\frac{\alpha}{\alpha-1}\right)^j \varepsilon_{t+j} \tag{16}$$

$$= S_t + B_t - v_t - \left(\frac{1}{1-\alpha}\right)\sum_{j=0}^{\infty}\left(\frac{\alpha}{\alpha-1}\right)^j \varepsilon_{t+j}. \tag{17}$$

The projection of $P_t - P_t'$ onto $L_t(x)$ results in a time series that contains components corresponding to the misspecification, bubble and omitted portfolio disturbance terms. The forecast error v_t is orthogonal to $L_t(x)$ by construction. However, the term $-(1/1-\alpha)\sum_{j=0}^{\infty} (\alpha/\alpha-1)^j \varepsilon_{t+j}$ may be correlated with variables observable at t when the portfolio balance disturbances are allowed to have arbitrary structure. Therefore, we place identifying restrictions on the portfolio disturbance to determine whether a nonzero $P_t - P_t'$ projection may be interpreted as evidence that the S_t and B_t terms are nonzero — i.e. prices do not obey the fundamental solution. We consider two common assumptions in the literature.

Case 1: Exact Cagan model

We first consider the case where the Cagan model holds exactly, i.e. $\varepsilon_t = 0$ for all t. This is a natural case to begin with because it represents the direct analog to the constant discount dividend stock price model. Goodfriend (1982) estimated the Cagan model using this assumption and provided some justification for it. In this case, $\varepsilon_t = 0\ \forall\ t$, and so $P_t - P_t'$ is equal to $P_t - P_t^*$. Durlauf and Hall (1989*c*), extending the seminal work of Hansen and Sargent (1991) on the testable implications of rational expectations models to the case of nonstationary deviations from a fundamental solution, provides all testable restrictions of the model which we summarise in

Theorem 1: *Testable restrictions on fundamental solution to the exact Cagan hyperinflation model.*

A) When model noise is stationary[3], *all testable implications of* H_0^f *are contained in the requirement that the projection of* $P_t - P_t'$ *onto* $L_t(x)$ *is zero.*

B) When model noise is nonstationary and M_t *and* P_t *are elements of* x_t, *then the projection of* $P_t - P_t'$ *onto* $L_t(x)$ *is well defined. All testable implications of* H_0^f *are contained in the requirement that the projection of* $P_t - P_t'$ *onto* $L_t(x)$ *is zero.*

Again, if model noise is nonstationary, the projection of $P_t - P_t'$ onto $L_t(x)$ is not always well defined as we are not working with L^2 processes. This is why it is important to include the history of both money and prices in the projection to ensure that nonstationary noise is recovered. If $P_t - P_t'$ is not orthogonal to $L_t(x)$, then at least one of the time series B_t or S_t is nonzero.

Case 2: Random walk portfolio disturbances

The most common assumption in empirical work on the Cagan model has been that the portfolio disturbances are a random walk:

$$\varepsilon_{t+1} = \varepsilon_t + u_{t+1} \tag{18}$$

where

$$\mathrm{E}\left(u_{t+1} \middle| \Im_t\right) = 0. \tag{19}$$

This assumption is made by Sargent and Wallace (1973), Salemi and Sargent (1979), Flood and Garber (1980), Burmeister and Wall (1982, 1987), and Casella (1989) among others. The empirical literature provides some support for the random walk specification in the sense that the goodness of fit of the Cagan model is improved when this specification is imposed. Furthermore, Khan (1975) calculated Durbin-Watson statistics for Cagan's regressions and found a pattern of substantial serial correlation; the Durbin-Watson statistic for the German case is 0.33.

When ε_t follows a random walk, $P_t - P_t'$ is no longer orthogonal to the history $L_t(x)$ since $\mathrm{E}\left(\varepsilon_{t+1} \middle| \Im_t\right) = \varepsilon_t \neq 0$ for $j \geq 1$. However, the problem may be solved by differencing. Consider

$$\Delta P_t - \Delta P_t' = -\Delta v_t + \Delta B_t + \Delta S_t - \left(\frac{1}{1-\alpha}\right)\sum_{j=0}^{\infty}\left(\frac{\alpha}{\alpha-1}\right)^j u_{t+j}. \tag{20}$$

[3] When we say that model noise is stationary, we mean that it is an L^2 process. When we say that it is nonstationary, we mean that it is either explosive or integrated of some order.

A generalisation of Theorem 1 verifies

Theorem 2: *Testable restrictions on fundamental model specification of the Cagan hyperinflation model with random walk portfolio disturbances.*

A) When model noise is stationary, all testable implications of H_0^f *are contained in the requirement that the projection of* $\Delta P_t - \Delta P_t'$ *onto* $L_{t-1}(x)$ *is zero.*

B) When model noise is nonstationary and M_t *and* P_t *are elements of* x_t*, then the projection of* $\Delta P_t - \Delta P_t'$ *onto* $L_{t-1}(x)$ *is well defined. All testable implications of* H_0^f *are contained in the requirement that the projection of* $\Delta P_t - \Delta P_t'$ *onto* $L_{t-1}(x)$ *is zero.*

Theorems 1 and 2 imply consistent tests for H_0^f. However, unless we have reason to believe that noise relative to the general model is negligible, these Theorems are not interpretable as tests for bubbles. We therefore consider tests which exclusively focus on possible misspecification of the general model.

4. SPECIFICATION RESTRICTIONS ON THE GENERAL SOLUTION

As described in Section 2, the variable of interest for testing H_0^g is r_{t+1}. In terms of the variables of the model,

$$r_{t+1} = \frac{\alpha}{\alpha-1}P_{t+1} - P_t + \frac{1}{1-\alpha}M_t - \frac{1}{1-\alpha}\varepsilon_t. \tag{21}$$

It is apparent, since the portfolio disturbances are not observable, that r_{t+1} is not an element of $L_\infty(x)$ and hence cannot be recovered from the data.

The observable component of r_{t+1} is h_{t+1}, representable as

$$\begin{aligned} h_{t+1} &= -\left[1-\left(\frac{\alpha}{\alpha-1}\right)L^{-1}\right]\left[P_t - P_t^* + \left(\frac{1}{1-\alpha}\right)\sum_{j=0}^{\infty}\left(\frac{\alpha}{\alpha-1}\right)^j \varepsilon_{t+j}\right] \\ &= r_{t+1} + \frac{1}{1-\alpha}\varepsilon_t \\ &= \frac{\alpha}{\alpha-1}P_{t+1} - P_t + \frac{1}{1-\alpha}M_t. \end{aligned} \tag{22}$$

The time series h_{t+1} represents the excess holding return on the money supply when there are no portfolio disturbances. Otherwise, h_{t+1} equals a mismeasured excess holding return and will not necessarily be orthogonal to variables observable at t.

The ability to distinguish misspecification from the presence of bubbles is a function of the structure of the spaces $L_t(P-P')$ and $L_t(h)$, which is the linear history of current and past h_t's. Mathematically, the analysis exploits the fact that since B_t is not an L^2 process, $L_t(P-P')$ is not a Hilbert space and hence $L_t(P-P')$ and $L_t(h)$ do not span the same random variables. If $P_t - P_t'$ were an L^2 process, then $L_t(P-P')$ and $L_t(h)$ would be equivalent Hilbert spaces. There is an information loss in analysing excess holding returns rather than forecast error levels which can be exploited to construct tests which are insensitive to the presence of bubbles.

As before, the appropriate transformation applied to h_{t+1} to retain the orthogonality properties of r_{t+1} will depend on the specific assumptions made concerning the structure of the portfolio disturbance ε_t.

Case 1: Exact Cagan model

When there is no portfolio disturbance, then the fundamental solution to prices is a geometrically weighted average of forecasts of current and future money. More importantly, $h_{t+1} = r_{t+1}$, so that the excess holding return on money is now observable. The following Theorem characterises the set of possible deviations generated exclusively by the model misspecification term.

Theorem 3: *Testable restrictions on model specification of the exact Cagan hyperinflation model.*

A) When model noise is stationary, all testable implications of H_0^g *are contained in the requirement that the projection of* h_{t+1} *onto* $L_t(x)$ *is zero.*

B) When model noise is nonstationary and M_t *and* P_t *are elements of* x_t, *then the projection of* h_{t+1} *onto* $L_t(x)$ *is well defined. All testable implications of* H_0^g *are contained in the requirement that the projection of* h_{t+1} *onto* $L_t(x)$ *is zero.*

Proof: We prove sufficiency for (A); necessity for (A) and (B) are verified in the Appendix.

Observe that if prices equal the fundamental solution, v_t is AR(1) since

$$\zeta_{t+1} = -\left[1-\left(\frac{\alpha}{\alpha-1}\right)L^{-1}\right]v_t$$

$$= \left(\frac{1}{1-\alpha}\right)\sum_{j=1}^{\infty}\left(\frac{\alpha}{\alpha-1}\right)^j \left\{\mathrm{E}(M_{t+j}|\Im_{t+1}) - \mathrm{E}(M_{t+j}|\Im_t) - \left[\mathrm{E}(\varepsilon_{t+j}|\Im_{t+1}) - \mathrm{E}(\varepsilon_{t+j}|\Im_t)\right]\right\}$$

(23)

which must be a white noise process, since by assumption, $L_t(\varepsilon)$ and $L_t(M)$ are both subsets of $L_t(x)$.

For the exact Cagan model,

$$h_{t+1} = \zeta_{t+1} + \frac{\alpha}{\alpha - 1}\xi_{t+1} \tag{24}$$

under H_0^g. In this case, ζ_{t+1} is a time $t+1$ information innovation with respect to money supply only; it is orthogonal to $\Im_t$ by the law of iterated expectations. The second term ξ_{t+1}, induced by a possible bubble, is orthogonal to $\Im_t$ by assumption. Further, since h_t and ξ_t are elements of $\Im_t$, this implies that ζ_t and ξ_t are elements of $\Im_{t+1}$ as well. As a result, h_{t+1} is the sum of two white noise terms which are uncorrelated at all leads and lags and therefore must be white noise. Since $L_t(x) \subseteq \Im_t$, sufficiency must hold.

If one treats h_{t+1} as the excess holding return on money then the Theorem really says that in the general solution to prices, the excess holding return must fulfil the zero expected profit condition (1). In this respect, the Cagan model is equivalent to the arbitrage condition in the constant discount dividend stock price model.

Case 2: Random walk portfolio disturbances

When the portfolio disturbance obeys a random walk, prices can no longer be directly constructed as a weighted average of expectations of current and future money supply levels. The observable innovation h_{t+1} will equal

$$h_{t+1} = \zeta_{t+1} + \frac{\alpha}{\alpha - 1}\xi_{t+1} + \frac{1}{1-\alpha}\varepsilon_t \tag{25}$$

under H_0^g. When ε_t is a random walk, h_{t+1} is no longer orthogonal to the history $L_t(x)$. Orthogonality breaks down since part of the current portfolio disturbance is measurable before t and is incorporated into prices dated t and earlier. As before, this problem can be solved by differencing h_{t+1}.

$$\Delta h_{t+1} = \zeta_{t+1} - \zeta_t + \frac{\alpha}{\alpha - 1}\xi_{t+1} - \frac{\alpha}{\alpha - 1}\xi_t + \frac{1}{1-\alpha}u_t \tag{26}$$

Δh_{t+1} is orthogonal to $\Im_{t-1}$ and hence $L_{t-1}(x)$ since the components of Δh_{t+1} in (26) are all white noise processes with respect to the econometrician's information set. Further, the model places no restrictions on the contemporaneous covariances between ζ_t, ξ_t and u_t. Consequently, Δh_{t+1} is an MA(1) process under the null hypothesis. This leads to

Theorem 4: *Testable restrictions on model specification of the Cagan hyperinflation model with random walk portfolio disturbances.*

A) When model noise is stationary, all testable implications of H_0^g *are contained in the requirement that the projection of* Δh_{t+1} *onto* $L_{t-1}(x)$ *is zero.*

B) When model noise is nonstationary and M_t *and* P_t *are elements of* x_t*, then the projection of* Δh_{t+1} *onto* $L_{t-1}(x)$ *is well defined. All testable implications of* H_0^g *are contained in the requirement that the projection of* Δh_{t+1} *onto* $L_{t-1}(x)$ *is zero.*

Proof: See Appendix.

5. RELATIONSHIP TO OTHER WORK

The signal extraction approach described above differs considerably from most analyses of the Cagan model. In order to facilitate comparisons, this section relates our approach and its estimation and testing procedure to some others which have been commonly used in the hyperinflation and other literatures.

The stock orthogonality conditions that we use, involving $P_t - P_t'$ and $\Delta P_t - \Delta P_t'$, have been principally employed in the finance literature on testing the constant discount dividend stock price model. The idea, that the difference between a perfect foresight price and a realised price must under the fundamental null be a forecast error, was first exploited by Shiller (1981) and LeRoy and Porter (1981) in deriving variance bounds tests. This idea was extended by Durlauf and Hall (1989*a*, 1989*b*) who showed how to obtain estimates of model noise relative to a fundamental price.

The standard approach to bubble testing has been to explicitly parameterise both a money supply equation and a bubble and then to estimate the bubble (or its coefficient or variance) and perhaps the model parameters as well. For instance, authors such as Flood and Garber (1980), Flood, Garber and Scott (1984), and Burmeister and Wall (1982) specify exogenous, autoregressive money supply equations; the first two use a deterministic and the third an explicit stochastic bubble specification. Burmeister and Wall (1987) and İmrohoroğlu (1989) allow the money supply equation to be endogenous, although its parameterisation is still explicit as are those of the bubbles. Casella (1989) also allows for endogenous money supply; furthermore, she exploits the relationship between the stock and flow representations through testing for a bubble by comparing the instrumental variables estimate of α from (28) and (29) below with one obtained by solving for the fundamental price. None of these approaches can distinguish between rejections due to bubbles and rejections due to misspecification, however, because under the misspecification alternative, the estimates of α in Casella's case (and of the unobserved component representing a bubble in the other papers) have complicated distributions which may or may not depend on bubble components. Testing our flow orthogonality condition and then comparing it to the stock condition does distinguish between model rejection due to misspecification and that due to bubbles.

Further, we observe that it is possible to exploit the properties of $P_t - P_t'$ and $\Delta P_t - \Delta P_t'$ in order to test for bubbles by exploiting the stationarity or nonstationarity of these time series in a way analogous to procedures proposed by Hamilton and Whiteman (1985) and Diba and Grossman (1988). In the exact case, $P_t - P_t'$ is uncorrelated with the money process under H_0^f. Therefore any nonstationarity in money will not be transmitted to $P_t - P_t'$. This series can be analysed under the exact specification to see whether it is a stationary process. Second, any nonstationarity induced in $P_t - P_t'$ by the random walk specification of the portfolio disturbance is eliminated by first differencing since $\Delta P_t - \Delta P_t'$ is stationary under the fundamental null. This second series therefore is appropriate to analyse when the portfolio disturbance is a random walk. By seeing whether $P_t - P_t'$ and $\Delta P_t - \Delta P_t'$ obey stationary specifications, one can directly test whether flow-based model rejections can be attributed to bubbles. As before, stationarity tests of this sort do not require specification of the stochastic process for money.

The flow orthogonality conditions that we use, involving h_{t+1} and Δh_{t+1}, have been employed previously by researchers estimating the Cagan equation. For instance, Goodfriend (1982) noted that under the exactness assumption, least squares yields consistent estimates of the parameters α and β under the assumption that u_{t+1} in

$$P_{t+1} = \frac{\beta}{\alpha} - \frac{\alpha - 1}{\alpha} P_t + \frac{1}{\alpha} M_t + u_{t+1} \tag{27}$$

is a forecast error and is therefore orthogonal to P_t and M_t. Dividing through by the constant factor $\alpha - 1$ and rearranging shows that this equation is equivalent to our h_{t+1}. Casella (1989) made use of the random walk flow orthogonality condition in estimating

$$(M_t - P_t) - (M_{t-1} - P_{t-1}) = \alpha[P_{t+1} - P_t - (P_t - P_{t-1})] + u_t - \alpha(w_t - w_{t-1}) \tag{28}$$

where

$$w_t = P_{t+1} - \mathrm{E}(P_{t+1} | \Im_t) \tag{29}$$

using instrumental variables; similarly, multiplying by a constant and rearranging shows its equivalence to Δh_{t+1}.

We estimate the parameters α and β in order to construct h_{t+1} and Δh_{t+1} and $P_t - P_t'$ and $\Delta P_t - \Delta P_t'$ as well (note that β does not enter the Δh_{t+1} equation). Since we are working with moment conditions, we use generalised method of moments (GMM) estimation. Using GMM essentially means that parameter values are chosen to minimise the variance of the noise component to prices, given a particular information set. By varying the information set, we obtain different parameter estimates and thus gain information on the robustness of the estimates; this will potentially shed light on the mixed empirical results currently in the literature. Since the estimators of Goodfriend and Casella can be obtained by particular choices of instrument sets and GMM weighting matrices, our results may be seen as complementary to and generalisations of their estimates.

6. EMPIRICAL ANALYSIS

In this section, we implement the projections of $P_t - P_t'$, $\Delta P_t - \Delta P_t'$, h_{t+1}, and Δh_{t+1} on to different information sets $L_t(x)$ and $L_{t-1}(x)$ for the German hyperinflation. The values of α and β used in constructing the variables $P_t - P_t'$, $\Delta P_t - \Delta P_t'$ are obtained from the generalised method of moments (GMM) estimation based on the orthogonality restrictions associated with the h_{t+1} and Δh_{t+1} equations. For each information set, GMM is applied to h_{t+1} and Δh_{t+1}, imposing the orthogonality restriction of the general null, and thereby producing a pair of α and β estimates. These estimates are then used to construct the $P_t - P_t'$ and $\Delta P_t - \Delta P_t'$ series whose orthogonality to the same information set can then be tested.[4] The various null hypotheses are tested through the construction of Wald statistics using the Newey-West variant of White's heteroskedasticity and autocorrelation correction. We discuss the use of this data in light of regime shifts and other issues before turning to the empirical results.

All reported regressions use monthly data sampled from 1920(1) to 1923(8) which is consistent with most empirical work on the German hyperinflation. The reason for truncating the sample at August 1923 is that a successful monetary reform was implemented which changed the data generating process (and

[4] One interesting feature of the quasi-difference flow regressions is the behavior of α. As predicted in the previous section, our estimates of α are remarkably similar to those of Goodfriend (1982) in the exact case and Casella (1989) in the random walk one. Goodfriend's estimate for Germany is –5.3, whereas many of our exact case estimates are around –4.3. Casella estimates α to equal –0.8, whereas under the random walk specification, our estimates generally range between –0.7 and –1.3. Complete results are available upon request.

invalidates most of the statistical theory underlying model estimation and testing) soon after that time.[5] The data we employ for this sample consist of the money series constructed by Flood and Garber (1980) and price data from the Statistisches Reichsamt (1925); both are described in appendices in Flood and Garber (1980) and Casella (1989). The P_t' series requires out-of-sample data which we take from LaHaye (1985). (LaHaye employs the same within-sample data as we do.)

In constructing an empirical representation for P_t', it is necessary to impose truncation and terminal values on the infinite sum. We employ the formula, due to Shiller (1981),

$$P_t' = -\beta\left[1-\left(\frac{\alpha}{\alpha-1}\right)^{T-t}\right]+\left(\frac{1}{1-\alpha}\right)\left[\sum_{j=0}^{T-t-1}\left(\frac{\alpha}{\alpha-1}\right)^{j} M_{t+j}+\left(\frac{\alpha}{\alpha-1}\right)^{T-t} P_T\right]. \quad (30)$$

There are three potential problems associated with the use of P_t' in our tests. First, Flood and Hodrick (1986) demonstrate that if the terminal value in the Shiller approximation contains a bubble, then it will be exactly cancelled out in the construction of P_t' and the stock projections will never reveal the bubble. By setting the terminal value equal to December 1926, which occurs substantially after the end of the hyperinflation, we avoid this problem.

The other two difficulties are both functions of the accuracy of the within-sample approximation of $P_t - P_t'$ to actual forecast errors. One, examined by Kim (1989), is that in some circumstances this approximation leads to poor finite sample properties for $P_t - P_t'$ projections. The other is that data after August 1923 incorporates the influence of nonzero reform probabilities and the reform itself which may lead to a bias in the results due to the 'peso problem'.[6]

The principal reason for the poor sampling results found in Kim (1989) is that in his application (involving consumption data) the discount rate is very small, about 0.005 on a monthly basis. In the present case, finite sample problems are likely to be mitigated by much larger discount rates of 0.15 to 0.35 for the α values we estimate and which are frequently found in the literature. At these rates, the difference between $\sum_{j=T-t+1}^{\infty}(\frac{\alpha}{\alpha-1})^j M_{t+j}$ and $(\frac{\alpha}{\alpha-1})^{T-t} P_T$, which represents the approximation error, is negligible for all but the final few observations.

[5] Researchers from Cagan on have found that model fit deteriorates rapidly as more data is included after August 1923. LaHaye (1985) presents a formal justification for this break and argues that regime shift probabilities were roughly zero before that time.

[6] If the end of the hyperinflation was uncertain from the perspective of the economy, then forecast errors will look irrational because the *ex post* rational price is far below the public's expectations. Put differently, the forecast error induced by the termination of the hyperinflation will not be offset by other forecast errors in the sample. (The law of large numbers cannot be invoked for singular events!)

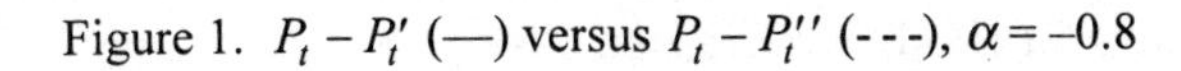

Figure 1. $P_t - P_t'$ (—) versus $P_t - P_t''$ (- - -), $\alpha = -0.8$

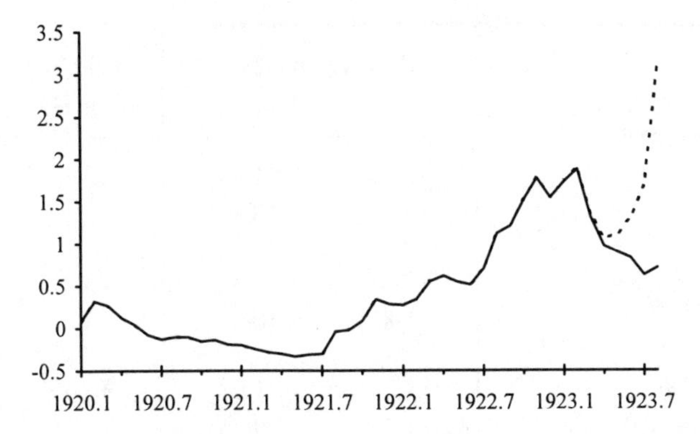

Figure 2. $P_t - P_t'$ (—) versus $P_t - P_t''$ (- - -), $\alpha = -4.5$

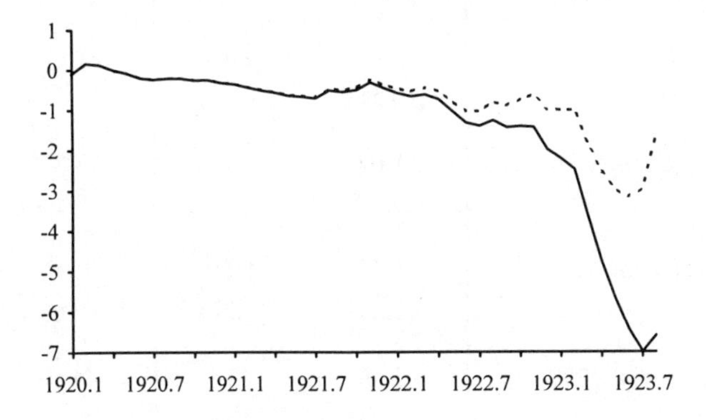

With regard to the regime issue, high discount rates increase the weight put on within-sample data in constructing P_t'. To verify that our P_t' variable is robust to shifts in the M_t regime, we performed the following experiment. We compared the historical $P_t - P_t'$ series with a synthetic series $P_t - P_t''$ in which money growth after August 1923 was replaced with forecasts of money growth based upon behaviour during the hyperinflation.[7] Figure 1 displays the two series when $\alpha = -0.8$. Until March 1923, the two series nearly coincide. For differing values of α, the fits vary, but the qualitative result is similar. Figure 2 displays the two series when $\alpha = -4.5$, approximately the worst case for our estimated values. Even

[7] Specifically, a fourth order polynomial time trend was fit to money growth between January 1920 and August 1923. The predictions of this trend after August 1923 were then employed in the computation of P_t''.

Table 1. Projection Results for H_0^f, Exact Equation Case

Elements of $L_t(x)$	GMM Estimates of α	and β	Wald Stat† for H_0^f
$\Delta m_t, \Delta p_t$	−2.71 (1.54)	1.44 (0.42)	63.13*
$\Delta m_t, \Delta m_{t-1}, \Delta p_t, \Delta p_{t-1}$	−4.21 (1.83)	1.59 (0.35)	58.36*
$\Delta m_t, \Delta m_{t-1}, \Delta m_{t-2}$, $\Delta p_t, \Delta p_{t-1}, \Delta p_{t-2}$	−4.20 (1.76)	1.62 (0.36)	85.49*
$\Delta m_t, \Delta m_{t-1}, \Delta m_{t-2}, \Delta m_{t-3}$, $\Delta p_t, \Delta p_{t-1}, \Delta p_{t-2}, \Delta p_{t-3}$	−4.36 (1.86)	1.68 (0.38)	77.64*
$\Delta^2 m_t, \Delta^2 p_t$	−1.09 (0.51)	1.16 (0.26)	21.60*
$\Delta^2 m_t, \Delta^2 m_{t-1}, \Delta^2 p_t, \Delta^2 p_{t-1}$	−1.07 (0.57)	1.16 (0.27)	48.75*
$\Delta^2 m_t, \Delta^2 m_{t-1}, \Delta^2 m_{t-2}$, $\Delta^2 p_t, \Delta^2 p_{t-1}, \Delta^2 p_{t-2}$	−1.45 (0.69)	(1.15) (0.28)	170.93*
$\Delta^2 m_t, \Delta^2 m_{t-1}, \Delta^2 m_{t-2}, \Delta^2 m_{t-3}$, $\Delta^2 p_t, \Delta^2 p_{t-1}, \Delta^2 p_{t-2}, \Delta^2 p_{t-3}$	−2.26 (1.08)	1.15 (0.31)	175.96*

†Asymptotic standard errors in parentheses. The regressions run are $P_t - P_t'$ on elements of $L_t(x)$; Wald statistics are for the hypothesis that the coefficients of all nonconstant regressors equal zero.

* Significant at the 5% level.

here, with the exception of the last few observations, the two series track closely. Given the information above, we conduct $P_t - P_t'$ and $\Delta P_t - \Delta P_t'$ projections by deleting the last seven observations, which appears to render the results robust to regime and sampling issues.

Table 1 reports Wald statistics for the exact Cagan model using first- and second-differenced combinations of money and prices regressed against $P_t - P_t'$. The results here show rejections for seven out of eight information sets. In Table 2, where $\Delta P_t - \Delta P_t'$ is the left hand side variable, the rejections are fewer than in

Table 2. Projection Results for H_0^f, Random Walk Case

Elements of $L_{t-1}(x)$	GMM Estimate of α	Wald Stat† for H_0^f
$\Delta m_{t-1}, \Delta p_{t-1}$	−1.48 (1.22)	8.64*
$\Delta m_{t-1}, \Delta m_{t-2}, \Delta p_{t-1}, \Delta p_{t-2}$	−1.30 (0.96)	12.48*
$\Delta m_{t-1}, \Delta m_{t-2}, \Delta m_{t-3}, \Delta p_{t-1}, \Delta p_{t-2}, \Delta p_{t-3}$	−0.79 (0.24)	24.15*
$\Delta m_{t-1}, \Delta m_{t-2}, \Delta m_{t-3}, \Delta m_{t-4}, \Delta p_{t-1}, \Delta p_{t-2}, \Delta p_{t-3}, \Delta p_{t-4}$	−0.80 (0.18)	24.57*
$\Delta^2 m_{t-1}, \Delta^2 p_{t-1}$	−1.28 (1.08)	6.11*
$\Delta^2 m_{t-1}, \Delta^2 m_{t-2}, \Delta^2 p_{t-1}, \Delta^2 p_{t-2}$	−0.75 (0.27)	6.63
$\Delta^2 m_{t-1}, \Delta^2 m_{t-2}, \Delta^2 m_{t-3}, \Delta^2 p_{t-1}, \Delta^2 p_{t-2}, \Delta^2 p_{t-3}$	−0.74 (0.25)	12.81*
$\Delta^2 m_{t-1}, \Delta^2 m_{t-2}, \Delta^2 m_{t-3}, \Delta^2 m_{t-4}, \Delta^2 p_{t-1}, \Delta^2 p_{t-2}, \Delta^2 p_{t-3}, \Delta^2 p_{t-4}$	−1.03 (0.21)	56.84*

† Asymptotic standard errors in parentheses. The regressions run are $\Delta P_t - \Delta P_t'$ on elements of $L_{t-1}(x)$; Wald statistics are for the hypothesis that the coefficients of all nonconstant regressors equal zero.

* Significant at the 5% level.

Table 1 but are still substantial: seven out of eight of the joint P_t and M_t information sets are significantly correlated with $\Delta P_t - \Delta P_t'$. These results provide very strong evidence against the null hypothesis that the German hyperinflation data are consistent with the fundamental solution to the Cagan model.

Table 3 presents specification tests of the exact Cagan model based upon the projection of h_{t+1} onto combinations of first and second differences of current and lagged prices and money supplies. The results in this table provide strong evidence against the general solution of the exact Cagan model: all joint first

Table 3. Projection Results for H_0^g, Exact Equation Case

Elements of $L_t(x)$	Wald Stat† for H_0^g	Elements of $L_t(x)$	Wald Stat† for H_0^g
$\Delta m_t, \Delta p_t$	7.06*	$\Delta^2 m_t, \Delta^2 p_t$	0.53
$\Delta m_t, \Delta m_{t-1}, \Delta p_t, \Delta p_{t-1}$	15.51*	$\Delta^2 m_t, \Delta^2 m_{t-1}, \Delta^2 p_t, \Delta^2 p_{t-1}$	0.66
$\Delta m_t, \Delta m_{t-1}, \Delta m_{t-2}, \Delta p_t, \Delta p_{t-1}, \Delta p_{t-2}$	27.74*	$\Delta^2 m_t, \Delta^2 m_{t-1}, \Delta^2 m_{t-2}, \Delta^2 p_t, \Delta^2 p_{t-1}, \Delta^2 p_{t-2}$	7.19
$\Delta m_t, \Delta m_{t-1}, \Delta m_{t-2}, \Delta m_{t-3}, \Delta p_t, \Delta p_{t-1}, \Delta p_{t-2}, \Delta p_{t-3}$	47.80*	$\Delta^2 m_t, \Delta^2 m_{t-1}, \Delta^2 m_{t-2}, \Delta^2 m_{t-3}, \Delta^2 p_t, \Delta^2 p_{t-1}, \Delta^2 p_{t-2}, \Delta^2 p_{t-3}$	32.64*

† The regressions run are h_{t+1} on elements of $L_t(x)$; Wald statistics are for the hypothesis that the coefficients of all nonconstant regressors equal zero.
* Significant at the 5% level.

differences of money and prices are statistically significant at the 5 percent level, and the largest second difference information set is significant as well. Table 4 summarises the results for the random walk Cagan model based on projecting Δh_{t+1}. Again we find rejections of the general solution for several of the information sets, particularly those employing longer lags. Together, these results indicate a rejection of the random walk specification as a description of the data. We therefore conclude that neither the exact Cagan model nor its random walk counterpart are consistent with the null hypothesis that $P_t = P_t^g$.

In comparing Tables 1–2 and Tables 3–4 we do find some decrease in the number of rejections of the model when one moves from forecast error levels to quasi-forward differences. For the exact Cagan model, we reject all eight times for the stock case and five times for the flow case. For the random walk specification, we reject seven times for the stock case and five times for the flow case. However, since we find that the stock and flow specifications are both consistently rejected for larger information sets, the rejections of both the exact and random walk fundamental nulls cannot necessarily be attributed to the presence of a rational bubble. In fact, both types of phenomena may be present — for example if the Δh_{t+1} projections are detecting a term which should be added to the fundamental solution, such an augmented Cagan equation will still admit linear bubble solutions.

Therefore, we attempt to see whether the fundamental model rejections can be attributed to a nonstationary noise component, as would be the case if a linear

Table 4. Projection Results for H_0^g, Random Walk Case

Elements of $L_{t-1}(x)$	Wald Stat† for H_0^g	Elements of $L_{t-1}(x)$	Wald Stat† for H_0^g
$\Delta m_{t-1}, \Delta p_{t-1}$	4.95	$\Delta^2 m_{t-1}, \Delta^2 p_{t-1}$	3.05
$\Delta m_{t-1}, \Delta m_{t-2}, \Delta p_{t-1}, \Delta p_{t-2}$	8.05	$\Delta^2 m_{t-1}, \Delta^2 m_{t-2}, \Delta^2 p_{t-1}, \Delta^2 p_{t-2}$	14.12*
$\Delta m_{t-1}, \Delta m_{t-2}, \Delta m_{t-3}, \Delta p_{t-1}, \Delta p_{t-2}, \Delta p_{t-3}$	28.46*	$\Delta^2 m_{t-1}, \Delta^2 m_{t-2}, \Delta^2 m_{t-3}, \Delta^2 p_{t-1}, \Delta^2 p_{t-2}, \Delta^2 p_{t-3}$	14.29*
$\Delta m_{t-1}, \Delta m_{t-2}, \Delta m_{t-3}, \Delta m_{t-4}, \Delta p_{t-1}, \Delta p_{t-2}, \Delta p_{t-3}, \Delta p_{t-4}$	30.30*	$\Delta^2 m_{t-1}, \Delta^2 m_{t-2}, \Delta^2 m_{t-3}, \Delta^2 m_{t-4}$ $\Delta^2 p_{t-1}, \Delta^2 p_{t-2}, \Delta^2 p_{t-3}, \Delta^2 p_{t-4}$	26.49*

† The regressions run are Δh_{t+1} on elements of $L_{t-1}(x)$; Wald statistics are for the hypothesis that all nonconstant regressors equal zero.
* Significant at the 5% level.

bubble term is present. Specifically, we examine the stationarity of $P_t - P_t'$ and $\Delta P_t - \Delta P_t'$ to directly test for the presence of bubbles. Our rejection of the null hypothesis that prices obey the general solution to the Cagan model, based on Tables 3 and 4, does not imply that bubbles are not present in the price data; rather it implies that the rejection of the fundamental solution which is reported in Tables 1 and 2 cannot necessarily be attributed to the presence of bubbles, as previous authors have done. In order to see whether bubbles are present in addition to model misspecification, we study the autoregressive structure of $P_t - P_t'$ and $\Delta P_t - \Delta P_t'$. Table 5 contains coefficients and standard errors from first order autoregressions for $P_t - P_t'$ and $\Delta P_t - \Delta P_t'$ for various values of α.[8] Recall that both series should be explosive, i.e. possess AR(1) coefficients outside the unit circle, if a bubble component exists for prices, regardless of the estimate of α used to construct the P_t' series.

The evidence in the table does not support the hypothesis that the source of model rejection is the presence of a bubble. The coefficients in the exact specification are close to one for all choices. Dickey-Fuller t-statistics, constructed using the Phillips-Perron correction [9], accept the null hypothesis that each $P_t - P_t'$

[8] The ability of these tests to detect bubbles does not depend upon the consistency of the estimates of α from the flow regressions. Even if H_0^g does not hold and the α estimates are inconsistent, the presence of an explosive bubble will still cause $P_t - P_t'$ and $\Delta P_t - \Delta P_t'$ to be nonstationary.

[9] All Phillips-Perron corrections employed Bartlett windows of length 10. See Fuller (1976, p. 373) for significance levels of the test statistics.

Table 5. First Order Autoregressions†

α	Bubble Growth Factor	$P_t - P_t'$	$\Delta P_t - \Delta P_t'$
–0.8	2.25	1.06 (0.05) [–2.44]*	0.29 (0.18) [–9.42]
–1.2	1.83	1.04 (0.05) [–2.66]*	0.25 (0.18) [–9.20]
–2.0	1.50	0.99 (0.07) [–3.28]*	0.15 (0.20) [–7.80]
–2.5	1.40	0.88 (0.07) [–3.28]*	0.15 (0.20) [–7.80]
–3.5	1.29	0.78 (0.12) [–2.44]*	0.04 (0.23) [–5.33]
–4.5	1.22	1.05 (0.06) [–0.55]*	0.12 (0.23) [–4.86]

† The third and fourth columns contain OLS estimates of the first-order autoregressive coefficients using a constant; OLS standard errors are in parentheses and (Phillips-Perron corrected) Dickey-Fuller Statistics are in brackets.

* Cannot reject the null hypothesis that the coefficient equals unity at the 5% level.

series contains a unit root. On the other hand, the first difference autoregressions generate small, statistically insignificant coefficients. Our tests reject the unit root null in all cases. There is little evidence of nonstationarity in either the $P_t - P_t'$ or $\Delta P_t - \Delta P_t'$ series. One explanation for the unit roots in the levels regressions that is consistent with the smaller noise magnitudes in the random walk specification relative to the exact specification is that ε_t is some ARIMA process other than a random walk. Our overall results therefore suggest that the rejections reported in Tables 1–4 are best regarded as evidence of model misspecification. At a minimum, if a nonfundamental solution to the model is present, its time series properties are very different from those of the linear bubbles which have been the object of so much of the empirical literature on hyperinflations.

7. SUMMARY AND CONCLUSIONS

This paper has developed a general testing methodology for the Cagan hyperinflation model under rational expectations. We have shown how to develop all testable implications of both fundamental and general solutions for prices in the model conditional on different specifications of money demand disturbances. Application of this methodology to the German hyperinflation reveals that the Cagan model is misspecified for the cases where money demand disturbances are either zero or obey a random walk. The rejections of the model do not imply that the source of the rejection is the presence of a rational bubble.

The Cagan model has provided important insights into the dynamics of hyperinflation and the interactions of expectations with inflation in general. These insights are in no way diminished by our results. We have explored the behaviour of a precise linear articulation of the Cagan model; it is perhaps not surprising that the complete set of restrictions implied by the formulation are inconsistent with the data. However, it is necessary to have a precisely specified model and test both fundamental and general solutions in order to be able to interpret rejections as evidence of bubbles. Our rejection of the general specification of the Cagan model therefore calls into question the claims by other authors to have found support for the presence of rational bubbles.

One implication of our paper is that it would be useful to explore alternative representations of the Cagan model which do not yield the strict orthogonality conditions exploited above. Two examples explored in Hooker (1990) are hidden state variables (excluding some variables from agents' information sets $\Im_t$) and serially correlated measurement error.

8. TECHNICAL APPENDIX

Proof of Theorem 1 and Theorem 3

Defining $N_t = B_t + S_t$ as model noise, let $N_{t|t}(x)$ equal the projection of $P_t - P_t'$, which is $B_t + S_t - v_t$ in the exact case, onto $L_t(x)$. Notice that since v_t is orthogonal to $L_t(x)$, $N_{t|t}(x)$ is the projection of $B_t + S_t$ onto $L_t(x)$. This projection in turn allows us to define two additional time series, $v_{t|t}(x)$ and $P^f_{t|t}(x)$ through the identities

$$v_{t|t}(x) = N_{t|t}(x) - P_t + P_t' \tag{1}$$

and

$$P^f_{t|t}(x) = P_t - N_{t|t}(x). \tag{2}$$

We can rewrite these equations as

$$P_t - P_t' = N_{t|t}(x) - v_{t|t}(x) \tag{3}$$

$$P_t' = P_{t|t}^f(x) + v_{t|t}(x). \tag{4}$$

This is the pair of equations analysed in Durlauf and Hall (1989*a*,*c*). Theorem 1.3 in Durlauf and Hall (1989*c*) verifies that when the null is the exact Cagan model, there exists some information set $\Im_t$, i.e. $L_t(x)$, such that this system exists as a possible structural representation for the Cagan model plus misspecification and noise. This means that the second moments of the data are consistent with this structural representation. Hence if $N_{t|t}(x) = 0$, then the data are consistent with the null hypothesis that the fundamental price process is the correct specification. All testable restrictions of the null model are correspondingly embodied in testing whether the projection $N_{t|t}(x)$ is zero. One possible problem with this projection is that when model noise N_t is nonstationary, then the projection of $N_{t|t}(x)$ may not be well defined, in the sense that there does not exist any $N_{t|t}(x) \in L_t(x)$ such that $P_t - P_t' - N_{t|t}(x)$ is stationary. Durlauf and Hall (1989*c*) also show that if current and lagged P_t's and M_t's are used in the $P_t - P_t'$ projection, then the explosive part of noise will be recovered, i.e. there is an estimate $N_{t|t}(x)$ constructed using some combination of current and lagged P_t's and M_t's such that any residual in the projection of $P_t - P_t'$ is stationary. These arguments verify Theorem 1.

Now consider the difference between projecting $P_t - P_t'$ versus h_{t+1} onto $L_t(x)$. If $P_t - P_t'$ is stationary, which means that the bubble is zero and noise is stationary, then the projection of $P_t - P_t'$ can equal zero if and only if the projection of h_{t+1} is zero. Letting $M_t(x)$ denote the projection operator onto $L_t(x)$, this follows first from observing that under H_0^g, current and future values of h_t are all orthogonal to the current information set, since $h_t \perp L_{t-i}(x) \Rightarrow h_{t+i} \perp L_t(x)\ \forall\, i \geq 1$ and second, from observing that the projection of $P_t - P_t'$ onto $L_t(x)$ is equivalent to a weighted average of current and future h_t projections, as $(P_t - P_t')M_t(x) = \sum_{j=0}^{\infty}(\frac{\alpha}{\alpha-1})^j h_{t+j+1}M_t(x)$ Formally, this occurs because $L_\infty(h)$ and $L_\infty(P-P')$ are equivalent Hilbert spaces. Hence any stationary noise will be detected.

Under a nonstationary alternative, $P_t - P_t'$ will no longer necessarily be representable as an element of the space spanned by $h_{t+1}, h_{t+2}, \ldots$. Three relevant cases exist. In case 1, $N_t = (\frac{\alpha}{\alpha-1})N_{t-1} - \zeta_t$, where $\mathrm{E}(\zeta_t | \Im_{t-1}) = 0$. In this case, the h_{t+1} projection is zero. This means that the h_{t+1} projection will not detect a rational bubble. In case 2, $N_t = (\frac{\alpha}{\alpha-1})N_{t-1} - \zeta_t$, but $\mathrm{E}(\zeta_t | \Im_{t-1}) \neq 0$. In this case, h_{t+1} is not white noise and noise will be detected. In case 3, $N_t = \pi N_{t-1} - \zeta_t$, $\pi \neq \frac{\alpha-1}{\alpha}$, $\pi \geq 1$ and ζ_t is stationary. In this case $-(1 - \frac{\alpha}{\alpha-1}L^{-1})(P_t - P_t')$ is still nonstationary. Again, following Durlauf and Hall (1989*c*), if current and lagged P_t's and M_t's are used in the h_{t+1} projection, then the explosive part of the filtered noise will be recovered. However, this means that the h_{t+1} projection itself is nonzero. Therefore, the h_{t+1} projection will detect a deviation from the null model for any type of nonstationary noise except a rational bubble. Theorem 3 is therefore verified.

Proof of Theorem 2 and Theorem 4

Theorem 2 is an implication of Theorem 1 and can be verified by a straightforward generalisation of Theorem 1.3 in Durlauf and Hall (1989*c*). As before, all projections are well defined so long as the histories of P_t and M_t are included in $L_t(x)$.

To verify Theorem 4, observe that when prices obey H_0^g, Δh_{t+1} is an MA(1) process. The MA(1) coefficient is unrestricted as the model contains no implications for the variance/covariance of ζ_t, ξ_t, and u_t. By a straightforward generalisation of Theorem 1, Δh_{t+1} must be orthogonal to $L_{t-1}(x)$.

Under the misspecification alternative, Δh_{t+1} equals

$$\Delta h_{t+1} = \zeta_{t+1} - \zeta_t + \frac{1}{1-\alpha} u_t - (1-L)\left[1 - \left(\frac{\alpha}{\alpha-1}\right) L^{-1}\right] N_t. \qquad (5)$$

Following the argument given in the proof of Theorem 1, the only case where the projection of Δh_{t+1} onto $L_{t-1}(x)$ is zero while the projection of $(\Delta P_t - \Delta P_t')$ onto $L_{t-1}(x)$ is nonzero occurs when N_t is a rational bubble. This verifies Theorem 4.

REFERENCES

BURMEISTER, E. and K. D. WALL (1982), 'Kalman Filtering Estimation of Unobserved Rational Expectations with an Application to the German Hyperinflation', *Journal of Econometrics*, 20, pp. 255–84.

—— (1987), 'Unobserved Rational Expectations and the German Hyperinflation with Endogenous Money Supply', *International Economic Review*, 28, pp. 15–32.

CAGAN, P. (1956), 'The Monetary Dynamics of Hyperinflation' in Friedman, M. (ed.), *Studies in the Quantity Theory of Money*, University of Chicago Press, Chicago.

CASELLA, A. (1989), 'Testing for Price Level Bubbles with Exogenous or Endogenous Fundamentals: the German Hyperinflation Once More', *Journal of Monetary Economics*, 24, pp. 109–22.

DIBA, B. and H. GROSSMAN (1988), 'Explosive Bubbles in Stock Prices?', *American Economic Review*,78, pp. 520–30.

DURLAUF, S. N. and R. E. HALL (1989*a*), 'Bounds on the Variances of Specification Errors in Models with Expectations', Working Paper no. 2936, NBER.

—— (1989*b*), 'Measuring Noise in Stock Prices', Working Paper, Stanford University.

—— (1989*c*), 'A Signal Extraction Approach to Recovering Noise in Expectations Based Models', Working Paper, Stanford University.

FLOOD, R. P. and P. M. GARBER (1980), 'Market Fundamentals Versus Price-level Bubbles: the First Tests', *Journal of Political Economy*, 88, pp. 745–70.

FLOOD, R. P., P. M. GARBER and L. O. SCOTT (1984), 'Multi-country Tests for Price Level Bubbles', *Journal of Economic Dynamics and Control*, 8, 329–40.

FLOOD, R. P. and R. J. HODRICK (1986), 'Asset Volatility, Bubbles, and Process Switching', *Journal of Finance*, XLI, pp. 831–42.

FULLER, W. (1976), *Introduction to Statistical Time Series*, John Wiley, New York.

GOODFRIEND, M. S. (1982), 'An Alternative Method of Estimating the Cagan Money Demand Function in Hyperinflation Under Rational Expectations', *Journal of Monetary Economics*, 9, pp. 43–57.

HAMILTON, J. D. and C. H. WHITEMAN (1985), 'The Observable Implications of Self-fulfilling Expectations', *Journal of Monetary Economics*, 16, pp. 353–73.

HANSEN, L. P. and T. J. SARGENT (1991), 'Exact Linear Rational Expectations Models: Specification and Estimation' in Hansen, L. P. and T. J. Sargent (eds.), *Rational Expectations Econometrics*, Westview Press, Boulder Colorado.

HOOKER, M. A. (1990), 'Dynamics of Hyperinflation with Hidden State Variables and Measurement Error', Working Paper, Dartmouth College.

İMROHOROĞLU, S. (1989), 'Testing for Sunspot Equilibria in the German Hyperinflation', Working Paper, University of Southern California.

KHAN, M. S. (1975), 'The Monetary Dynamics of Hyperinflation: a Note', *Journal of Monetary Economics*, 1, pp. 355–62.

KIM, C. (1989), 'Measuring Noise in the Permanent Income Hypothesis by Noise-consumption Ratio', Working Paper, Rutgers University.

LAHAYE, L. (1985), 'Inflation and Currency Reform', *Journal of Political Economy*, 93, pp. 537–60.

LEROY, S. and R. PORTER (1981), 'The Present Value Relation: Tests Based on Implied Variance Bounds', *Econometrica*, LXIX, pp. 555–74.

PHILLIPS, P. C. B. and P. PERRON (1988), 'Testing for a Unit Root in Time Series Regression', *Biometrika*, 75, pp. 335–46.

SALEMI, M. and T. J. SARGENT (1979), 'The Demand for Money During Hyperinflations Under Rational Expectations: II', *International Economic Review*, 20, pp. 741–58.

SARGENT, T. J. (1977), 'The Demand for Money During Hyperinflations Under Rational Expectations', *International Economic Review*, 18, pp. 59–82.

SARGENT, T. J. and N. WALLACE (1973), 'Rational Expectations and the Dynamics of Hyperinflation', *International Economic Review*, 14, pp. 328–50.

SHILLER, R. J. (1981), 'Do Stock Prices Move Too Much to be Justified by Subsequent Changes in Dividends?', *American Economic Review*, LXXI, pp. 421–36.

STATISTISCHES REICHSAMT (1925), *Zahlen zur geldenwertung in Deutschland 1914 bis 1923,* Reimar Hobbing, Berlin.

WEBB, S. B. (1985), 'Government Debt and Inflationary Expectations as Determinants of the Money Supply in Germany: 1919–23', *Journal of Money, Credit and Banking*, 17, pp. 479–92.

10

Regime switching with time-varying transition probabilities

Francis X. Diebold, Joon-Haeng Lee and Gretchen C. Weinbach[*]

The Markov switching model is useful because of the potential it offers for capturing occasional but recurrent regime shifts in a simple dynamic econometric model. Existing treatments, however, restrict the transition probabilities to be constant over time; that is, the probability of switching from one regime to the other cannot depend on the behaviour of underlying economic fundamentals. In contrast, we propose a class of Markov switching models in which the transition probabilities can vary with fundamentals. We develop an EM algorithm for estimation of the model and we illustrate it with a simulation example. We conclude with a discussion of directions for future research, including application to exchange rate and business cycle modelling.

1. INTRODUCTION

Models incorporating nonlinearities associated with regime switching have a long tradition in empirical macroeconomics and dynamic econometrics.[1] Key methodological contributions include the early work of Quandt (1958) and Goldfeld and Quandt (1973), and the more recent work of Hamilton (1990). Recent substantive applications include Hamilton (1988) (interest rates), Hamilton

* Financial support from the U.K. Economic and Social Research Council under grant R000233447 is gratefully acknowledged by both authors. We are indebted to Neil Ericsson for helpful comments.

[1] For a survey of the nonlinear tradition in empirical macroeconomics, with particular attention paid to regime switching, see Diebold and Rudebusch (1993).

(1989) (aggregate output), Cecchetti, Lam and Mark (1990) and Abel (1992) (stock returns), and Engel and Hamilton (1990) (exchange rates), among many others.

Our attention here focuses on Hamilton's Markov switching model, which has become very popular. In Hamilton's model, time-series dynamics are governed by a finite-dimensional parameter vector, which switches (potentially each period) depending upon which of two states is realised, with state transitions governed by a first-order Markov process with constant transition probabilities.

Although the popularity of Hamilton's model is well deserved, it nevertheless incorporates a potentially severely binding constraint, the constancy of state transition probabilities. Economic considerations suggest the desirability of allowing the transition probabilities to vary. As an example, consider the process of exchange rate revaluation. It is plausible that the likelihood of exchange rate revaluation increases under progressively more severe over- or undervaluation on the basis of economic fundamentals, and certainly, one would not want to exclude that possibility from the outset.

We therefore propose in this paper a class of Markov switching models in which the transition probabilities are endogenous.[2] We discuss the model in Section 2, develop an EM algorithm for parameter estimation in Section 3, and illustrate the methodology with a simulation example in Section 4. We conclude with a discussion of directions for future research in Section 5.

2. THE MODEL

Let $\{s_t\}_{t=1}^{T}$ be the sample path of a first-order, two-state Markov process with transition probability matrix illustrated in Figure 1. As is apparent in the figure, the two transition probabilities are time-varying, evolving as logistic functions of $x'_{t-1}\beta_i, i=0,1$, where the $(k \times 1)$ conditioning vector x_{t-1} contains economic variables that affect the state transition probabilities. It will be convenient to stack the two sets of parameters governing the transition probabilities into a $(2k \times 1)$ vector, $\beta = (\beta'_0, \beta'_1)'$.

It is obvious, but worth noting, that when the last $(k-1)$ terms of the $(1 \times k)$ transition probability parameter vectors, β_0 and β_1, are set to zero, the transition probability functions are time-invariant so that p_t^{00} and p_t^{11} are simply constants; our model collapses to that of Hamilton (1990).

---

[2] The first work in this area is Lee (1991), from which this paper draws. Related subsequent literature includes Fillardo (1991), who considers Markov-switching business-cycle models with transition probabilities that change with movements in an index of leading indicators, as well as Ghysels (1992) and De Toldi, Gourieroux and Monfort (1992), who consider duration models with hazard rates that vary across seasons.

Figure 1. Transition Probability Matrix

		Time t	
		State 0	**State 1**
Time $t-1$	**State 0**	p_t^{00} $P(s_t = 0 \| s_{t-1} = 0, x_{t-1}; \beta_0)$ $\frac{\exp(x'_{t-1}\beta_0)}{1+\exp(x'_{t-1}\beta_0)}$	$p_t^{01} = (1 - p_t^{00})$ $P(s_t = 1 \| s_{t-1} = 0, x_{t-1}; \beta_0)$ $1 - \frac{\exp(x'_{t-1}\beta_0)}{1+\exp(x'_{t-1}\beta_0)}$
	State 1	$p_t^{10} = (1 - p_t^{11})$ $P(s_t = 0 \| s_{t-1} = 1, x_{t-1}; \beta_1)$ $1 - \frac{\exp(x'_{t-1}\beta_1)}{1+\exp(x'_{t-1}\beta_1)}$	p_t^{11} $P(s_t = 1 \| s_{t-1} = 1, x_{t-1}; \beta_1)$ $\frac{\exp(x'_{t-1}\beta_1)}{1+\exp(x'_{t-1}\beta_1)}$

Note to Figure: $x_{t-1} = (1, x_{1,t-1}, \ldots, x_{(k-1),t-1})'$ and $\beta_i = (\beta_{i0}, \beta_{i1}, \ldots, \beta_{i(k-1)})'$, $i = 0, 1$.

Let $\{y_t\}_{t=1}^T$ be the sample path of a time series that depends on $\{s_t\}_{t=1}^T$ as follows:

$$\left(y_t \middle| s_t = i; \quad \alpha_i\right) \overset{iid}{\sim} \mathrm{N}\left(\mu_i, \sigma_i^2\right), \tag{1}$$

where $\alpha_i = (\mu_i, \sigma_i^2)'$, $i = 0, 1$.[3] Thus, the density of y_t conditional upon s_t is:

$$f\left(y_t \middle| s_t = i; \alpha_i\right) = \frac{1}{\sqrt{2\pi}\sigma_i} \exp\left(\frac{-(y_t - \mu_i)^2}{2\sigma_i^2}\right), \tag{2}$$

$i = 0, 1$. It will be convenient to stack the two sets of parameters governing the densities into a (4×1) vector, $\alpha = (\alpha_0', \alpha_1')'$.

As we shall see, a quantity of particular interest in the likelihood function is $P(s_1)$, which denotes $P(S_1 = s_1)$. Regarding x_t, there are two cases to consider,

[3] Generalisations to allow for more than two states and/or intra-state dynamics are straightforward but tedious, so we shall not consider them here.

stationary and nonstationary. In the stationary case,

$$P(s_1) = P(s_1 | \underline{x}_T; \theta) = P(s_1; \beta).$$

That is, $P(s_1)$ is simply the long-run probability of $S_1 = s_1$, which in turn is determined by β. In the nonstationary case, the long-run probability does not exist, and so $P(S_1 = s_1)$ must be treated as an additional parameter to be estimated. It turns out, as we show subsequently, that $P(S_1 = 1)$ is all that is needed to construct the first likelihood term. We shall call this quantity 'ρ' in both the stationary and nonstationary cases, remembering that in the stationary case ρ is *not* an additional parameter to be estimated, but rather is determined by β, while the nonstationary case ρ *is* an additional parameter to be estimated.[4] In certain situations, computation of ρ in the stationary case may be done via simulation; see Diebold and Schuermann (1992).

Let $\theta = (\alpha', \beta', \rho)'$ be the $(2k + 5 \times 1)$ vector of all model parameters. The complete-data likelihood is then[5]

$$\begin{aligned} f(\underline{y}_T, \underline{s}_T | \underline{x}_T; \theta) &= f(y_1, s_1 | \underline{x}_T; \theta) \prod_{t=2}^{T} f(y_t, s_t | \underline{y}_{t-1}, \underline{s}_{t-1}, \underline{x}_T; \theta) \\ &= f(y_1 | s_1, \underline{x}_T; \theta) P(s_1) \prod_{t=2}^{T} f(y_t | s_t, \underline{y}_{t-1}, \underline{s}_{t-1}, \underline{x}_T; \theta) P(s_t | \underline{y}_{t-1}, \underline{s}_{t-1}, \underline{x}_t; \theta) \\ &= f(y_1 | s_1; \alpha) P(s_1) \prod_{t=2}^{T} f(y_t | s_t; \alpha) P(s_t | s_{t-1}, x_{t-1}; \beta), \end{aligned}$$

here f denotes any density and underlining denotes past history of the variable from $t = 1$ to the variable subscript.

It will prove convenient to write the complete-data likelihood in terms of indicator functions,

$$\begin{aligned} f(\underline{y}_T, \underline{s}_T | \underline{x}_T; \theta) = {} & [\mathrm{I}(s_1 = 1) f(y_1 | s_1 = 1; \alpha_1) \rho + \mathrm{I}(s_1 = 0) f(y_1 | s_1 = 0; \alpha_0)(1 - \rho)] \\ & \times \prod_{t=2}^{T} \{ \mathrm{I}(s_t = 1, s_{t-1} = 1) f(y_t | s_t = 1; \alpha_1) p_t^{11} \\ & + \mathrm{I}(s_t = 0, s_{t-1} = 1) f(y_t | s_t = 0; \alpha_0)(1 - p_t^{11}) \\ & + \mathrm{I}(s_t = 1, s_{t-1} = 0) f(y_t | s_t = 1; \alpha_1)(1 - p_t^{00}) \\ & + \mathrm{I}(s_t = 0, s_{t-1} = 0) f(y_t | s_t = 0; \alpha_0) p_t^{00} \}. \end{aligned}$$

Conversion to log form yields

---

[4] Alternatively, if prior information is available, ρ may be set accordingly. The issues are analogous to those that arise with initialisation of the Kalman filter in the nonstationary case.

[5] 'Complete-data' refers to the (hypothetical) assumption that both $\{y_t\}$ and $\{s_t\}$ are observed.

$$\log f(\underline{y}_T, \underline{s}_T | \underline{x}_T; \theta) = \mathrm{I}(s_1 = 1)\left[\log f(y_1 | s_1 = 1; \alpha_1) + \log \rho\right]$$
$$+ \mathrm{I}(s_1 = 0)\left[\log f(y_1 | s_1 = 0; \alpha_0) + \log(1-\rho)\right]$$
$$+ \sum_{t=2}^{T} \left\{ \mathrm{I}(s_t = 1) \log f(y_t | s_t = 1; \alpha_1) + \mathrm{I}(s_t = 0) \log f(y_t | s_t = 0; \alpha_0) \right.$$
$$+ \mathrm{I}(s_t = 1, s_{t-1} = 1) \log(p_t^{11}) + \mathrm{I}(s_t = 0, s_{t-1} = 1) \log(1 - p_t^{11})$$
$$\left. + \mathrm{I}(s_t = 1, s_{t-1} = 0) \log(1 - p_t^{00}) + \mathrm{I}(s_t = 0, s_{t-1} = 0) \log(p_t^{00}) \right\}.$$

The complete-data log likelihood cannot be constructed in practice, because the complete data are not observed. Conceptually, the fact that the states are unobserved is inconsequential, because the incomplete-data log likelihood may be obtained by summing over all possible state sequences,

$$\log f(\underline{y}_T | \underline{x}_T; \theta) = \log\left(\sum_{s_1=0}^{1} \sum_{s_2=0}^{1} \cdots \sum_{s_T=0}^{1} f(\underline{y}_T, \underline{s}_T | \underline{x}_T; \theta) \right),$$

and then maximised with respect to θ. In practice, however, construction and numerical maximisation of the incomplete-data log likelihood in this way is computationally intractable, as $\{s_t\}_{t=1}^{T}$ may be realised in 2^T ways. Therefore, following Hamilton's (1990) suggestion for the case of constant transition probabilities, we propose an EM algorithm for maximisation of the incomplete-data likelihood.

3. MODEL ESTIMATION: THE EM ALGORITHM

The EM algorithm is a stable and robust procedure for maximising the incomplete-data log likelihood via iterative maximisation of the expected complete-data log likelihood, conditional upon the observable data.[6] The procedure, shown schematically in Figure 2, amounts to the following:[7]

(1) Pick $\theta^{(0)}$.

(2) Get:

$$P(s_t = 1 | \underline{y}_T, \underline{x}_T; \theta^{(0)}) \quad \forall t,$$
$$P(s_t = 0 | \underline{y}_T, \underline{x}_T; \theta^{(0)}) \quad \forall t,$$
$$P(s_t = 1, s_{t-1} = 1 | \underline{y}_T, \underline{x}_T; \theta^{(0)}) \quad \forall t,$$
$$P(s_t = 0, s_{t-1} = 1 | \underline{y}_T, \underline{x}_T; \theta^{(0)}) \quad \forall t,$$

---

[6] Insightful discussions of the EM algorithm may be found in Dempster, Laird and Rubin (1977), Watson and Engle (1983), and Ruud (1991).

[7] Parameter superscripts count iterations.

$$P\left(s_t = 1, s_{t-1} = 0 \middle| \underline{y}_T, \underline{x}_T; \theta^{(0)}\right) \quad \forall t,$$
$$P\left(s_t = 0, s_{t-1} = 0 \middle| \underline{y}_T, \underline{x}_T; \theta^{(0)}\right) \quad \forall t;$$

construct $\mathrm{E} \log f(\underline{y}_T, \underline{s}_T | \underline{x}_T; \theta^{(0)})$ by replacing I's with P's.

(3) Set $\theta^{(1)} = \arg\max_{\theta} \mathrm{E}\left[\log f(\underline{y}_T, \underline{s}_T | \underline{x}_T; \theta^{(0)})\right]$.

(4) Iterate to convergence.

Step (1) simply assigns an initial guess to the parameter vector, $\theta^{(0)}$, in order to start the EM algorithm. Step (2) is the 'E' (expectation) part of the algorithm, which produces smoothed state probabilities conditional upon $\theta^{(0)}$, while step (3) is the 'M' (maximisation) part, which produces an updated parameter estimate, $\theta^{(1)}$, conditional upon the smoothed state probabilities obtained in step (2). The convergence criterion adopted in (4) may be based upon various standard criteria, such as the change in the log likelihood from one iteration to the next, the value of the gradient vector, or $\|\theta^{(j)} - \theta^{(j-1)}\|$, for various norms $\|\bullet\|$.

3.1. The expectation step

We wish to take expectations of the complete-data log likelihood, conditional upon the observed data. As in Hamilton (1990), this amounts to substitution of smoothed state probabilities (to be derived below) for indicator functions in the complete-data log likelihood,

$$\begin{aligned}
\mathrm{E}\left[\log f\left(\underline{y}_T, \underline{s}_T \middle| \underline{x}_T; \theta^{(j-1)}\right)\right] = {} & \rho^{(j-1)}\left[\log f\left(y_1 \middle| s_1 = 1; \alpha_1^{(j-1)}\right) + \log \rho^{(j-1)}\right] \\
& + \left(1 - \rho^{(j-1)}\right)\left[\log f\left(y_1 \middle| s_1 = 0; \alpha_0^{(j-1)}\right) + \log\left(1 - \rho^{(j-1)}\right)\right] \\
& + \sum_{t=2}^{T} \Big\{ P\left(s_t = 1 \middle| \underline{y}_T, \underline{x}_T; \theta^{(j-1)}\right) \log f\left(y_t \middle| s_t = 1; \alpha_1^{(j-1)}\right) \\
& + P\left(s_t = 0 \middle| \underline{y}_T, \underline{x}_T; \theta^{(j-1)}\right) \log f\left(y_t \middle| s_t = 0; \theta_0^{(j-1)}\right) \\
& + P\left(s_t = 1, s_{t-1} = 1 \middle| \underline{y}_T, \underline{x}_T; \theta^{(j-1)}\right) \log\left(p_t^{11}\right) \\
& + P\left(s_t = 0, s_{t-1} = 1 \middle| \underline{y}_T, \underline{x}_T; \theta^{(j-1)}\right) \log\left(1 - p_t^{11}\right) \\
& + P\left(s_t = 1, s_{t-1} = 0 \middle| \underline{y}_T, \underline{x}_T; \theta^{(j-1)}\right) \log\left(1 - p_t^{00}\right) \\
& + P\left(s_t = 0, s_{t-1} = 0 \middle| \underline{y}_T, \underline{x}_T; \theta^{(j-1)}\right) \log\left(p_t^{00}\right) \Big\},
\end{aligned} \tag{3}$$

where the smoothed state probabilities are obtained from the optimal nonlinear smoother, conditional upon the current 'best guess' of θ, $\theta^{(j-1)}$.

Given $\theta^{(j-1)}$, $\underline{y}_T$, and $\underline{x}_T$, the algorithm for calculating the smoothed state probabilities for iteration j is as follows:

Figure 2. The EM algorithm

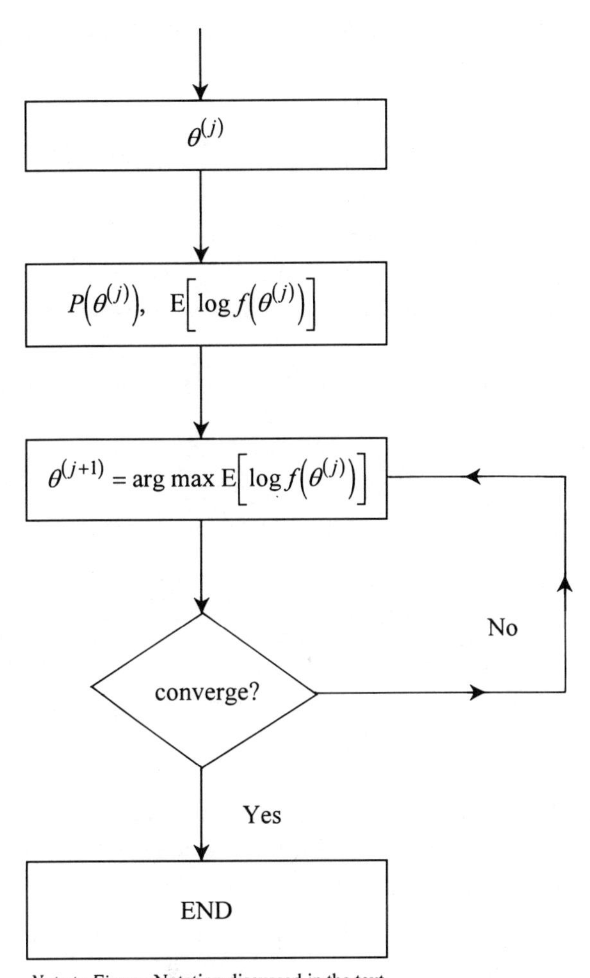

Note to Figure: Notation discussed in the text

1. Calculate the sequence of conditional densities of y_t given by (2.2) (a $(T \times 2)$ matrix), and transition probabilities given by Figure 1 (a $(T-1 \times 4)$ matrix).

2. Calculate filtered joint state probabilities (a $(T-1 \times 4)$ matrix) by iterating on steps 2a–2d below for $t = 2, ..., T$:

2a. Calculate the joint conditional distribution of (y_t, s_t, s_{t-1}) given $\underline{y}_{t-1}$ and $\underline{x}_{t-1}$ (four numbers): For $t = 2$, the joint conditional distribution is given by

$$f(y_2, s_2, s_1 | y_1, x_1; \theta^{(j-1)}) = f(y_2 | s_2; \alpha^{(j-1)}) P(s_2 | s_1, x_1; \beta^{(j-1)}) P(s_1).$$

For subsequent time t, the joint conditional distribution is

$$f(y_t, s_t, s_{t-1} | \underline{y}_{t-1}, \underline{x}_{t-1}; \theta^{(j-1)}) =$$

$$\sum_{s_{t-2}=0}^{1} f(y_t | s_t; \alpha^{(j-1)}) P(s_t | s_{t-1}, x_{t-1}; \beta^{(j-1)}) P(s_{t-1}, s_{t-2} | \underline{y}_{t-1}, \underline{x}_{t-1}; \theta^{(j-1)})$$

where the conditional density $f(y_t | s_t; \alpha^{(j-1)})$ and transition probabilities $P(s_t | s_{t-1}, x_{t-1}; \beta^{(j-1)})$ are given by step 1, and $P(s_{t-1}, s_{t-2} | \underline{y}_{t-1}, \underline{x}_{t-1}; \theta^{(j-1)})$ is the filtered probability resulting from execution of step 2 for the previous t value.

2b. Calculate the conditional likelihood of y_t (one number):

$$f(y_t | \underline{y}_{t-1}, \underline{x}_{t-1}; \theta^{(j-1)}) = \sum_{s_t=0}^{1} \sum_{s_{t-1}=0}^{1} f(y_t, s_t, s_{t-1} | \underline{y}_{t-1}, \underline{x}_{t-1}; \theta^{(j-1)}).$$

2c. Calculate the time-t filtered state probabilities (four numbers):

$$P(s_t, s_{t-1} | \underline{y}_t, \underline{x}_t; \theta^{(j-1)}) = \frac{f(y_t, s_t, s_{t-1} | \underline{y}_t, \underline{x}_t; \theta^{(j-1)})}{f(y_t | \underline{y}_t, \underline{x}_t; \theta^{(j-1)})},$$

where the numerator is the joint conditional distribution of (y_t, s_t, s_{t-1}) from step 2a and the denominator is the conditional likelihood of y_t from step 2b above.

2d. These four filtered probabilities are used as input for step 2a to calculate the filtered probabilities for the next time period, and steps 2a–2d are repeated $(T-2)$ times.

3. Calculate the smoothed joint state probabilities as follows (a $(T-1 \times 6)$ matrix):

3a. For $t = 2$ and a given valuation of (s_t, s_{t-1}), sequentially calculate the joint probability of $(s_\tau, s_{\tau-1}, s_t, s_{t-1})$ given $\underline{y}_\tau$ and $\underline{x}_\tau$, for $\tau = t+2, t+3, \ldots, T$:[8]

$$P(s_\tau, s_{\tau-1}, s_t, s_{t-1} | \underline{y}_\tau, \underline{x}_\tau; \theta^{(j-1)}) =$$

$$\frac{\sum_{s_{\tau-2}=0}^{1} f(y_\tau | s_\tau; \alpha^{(j-1)}) P(s_\tau | s_{\tau-1}, x_{\tau-1}; \beta^{(j-1)}) P(s_{\tau-1}, s_{\tau-2}, s_t, s_{t-1} | \underline{y}_{\tau-1}, \underline{x}_{\tau-1}; \theta^{(j-1)})}{f(y_\tau | \underline{y}_{\tau-1}, \underline{x}_{\tau-1}; \theta^{(j-1)})}$$

[8] There are of course four possible (s_t, s_{t-1}) sequences: (0,0), (0,1), (1,0) and (1,1).

where the first two terms in the numerator are given by step 1, the third by the previous step 3a computation, and the denominator by step 2b. When $\tau = t+2$, the third term in the numerator is initialised with the following expression:

$$P\left(s_{t+1}, s_t, s_{t-1} \middle| \underline{y}_{t+1}, \underline{x}_{t+1}; \theta^{(j-1)}\right) = \frac{f\left(y_{t+1} \middle| s_{t+1}; \alpha^{(j-1)}\right) P\left(s_{t+1} \middle| s_t, x_t; \beta^{(j-1)}\right) P\left(s_t\ s_{t-1} \middle| \underline{y}_t, \underline{x}_t; \theta^{(j-1)}\right)}{f\left(y_{t+1} \middle| \underline{y}_t, \underline{x}_t; \theta^{(j-1)}\right)}.$$

For each τ value we produce a (4×1) vector of probabilities corresponding to the four possible valuations of $(s_\tau, s_{\tau-1})$. Thus, upon reaching $\tau = T$, we have computed and saved a $(T-3) \times 4$ matrix, the last row of which is used in step 3b below.

3b. Upon reaching $\tau = T$, the smoothed joint state probability for time t and the chosen valuation of (s_t, s_{t-1}) is calculated as

$$P\left(s_t, s_{t-1} \middle| \underline{y}_T, \underline{x}_T; \theta^{(j-1)}\right) = \sum_{s_T=0}^{1} \sum_{s_{T-1}=0}^{1} P\left(s_T, s_{T-1}, s_t, s_{t-1} \middle| \underline{y}_T, \underline{x}_T; \theta^{(j-1)}\right).$$

3c. Steps 3a and 3b are repeated for all possible time t valuations (s_t, s_{t-1}), until a smoothed probability has been calculated for each of the four possible valuations. At this point we have a (1×4) vector of smoothed joint state probabilities for (s_t, s_{t-1}).

3d. Steps 3a–3c are repeated for t = 3, 4,...,T, yielding a total of $(T-1 \times 4)$ smoothed joint state probabilities.

4. Smoothed marginal state probabilities are found by summing over the smoothed joint state probabilities. For example,

$$\begin{aligned} P\left(s_t = 1 \middle| \underline{y}_T, \underline{x}_T; \theta^{(j-1)}\right) = & \; P\left(s_t = 1, s_{t-1} = 1 \middle| \underline{y}_T, \underline{x}_T; \theta^{(j-1)}\right) \\ & + P\left(s_t = 1, s_{t-1} = 0 \middle| \underline{y}_T, \underline{x}_T; \theta^{(j-1)}\right). \end{aligned}$$

These $(T-1 \times 6)$ smoothed state probabilities are used as input for the maximisation step, which we now describe.

3.2. The maximisation step

Given the smoothed state probabilities, the expected complete-data log likelihood, given by (3), is maximised directly with respect to the model parameters. The resulting $2k + 5$ first-order conditions are linear both in ρ and the conditional density parameter vector α, and nonlinear in the transition probability parameter vector β. Moreover, these two sets of parameters appear in distinctly different terms in the likelihood function. Due to this separability, five of the first-order conditions are linear in the parameters:

$$\sum_{t=1}^{T} P\left(s_t = i \middle| \underline{y}_T, \underline{x}_T; \theta^{(j-1)}\right)\left(y_t - \mu_i^{(j)}\right) = 0$$

$$\sum_{t=1}^{T} P\left(s_t = i \middle| \underline{y}_T, \underline{x}_T; \theta^{(j-1)}\right)\left(\frac{(y_t - \mu_i^{(j)})^2}{(\sigma_i^2)^{(j)}} - 1\right) = 0$$

$$P\left(s_1 = 1 \middle| \underline{y}_T, \underline{x}_T; \theta^{(j-1)}\right)\left(\frac{1}{\rho}\right) - 1 = 0,$$

and yield immediate closed-form expressions for the maximum likelihood estimators.

$$\mu_i^{(j)} = \frac{\sum_{t=1}^{T} y_t P\left(s_t = i \middle| \underline{y}_T, \underline{x}_T; \theta^{(j-1)}\right)}{\sum_{t=1}^{T} P\left(s_t = i \middle| \underline{y}_T, \underline{x}_T; \theta^{(j-1)}\right)}$$

$$\left(\sigma_i^2\right)^{(j)} = \frac{\sum_{t=1}^{T} \left(y_t - \mu_i^{(j)}\right)^2 P\left(s_t = i \middle| \underline{y}_T, \underline{x}_T; \theta^{(j-1)}\right)}{\sum_{t=1}^{T} P\left(s_t = i \middle| \underline{y}_T, \underline{x}_T; \theta^{(j-1)}\right)}$$

$$\rho^{(j)} = P\left(s_1 = 1 \middle| \underline{y}_T, \underline{x}_T; \theta^{(j-1)}\right),$$

$i = 0,1$.

However, given our use of logit transition probability functions, the remaining $2k$ first-order conditions are *nonlinear* in β, and are given by[9]

$$\sum_{t=2}^{T} x_{t-1}\left\{P\left(s_t = 0, s_{t-1} = 0 \middle| \underline{y}_T, \underline{x}_T; \theta^{(j-1)}\right) - p_t^{00} P\left(s_{t-1} = 0 \middle| \underline{y}_T, \underline{x}_T; \theta^{(j-1)}\right)\right\} = 0$$

$$\sum_{t=2}^{T} x_{t-1}\left\{P\left(s_t = 1, s_{t-1} = 1 \middle| \underline{y}_T, \underline{x}_T; \theta^{(j-1)}\right) - p_t^{11} P\left(s_{t-1} = 1 \middle| \underline{y}_T, \underline{x}_T; \theta^{(j-1)}\right)\right\} = 0.$$

Closed-form solutions are found by linearly approximating p_t^{00} and p_t^{11} using a first-order Taylor series expansion around $\beta_0^{(j-1)}$ and $\beta_1^{(j-1)}$, respectively. These linear approximations are given by

$$p_t^{00}\left(\beta_0^{(j-1)}\right) \approx p_t^{00}\left(\beta_0^{(j-1)}\right) + \left.\frac{\partial p_t^{00}(\beta_0)}{\partial \beta_0}\right|_{\beta_0 = \beta_0^{(j-1)}} \left(\beta_0 - \beta_0^{(j-1)}\right)$$

---

[9] A variety of alternative functional forms, in addition to the logit, are examined in Lee (1991).

$$p_t^{11}\left(\beta_1^{(j-1)}\right) \approx p_t^{11}\left(\beta_1^{(j-1)}\right) + \left.\frac{\partial p_t^{11}(\beta_1)}{\partial \beta_1}\right|_{\beta_1=\beta_1^{(j-1)}} \left(\beta_1 - \beta_1^{(j-1)}\right).$$

For simplicity, we adopt the following notation:

$$p_{it}^{00}\left(\beta_0^{(j-1)}\right) = \left.\frac{\partial p_t^{00}(\beta_0)}{\partial \beta_{i0}}\right|_{\beta_0=\beta_0^{(j-1)}}, \quad i = 0,\ldots,k-1$$

$$p_{it}^{11}\left(\beta_1^{(j-1)}\right) = \left.\frac{\partial p_t^{11}(\beta_1)}{\partial \beta_{i1}}\right|_{\beta_1=\beta_1^{(j-1)}}, \quad i = 0,\ldots,k-1,$$

so that the vectors of partials are $(1 \times k)$ row vectors given by

$$\left.\frac{\partial p_t^{00}(\beta_0)}{\partial \beta_0}\right|_{\beta_0=\beta_0^{(j-1)}} = \left(p_{0t}^{00}\left(\beta_0^{(j-1)}\right), p_{1t}^{00}\left(\beta_0^{(j-1)}\right),\ldots,p_{(k-1)t}^{00}\left(\beta_0^{(j-1)}\right)\right)$$

$$\left.\frac{\partial p_t^{11}(\beta_1)}{\partial \beta_1}\right|_{\beta_1=\beta_1^{(j-1)}} = \left(p_{0t}^{11}\left(\beta_1^{(j-1)}\right), p_{1t}^{11}\left(\beta_1^{(j-1)}\right),\ldots,p_{(k-1)t}^{11}\left(\beta_1^{(j-1)}\right)\right),$$

and the individual partials are given by

$$p_{it}^{00}\left(\beta_0^{(j-1)}\right) = x_{i,t-1}\left(p_t^{00}\left(\beta_0^{(j-1)}\right) - p_t^{00}\left(\beta_0^{(j-1)}\right)^2\right), \quad i = 0, \ldots, k-1$$

$$p_{it}^{11}\left(\beta_1^{(j-1)}\right) = x_{i,t-1}\left(p_t^{11}\left(\beta_1^{(j-1)}\right) - p_t^{11}\left(\beta_1^{(j-1)}\right)^2\right), \quad i = 0, \ldots, k-1,$$

where $x_{i,t-1}$ is the ith element of x_{t-1}.

Substituting these linear approximations for the transition probabilities into the $2k$ nonlinear first-order conditions results in $2k$ *linear* first-order conditions given by

$$\sum_{t=2}^{T} x_{t-1}\Bigg\{P\left(s_t = 0, s_{t-1} = 0 \middle| \underline{y}_T, \underline{x}_T;\theta^{(j-1)}\right)$$
$$-P\left(s_{t-1} = 0 \middle| \underline{y}_T, \underline{x}_T;\theta^{(j-1)}\right)\left[p_t^{00}\left(\beta_0^{(j-1)}\right) + \frac{\partial p_t^{00}(\beta_0)}{\partial \beta_0}\left(\beta_0 - \beta_0^{(j-1)}\right)\right]\Bigg\} = 0$$

$$\sum_{t=2}^{T} x_{t-1}\Bigg\{P\left(s_t = 1, s_{t-1} = 1 \middle| \underline{y}_T, \underline{x}_T;\theta^{(j-1)}\right)$$
$$-P\left(s_{t-1} = 1 \middle| \underline{y}_T, \underline{x}_T;\theta^{(j-1)}\right)\left[p_t^{11}\left(\beta_1^{(j-1)}\right) + \frac{\partial p_t^{11}(\beta_1)}{\partial \beta_1}\left(\beta_1 - \beta_1^{(j-1)}\right)\right]\Bigg\} = 0,$$

where all derivatives are understood to be evaluated at $\beta_0^{(j-1)}$ or $\beta_1^{(j-1)}$, as relevant.

Solving these, we obtain a closed-form solution for $\beta_0^{(j)}$,

$$\beta_0^{(j)} = \left(\sum_{t=2}^{T} x_{t-1} P\left(s_{t-1} = 0 \middle| \underline{y}_T, \underline{x}_T; \theta^{(j-1)}\right) \frac{\partial p_t^{00}(\beta_0)}{\partial \beta_0} \right)^{-1}$$

$$\times \left(\sum_{t=2}^{T} x_{t-1} \left\{ \begin{array}{l} P\left(s_t = 0, s_{t-1} = 0 \middle| \underline{y}_T, \underline{x}_T; \theta^{(j-1)}\right) - \\ \quad P\left(s_{t-1} = 0 \middle| \underline{y}_T, \underline{x}_T; \theta^{(j-1)}\right) \left[p_t^{00}\left(\beta_0^{(j-1)}\right) - \frac{\partial p_t^{00}(\beta_0)}{\partial \beta_0} \beta_0^{(j-1)} \right] \end{array} \right\} \right).$$

Similarly the closed-form solution for $\beta_1^{(j)}$ is

$$\beta_1^{(j)} = \left(\sum_{t=2}^{T} x_{t-1} P\left(s_{t-1} = 1 \middle| \underline{y}_T, \underline{x}_T; \theta^{(j-1)}\right) \frac{\partial p_t^{11}(\beta_1)}{\partial \beta_1} \right)^{-1}$$

$$\times \left(\sum_{t=2}^{T} x_{t-1} \left\{ \begin{array}{l} P\left(s_t = 1, s_{t-1} = 1 \middle| \underline{y}_T, \underline{x}_T; \theta^{(j-1)}\right) - \\ \quad P\left(s_{t-1} = 1 \middle| \underline{y}_T, \underline{x}_T; \theta^{(j-1)}\right) \left[p_t^{11}\left(\beta_1^{(j-1)}\right) - \frac{\partial p_t^{11}(\beta_1)}{\partial \beta_1} \beta_1^{(j-1)} \right] \end{array} \right\} \right).$$

The cases of $k = 2$ and $k = 3$ are of particular interest in applied work. For this reason, we catalog explicit expressions for $\beta_0^{(j)}$ and $\beta_1^{(j)}$ in those cases in the appendix.

4. SIMULATION RESULTS

In order to demonstrate the methodology, we present the results of a simulation exercise. Sample size is 100. We set $k = 2$ so that the time-varying transition probabilities are driven by one x series. The transition probability parameters, α, are chosen and the x series constructed so that the (true) probabilities of staying in state, p_t^{00} and p_t^{11}, each alternate between 0.40 and 0.90 over successive sets of twenty sample observations, beginning with $p_t^{00} = 0.40$ and $p_t^{11} = 0.90$. The chosen parameter values are shown in Table 1, the simulated y and s sequences are shown in Figure 3, the x sequence is shown in Figure 4, and the resultant probabilities of staying in state are shown in Figure 7 (labeled 'actual').

Parameter estimation using the EM algorithm begins at the true parameter values. Convergence of the EM algorithm is checked as follows. Upon the calculation of each new parameter vector, say the jth, a comparison is make with the previous vector, the $(j-1)^{st}$. If the absolute value of the maximal difference in like elements of $\theta^{(j)}$ and $\theta^{(j-1)}$ is less then $10e^{-8}$, iteration is terminated. Convergence was obtained in 462 iterations, and the location of the likelihood maximum obtained was robust to a variety of alternative start up parameters. As

Table 1. Estimation Results

Parameter	θ_0	$\hat{\theta}$	$\hat{\theta}_H$
μ_0	-1.00	-1.62	-0.38
σ_0^2	4.00	2.83	3.73
μ_1	1.00	1.27	2.25
σ_1^2	4.00	2.97	2.24
β_{00}	0.79	-1.53	1.72
β_{01}	-2.00	-2.89	NA
β_{10}	1.00	0.94	0.49
β_{11}	2.00	2.07	NA
ρ	1.00	0.97	0.29
MSE		0.13	0.27
Iterations		462	208
lnL		-238.17	-247.53

Notes: θ_0 is the true parameter vector; $\hat{\theta}$ is the estimated parameter vector, and $\hat{\theta}_H$ is the estimated parameter vector for the Hamilton model, obtained by constraining β_{01} and β_{11} to be zero.

Figure 3. Y and S sequences

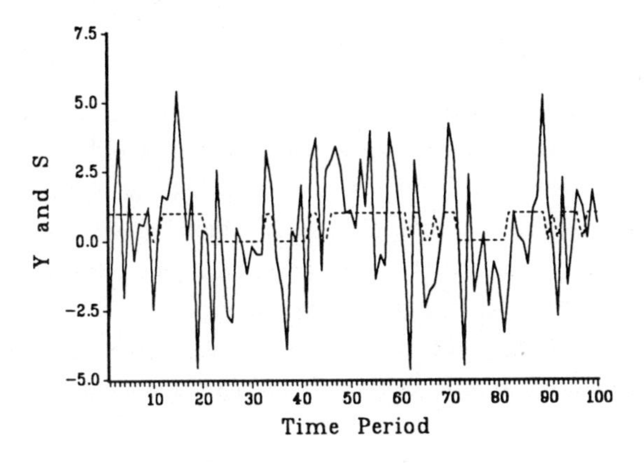

Figure 4. X sequence

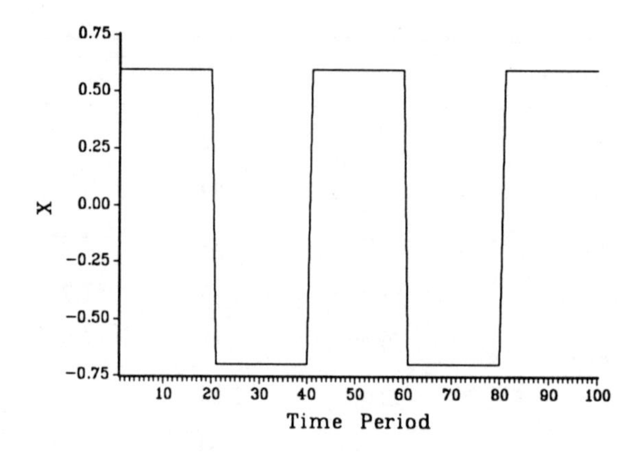

Figure 5. Sequence of log-likelihood values

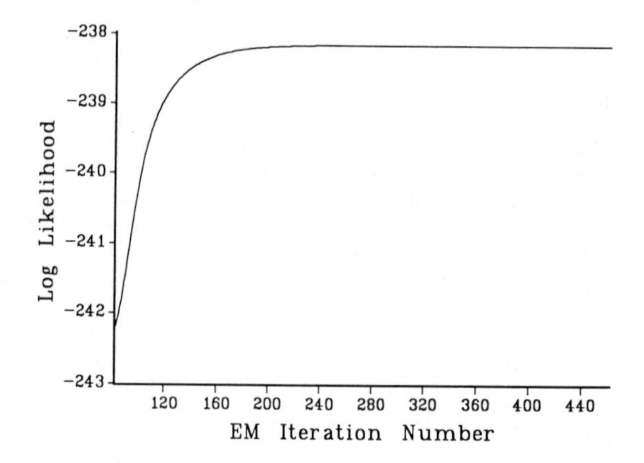

shown in Figure 5, the EM algorithm gets close to the likelihood maximum very quickly, but then takes more iterations to reach convergence.[10]

The resultant maximum likelihood parameter estimates are given in Table 1, labelled $\hat{\theta}$. Given the small sample size, the likelihood maximum is fairly close to the true parameter vector, the main exception being the estimate of β_{00}, which diverges from the true value by a rather large amount.

In Figure 6 we graph the time series of true states and smoothed state probabilities produced by the EM algorithm. The smoothed state probabilities,

[10] This behaviour is noted frequently in the literature.

Figure 6. Actual and smoothed state sequences

(a) Time-varying probabilities, MSE = 0.13

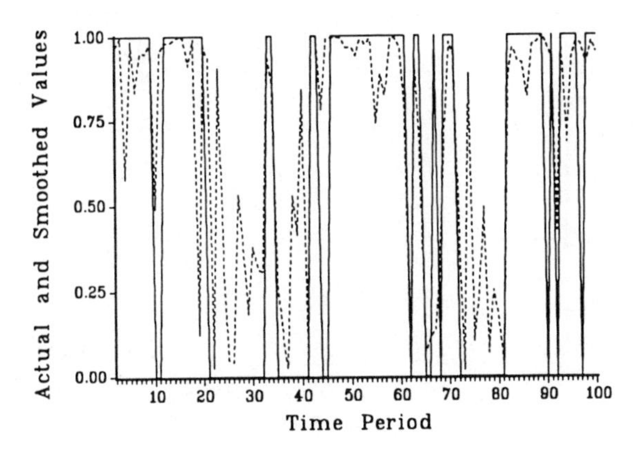

(b) Constant probabilities, MSE = 0.27

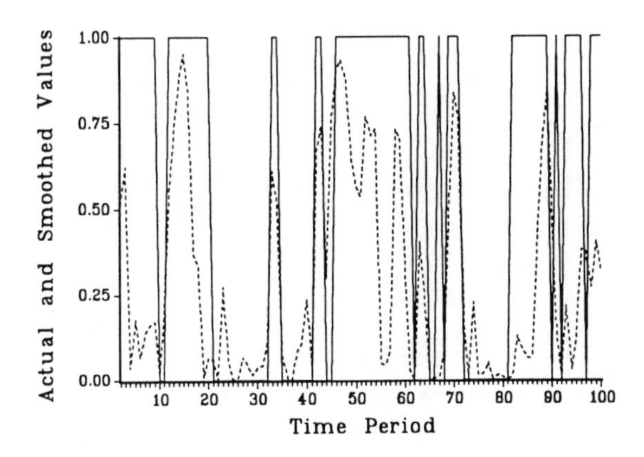

which are the EM algorithm's best guess at the state each period based on our time-varying transition probability model (Figure 6(a)), track the true states quite well. The mean-squared state extraction error using our model is 0.13.[11]

Next, we use the EM algorithm to fit a Hamilton model, which does not allow for time-varying transition probabilities, to the same dataset. The Hamilton model parameter estimates are given in Table 1, labelled $\hat{\theta}_H$. The likelihood ratio test statistic clearly rejects the null of constant transition probabilities. The time series

[11] The mean squared state extraction error using the *true* parameter values, which may be viewed as a lower bound, is 0.11.

Figure 7. Actual and fitted values

(a) p^{00}

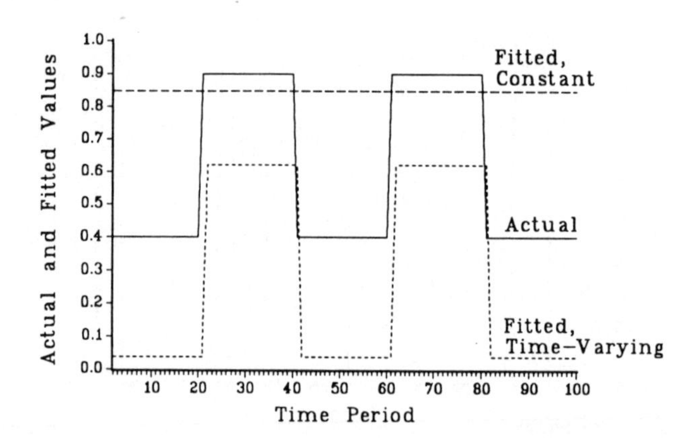

(b) p^{11}

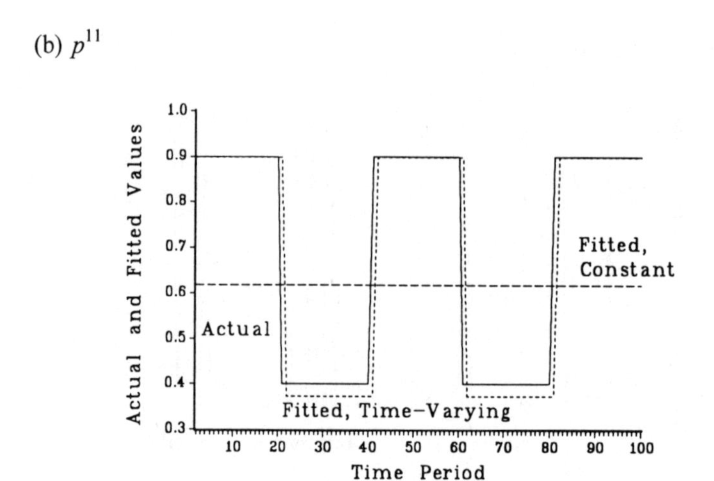

of true states and smoothed state probabilities that result from the fitted Hamilton model appear in Figure 6(b). The mean-squared state extraction error is 0.27, more than twice as large as that resulting from our time-varying transition probability model.

The fitted values of the transition probabilities for our model and the Hamilton model appear in Figures 7(a) and (b), along with the true probabilities. Our estimates do a reasonable job of tracking the time-varying probabilities, whereas, needless to say, the Hamilton estimates do not. As intuition suggests, the Hamilton estimates lie between the actual 'high' and 'low' values. Their restriction to constancy is responsible for the higher mean-squared state extraction error associated with the Hamilton model.

5. CONCLUDING REMARKS

This paper has been largely methodological, and numerous additional methodological issues are currently under investigation, including formal asymptotic distribution theory, elimination of the linear approximation employed in solving the first-order conditions, model specification tests, and analytic determination of ergodic probabilities. We shall not dwell on those issues here; instead, we shall briefly discuss two potentially fruitful areas of application.

The first concerns exchange rate dynamics. Engel and Hamilton (1990) have suggested that exchange rates may follow a switching process. We agree. But certainly, it is highly restrictive to require constancy of the transition probabilities. Rather, they should be allowed to vary with fundamentals, such as relative money supplies, relative real outputs, interest rate differentials, and so forth. Moreover, Mark (1992) produces useful indexes of fundamentals, which may be exploited to maintain parsimony. We shall provide a detailed report on this approach in a future paper.

The second concerns aggregate output dynamics. Diebold, Rudebusch and Sichel (1993) have found strong duration dependence in postwar US contractions. That is, the longer a contraction persists, the more likely it is to end. That suggests allowing the transition probabilities in a Markov switching model of aggregate output dynamics to depend on length-to-date of the current regime, which can readily be achieved by expanding the state space of the process.[12]

APPENDIX

The general form of the maximum expected complete-data likelihood estimators for the $2k$ transition probability function parameters, $\beta_0^{(j)}$ and $\beta_1^{(j)}$, is given in Section 3.2. Here we include the explicit expressions for the cases of $k = 2$ and $k = 3$, which are of particular interest in applied work. Due to space limitations, it is understood that in the expressions that follow all smoothed probabilities are conditional on $\underline{y}_T$ and $\underline{x}_T$ given $\theta^{(j-1)}$, and that transition probabilities p_t^{00}, p_t^{11} and their derivatives are evaluated at $\beta_0^{(j-1)}$ and $\beta_1^{(j-1)}$, respectively.

[12] We thank Atsushi Kajii and Jim Hamilton for pointing this out.

<u>$k = 2$</u>

$$\beta_0^{(j)} = \begin{pmatrix} \beta_{00}^{(j)} \\ \beta_{01}^{(j)} \end{pmatrix} = \begin{pmatrix} \sum_{t=2}^{T} x_{0,t-1} P(s_{t-1}=0) p_{0t}^{00} & \sum_{t=2}^{T} x_{0,t-1} P(s_{t-1}=0) p_{1t}^{00} \\ \sum_{t=2}^{T} x_{1,t-1} P(s_{t-1}=0) p_{0t}^{00} & \sum_{t=2}^{T} x_{1,t-1} P(s_{t-1}=0) p_{1t}^{00} \end{pmatrix}^{-1}$$

$$\times \begin{pmatrix} \sum_{t=2}^{T} x_{0,t-1} \left\{ P(s_t=0, s_{t-1}=0) - P(s_{t-1}=0) \left[p_t^{00} - \frac{\partial p_t^{00}}{\partial \beta_0} \beta_0^{(j-1)} \right] \right\} \\ \sum_{t=2}^{T} x_{1,t-1} \left\{ P(s_t=0, s_{t-1}=0) - P(s_{t-1}=0) \left[p_t^{00} - \frac{\partial p_t^{00}}{\partial \beta_0} \beta_0^{(j-1)} \right] \right\} \end{pmatrix}$$

$$\beta_1^{(j)} = \begin{pmatrix} \beta_{10}^{(j)} \\ \beta_{11}^{(j)} \end{pmatrix} = \begin{pmatrix} \sum_{t=2}^{T} x_{0,t-1} P(s_{t-1}=1) p_{0t}^{11} & \sum_{t=2}^{T} x_{0,t-1} P(s_{t-1}=1) p_{1t}^{11} \\ \sum_{t=2}^{T} x_{1,t-1} P(s_{t-1}=1) p_{0t}^{11} & \sum_{t=2}^{T} x_{1,t-1} P(s_{t-1}=1) p_{1t}^{11} \end{pmatrix}^{-1}$$

$$\times \begin{pmatrix} \sum_{t=2}^{T} x_{0,t-1} \left\{ P(s_t=1, s_{t-1}=1) - P(s_{t-1}=1) \left[p_t^{11} - \frac{\partial p_t^{11}}{\partial \beta_1} \beta_1^{(j-1)} \right] \right\} \\ \sum_{t=2}^{T} x_{1,t-1} \left\{ P(s_t=1, s_{t-1}=1) - P(s_{t-1}=1) \left[p_t^{11} - \frac{\partial p_t^{11}}{\partial \beta_1} \beta_1^{(j-1)} \right] \right\} \end{pmatrix}$$

<u>$k = 3$</u>

$$\beta_0^{(j)} = \begin{pmatrix} \beta_{00}^{(j)} \\ \beta_{01}^{(j)} \\ \beta_{02}^{(j)} \end{pmatrix} =$$

$$\begin{pmatrix} \sum_{t=2}^{T} x_{0,t-1} P(s_{t-1}=0) p_{0t}^{00} & \sum_{t=2}^{T} x_{0,t-1} P(s_{t-1}=0) p_{1t}^{00} & \sum_{t=2}^{T} x_{0,t-1} P(s_{t-1}=0) p_{2t}^{00} \\ \sum_{t=2}^{T} x_{1,t-1} P(s_{t-1}=0) p_{0t}^{00} & \sum_{t=2}^{T} x_{1,t-1} P(s_{t-1}=0) p_{1t}^{00} & \sum_{t=2}^{T} x_{1,t-1} P(s_{t-1}=0) p_{2t}^{00} \\ \sum_{t=2}^{T} x_{2,t-1} P(s_{t-1}=0) p_{0t}^{00} & \sum_{t=2}^{T} x_{2,t-1} P(s_{t-1}=0) p_{1t}^{00} & \sum_{t=2}^{T} x_{2,t-1} P(s_{t-1}=0) p_{2t}^{00} \end{pmatrix}^{-1}$$

$$\times \begin{pmatrix} \sum_{t=2}^{T} x_{0,t-1} \left\{ P(s_t=0, s_{t-1}=0) - P(s_{t-1}=0) \left[p_t^{00} - \frac{\partial p_t^{00}}{\partial \beta_0} \beta_0^{(j-1)} \right] \right\} \\ \sum_{t=2}^{T} x_{1,t-1} \left\{ P(s_t=0, s_{t-1}=0) - P(s_{t-1}=0) \left[p_t^{00} - \frac{\partial p_t^{00}}{\partial \beta_0} \beta_0^{(j-1)} \right] \right\} \\ \sum_{t=2}^{T} x_{2,t-1} \left\{ P(s_t=0, s_{t-1}=0) - P(s_{t-1}=0) \left[p_t^{00} - \frac{\partial p_t^{00}}{\partial \beta_0} \beta_0^{(j-1)} \right] \right\} \end{pmatrix}$$

$$\beta_1^{(j)} = \begin{pmatrix} \beta_{10}^{(j)} \\ \beta_{11}^{(j)} \\ \beta_{12}^{(j)} \end{pmatrix} =$$

$$\begin{pmatrix} \sum_{t=2}^{T} x_{0,t-1} P(s_{t-1}=1) p_{0t}^{00} & \sum_{t=2}^{T} x_{0,t-1} P(s_{t-1}=1) p_{1t}^{00} & \sum_{t=2}^{T} x_{0,t-1} P(s_{t-1}=1) p_{2t}^{00} \\ \sum_{t=2}^{T} x_{1,t-1} P(s_{t-1}=1) p_{0t}^{00} & \sum_{t=2}^{T} x_{1,t-1} P(s_{t-1}=1) p_{1t}^{00} & \sum_{t=2}^{T} x_{1,t-1} P(s_{t-1}=1) p_{2t}^{00} \\ \sum_{t=2}^{T} x_{2,t-1} P(s_{t-1}=1) p_{0t}^{00} & \sum_{t=2}^{T} x_{2,t-1} P(s_{t-1}=1) p_{1t}^{00} & \sum_{t=2}^{T} x_{2,t-1} P(s_{t-1}=1) p_{2t}^{00} \end{pmatrix}^{-1}$$

$$\times \begin{pmatrix} \sum_{t=2}^{T} x_{0,t-1} \left\{ P(s_t=1, s_{t-1}=1) - P(s_{t-1}=1) \left[p_t^{11} - \frac{\partial p_t^{11}}{\partial \beta_1} \beta_1^{(j-1)} \right] \right\} \\ \sum_{t=2}^{T} x_{1,t-1} \left\{ P(s_t=1, s_{t-1}=1) - P(s_{t-1}=1) \left[p_t^{11} - \frac{\partial p_t^{11}}{\partial \beta_1} \beta_1^{(j-1)} \right] \right\} \\ \sum_{t=2}^{T} x_{2,t-1} \left\{ P(s_t=1, s_{t-1}=1) - P(s_{t-1}=1) \left[p_t^{11} - \frac{\partial p_t^{11}}{\partial \beta_1} \beta_1^{(j-1)} \right] \right\} \end{pmatrix}$$

REFERENCES

CECCHETTI, S. G., P. LAM and N. C. MARK (1990), 'Mean Reversion in Equilibrium Asset Prices', *American Economic Review*, 80, 3, pp. 398–418.

DEMPSTER, A. P., N. M. LAIRD and D. B. RUBIN (1977), 'Maximum Likelihood from Incomplete Data via the EM Algorithm', *Journal of the Royal Statistical Society*, 39, pp. 1–38.

DE TOLDI, M., C. GOURIEROUX and A. MONFORT (1992), 'On Seasonal Effects in Duration Models', Working Paper #9216, INSEE, Paris.

DIEBOLD, F. X. and G. D. RUDEBUSCH (1993), 'Measuring Business Cycles: A Modern Perspective', Manuscript, Dept. of Economics, University of Pennsylvania.

DIEBOLD, F. X., G. D. RUDEBUSCH and D. SICHEL (1993), 'Further Evidence on Business Cycle Duration Dependence', in J. H. Stock and M. W. Watson (eds.), *New Research on Business Cycles, Indicators and Forecasting*, University of Chicago Press for NBER, Chicago, pp. 255–84.

DIEBOLD, F. X. and T. SCHUERMANN (1992), 'Exact Maximum Likelihood Estimation of ARCH Models', Manuscript, Dept. of Economics, University of Pennsylvania.

ENGEL, C. and J. D. HAMILTON (1990), 'Long Swings in the Dollar: Are They in the Data and do Markets Know it?', *American Economic Review*, 80, 4, pp. 689–713.

FILLARDO, A. J. (1991), 'Business Cycle Phases and Their Transitions', Manuscript, Dept. of Economics, University of Chicago.

GHYSELS, E. (1992), 'A Time Series Models of Growth Cycles and Seasonals with Stochastic Regime Switches', Manuscript, CRDE, Dept. of Economics, University of Montreal.

GOLDFELD, S. M. and R. E. QUANDT (1973), 'A Markov Model for Switching Regressions', *Journal of Econometrics*, 1, pp. 3–16.

HAMILTON, J. D. (1988), 'Rational-Expectations Econometric Analysis of Changes in Regime: An Investigation of the Term Structure of Interest Rates', *Journal of Economic Dynamics and Control*, 12, pp. 385–423.

—— (1989), 'A New Approach to the Economic Analysis of Nonstationary Time Series and the Business Cycle', *Econometrica*, 57, pp. 357–84.

—— (1990), 'Analysis of Time Series Subject to Changes in Regime', *Journal of Econometrics*, 45, pp. 39–70.

LEE, J.-H. (1991), 'Nonstationary Markov Switching Models of Exchange Rates: The Pound-Dollar Exchange Rate', PhD Dissertation, University of Pennsylvania.

MARK, N. C. (1992), 'Exchange Rates and Fundamentals: Evidence on Long-Horizon Predictability and Overshooting', Manuscript, Dept. of Economics, Ohio State University.

QUANDT, R. E. (1958), 'The Estimation of Parameters of Linear Regression System Obeying Two Separate Regimes', *Journal of the American Statistical Association*, 55, pp. 873–80.

RUUD, P. A. (1991), 'Extension of Estimation Methods Using the EM Algorithm', *Journal of Econometrics*, 49, pp. 305–41.

WATSON, M. W. and R. F. ENGLE (1983), 'Alternative Algorithms for the Estimation of Dynamic Factor, Mimic and Varying Coefficient Models', *Journal of Econometrics*, 15, pp. 385–400.

Name Index

Abel, 283
Ahn, S. K. 88
Anderson, T. W. 90
Andrews, D. W. K. 7, 97
Andrews, M. K. 52
Artis, M. 12

Baba, Y., 194
Baillie, R. T. 18, 22, 48
Banerjee, A. 30, 107, 108, 184, 239
Bardsen, G. 88
Barsky, R. B. 172, 176
Bartelsman, E. J. 179n, 187
Baumol, W. J. 193
Beaulieu, J. J. 161, 165, 166, 172, 179n, 180n
Bell, W. R. 180n, 182
Bentley, B. 255
Bera, A. K. 206n
Bewley, R. A. 88, 90, 92, 105, 106, 110
Birchenhall, C. R. 166, 180n
Bishop, S. L. 214
Bjørnstad, J. F. 38
Bladen-Hovell, R. C. 220
Bock, M. E. 19
Bossaerts, P. 88
Boswijk, H. P. 175
Box, G. E. P. 12, 88, 90, 104, 154, 175, 206n
Brandner, P. 46
Breusch, T. 89
Brown, B. W. 10
Brown, R. L. 207
Burmeister, E. 258, 265, 269
Burnside, C. 225, 227, 232, 235
Burridge, P. 172n, 175, 182
Byron, R. P. 231

Cagan, P. 3, 258, 259, 272n
Calzolari, G. 22
Campbell, J. Y. 16n
Campos, J. 179n, 192n, 204, 211, 212, 216
Canova, F. 1, 3, 161, 166, 172, 226
Carvalho, J. L. 161
Casella, A. 258, 260, 265, 269, 270, 271, 272
Cecchetti, S. G. 283
Chambers, M. J. 46
Choi, B. 89
Chong, Y. Y. 18, 22, 38, 48, 190
Chow, G. C. 228, 242
Christiano, L. J. 235
Chui, A. P. L. 166
Clements, M. P. 1, 10, 11, 12, 18, 20, 31, 32, 35, 36, 40, 42, 44, 46, 47, 179n
Cleveland, W. P. 182, 187
Clews, R. 222
Coen, P. G. 12
Coghlan, R. T. 211
Cogley, T. 232
Coleman, J. 179n
Crutchfield, J. 176
Cuthbertson, K. 211

Davidson, J. E. H. 187, 211
Davidson, R. 246
Deistler, M. 56
DeJong, D. 179n
Dempster, A. P. 287n
DeToldi, M. 284n
Dhrymes, P. J. 38
Diba, B. 259, 270
Dickey, D. A. 89, 164, 165, 166, 180n, 194
Diebold, F. X. 3, 38, 283n, 299
Doan, T. 12

Dolado, J. J. 50
Doornik, J. 220
Dufour, J.-M. 207
Durbin, J. 182, 206n, 207
Durlauf, S. N. 1, 3, 88, 140, 225n, 263, 269, 280, 281

Eichenbaum, M. 225, 235
Engel, C. 284, 299
Engle, R. F. 8, 17, 33, 34, 40, 42, 43, 46, 87, 88, 89, 90, 92, 98, 99, 107, 108, 133, 135, 154, 165n, 169, 171, 172, 175, 179n, 184, 206n, 211, 239, 287n
Ericsson, N. R. 2, 9, 12, 18, 34, 35, 38, 40, 172n, 180, 181, 184, 190, 192n, 194, 201, 204, 211, 212, 214, 216, 283n
Ermini, L. 190
Evans, G. W. 232
Evans, J. M. 207

Faust, J. 179n
Favero, C. 28, 34, 193
Fillardo, A. J. 284n
Finn, M. 1, 3
Fisher, L. 88, 90, 105, 110
Fisher, P. G. 52
Flood, R. P. 265, 269, 272
Franses, P. H. 154n, 165, 166, 175, 180n
Friedman, M. 192n, 194
Fuller, W. A. 89, 164, 165, 166, 180n, 194, 277n

Galbraith, J. W. 50
Garber, P. M. 265, 269, 272
Gersowitz, M. 175
Ghysels, E. 166, 172, 176, 180n, 189, 284n
Godfrey, L. G. 206n
Goldfeld, S. M. 194, 207, 283
Gomme, E. D. 50
Goodfriend, M. S. 264, 270, 271
Goodhart, C. A. E. 194
Gourieroux, C. 284n
Granger, C. W. J. 8, 17, 34, 36, 39, 40, 42, 46, 87, 88, 89, 98, 107, 108, 133, 135, 154, 165n, 169, 179n, 182, 184, 186, 190, 239
Gregory, A. W. 89, 128, 226
Grether, D. M. 161, 182
Grossman, H. 259, 270

Hacche, G. 211
Hall, R. E. 263, 269, 280, 281
Hall, S. G. 39n, 211, 214
Hallman, J. 190
Hamilton, J. D. 259, 270, 283, 284, 287, 288, 299
Hannan, E. J. 55, 56
Hansen, B. E. 5, 7, 88, 89, 92, 94, 97, 108, 141, 144
Hansen, L. P. 161, 166, 172, 176, 264
Hargreaves, C. P. 2, 6, 7, 8, 87, 92, 103, 127, 129
Harris, D. 2
Harvey, A. C. 12, 175, 206n
Hasza, D. P. 164, 165, 166, 180n
Healey, J. 222
Helkie, W. 179n
Henderson, D. 179n
Hendry, D. F. 1, 2, 4, 10, 11, 12, 15n, 18, 20, 22, 25, 28, 31, 32, 34, 35, 36, 38, 40, 42, 44, 46, 47, 48, 87n 88, 89, 99, 172n, 180, 181, 184, 187, 190, 192n, 193, 194, 201, 204, 211, 212, 214, 231
Henry, S. G. B. 211
Hillier, G. H. 90, 136, 137, 138
Hillmer, S. C. 175
Hodrick, R. J. 272
Hogarth, G. 222
Hooker, M. A. 1, 3, 279
Hoover, K. D. 226
Hsiao, C. 130
Hylleberg, S. 2, 6, 12, 154, 161, 164, 165, 166n, 169, 171, 172, 175, 180n, 183

İmrohoroğlu, S. 259
Inder, B. 2, 135, 151

Jarque, C. M. 206n
Jenkins, J. M. 12, 154, 175
Jevons, W. S. 153
Johansen, S. 2, 4, 5, 6, 7, 8, 40, 41, 42, 88, 90, 92, 100, 103, 133, 166, 171n, 179n, 180, 184, 187, 193, 200, 204, 237, 238, 239n, 241
Joreskog, K. G. 4
Jørgensen, C. 161, 166n, 172, 180n
Judd, J. P. 207
Judge, G. G. 19
Juselius, K. 7, 103, 184, 187, 193

Kajii, A. 283n, 299n
Kalman, R. E. 12
Kendall, M. G. 50
Kenny, P. B. 182
Khan, M. S. 265
Kim, C. 272
King, M. L. 136, 137, 138
King, R. G. 225n, 226, 228, 237, 238n
Klein, L. R. 52
Kloek, T. 175, 180n
Koopmans, 5
Kremers, J. J. M. 239
Kunst, R. S. 46
Kwiatkowski, D. 2, 8,133, 134
Kydland, F. E. 225, 226, 233

LaHaye, L. 272
Laidler, D. E. W. 194
Laird, N. M. 287n
Lam, P. 283
Lee, H. S. 165n, 166, 171n, 180n, 189
Lee, J.-H. 3, 284n, 292n
Leeper, E. 179n
Lenum, 102
LeRoy, S. 269
Litterman, R. 12
Long, J. B. 225, 226, 227, 229
Longbottom, J. A. 52
Loretan, M. 5, 8, 88, 92, 108
Lovell, M. C. 182, 191n
Lu, M. 40
Lucas, R. E. 10, 32, 34, 233
Lütkepohl, H. 43

MacKinnon, J. G. 175, 198, 207, 246
Mankiw, N. G. 16n
Mann, C. 222
Maravall, A. 175, 183, 184
Marcet, A. 228
Mariano, R. S. 10
Mark, N. C. 283, 299
Marquez, J. R. 18, 179n
McGrattan, E. B. 226, 228
Mehra, J. 179n
Milbourne, R. 194
Miller, M. H. 194
Mills, J. A. 59
Miron, J. A. 161, 165, 166, 171, 172, 175, 176, 180n
Mizon, G. E. 12, 38, 40, 179n, 187, 190, 204n, 211, 231
Monahan, J. C. 97
Monfort, A. 231, 284n
Moss, S. 50
Muellbauer, J. 179n

Nason, J. M. 232
Nelson, C. 12, 32, 39
Nerlove, M. 161, 182
Neusser, K. 237
Newbold, P. 36, 39
Newey, W. K. 144
Nicholls, D. F. 206n, 207
Noh, J. 166, 180n

Ogaki, M. 131
O'Hanian, L. 283n
Orden, D. 88, 90, 92, 105, 106, 110
Ormerod, P. 50
Orr, D. 194
Osborn, D. R. 166, 175, 176, 179n, 180n, 184
Osterwald-Lenum, M. 102, 200
Ouliaris, S. 53n, 88, 89, 133, 149

Pagan, A. R. 1, 2, 3, 5, 6, 87n, 206n, 207, 283n
Palm, F. 232
Park, J. Y. 88, 89, 91, 92, 142
Parzen, 7, 97, 123
Paulsen, J. 56
Perron, P. 89, 189
Phillips, P. C. B. 2, 3, 4, 5, 6, 7, 8, 42, 55, 56, 57, 58, 59, 63, 87n, 88, 89, 90, 91, 92, 94, 103, 106, 108, 133, 140, 141, 142, 144, 147, 149, 184
Pierce, D. A. 175, 183, 206n
Ploberger, W. 2, 54, 55, 56, 57, 58, 59, 63
Plosser, C. I. 225, 226, 227, 228, 229
Porter, R. 269
Pötscher, B. M. 56
Prasad, K. 59
Prescott, E. C. 225, 226, 229, 233

Quandt, R. E. 283

Rabemananjara, R. 231
Ramsey, J. B. 206n

Rebelo, S. 225, 226, 228, 235
Reichlin, L. 16n
Reinsel, G. C. 88
Richard, J.-F. 38, 88, 99, 179n, 190, 211
Rissanen, J. 55, 59
Robertson, J. 225n
Rubin, D. B. 287n
Rudebusch, G. D. 283n, 299
Ruud, P. A. 287n

Said, S. E. 89
Saikonnen, P. 88
Salemi, M. 258, 265
Sargent, T. J. 176, 227, 233, 258, 259, 264, 265
Scadding, J. L. 207
Schmidt, P. C. B. 8, 18, 22, 48
Schuermann, T. 302
Schwarz, G. 56
Schwartz, A. J. 192n
Scott, L. O. 269
Shephard, N. 12
Shiller, R. J. 269, 272
Shin, Y. 8
Sichel, D. E. 194, 207, 299
Siklos, P. L. 189
Silvapulle, P. 135, 151
Sims, C. A. 12, 172, 180, 182
Smith, G. W. 194, 226
Smith, J. P. 166
Snyder, R. D. 136, 138
Sorbom, D. 8
Sørensen, N. K. 161, 166n, 172, 180n
Spanos, A. 231
Spitzer, J. J. 22
Srba, F. 220
Starr, R. M. 194
Stentoft, K. 153n
Stock, J. H. 12, 42, 88, 89

Tiao, G. C. 88, 90, 104, 175, 182
Tinbergen, J. 231
Tobin, J. 193
Todd, P. H. J. 175, 176
Todd, R. M. 12,
Topping, S. L. 214
Tran, H.-A. 2, 12, 172n, 181, 194, 201, 204, 211, 212, 216
Trivedi, P. K. 15n
Trundle, J. M. 211
Tsay, R. S. 56
Turner, D. S. 10

Wall, K. D. 258, 265, 269
Wallace, N. 259, 265
Wallis, K. F. 12, 26, 31, 33, 172n, 175, 179n, 180, 182, 184, 189, 190, 192n, 232
Watson, G. S. 206n
Watson, M. W. 12, 42, 88, 89, 136, 138, 233, 234, 287n
Webb, S. B. 259
Weinbach, G. C. 3
West, K. D. 17, 144
White, H. 206n, 207
Whiteman, C. H. 259, 270
Whitley, J. D. 12, 33
Wickens, M. R. 89
Wilcox, D. 179n, 180n
Wilcox, J. B. 211

Yeo, S. 220
Yoo, B. S. 17, 42, 43, 46, 88, 90, 92, 98, 99, 154, 169n, 171

Zeldes, S. P. 175
Zellner, A. 176, 232

Subject Index

accelerator mechanism, 244
ADF test, 194, 239, 241
ADL, 216
AIC, 56
analogue model, 182
AR(p) + trend(r), 63
ARMA(p,q) + trend(r), 63
augmentation, 169
Australian macroeconomic data, 54
autoregressive distributed lag, 205

band spectrum regression, 175
Bayes model measure, 54
BIC, 56
Box-Jenkins methods, 259
Brownian motion, 140
bubble, 269
Buys Ballot plot, 161

Cagan hyperinflation model, 257, 260, 265
calibration, 228
  calibrated models, 233
Chow statistics, 207
cointegration, 180, 184, 186, 200, 201, 230
  237, 238, 239
  cointegrating errors 240
  cointegrating restrictions 227, 243
  cointegrating space, 93
  cointegrating vectors, 237, 238, 240
common trends, 236
conditional model, 189, 204
conditional money demand, 180
constancy, 207

diagnostic statistics, 206
differences, 216
differential, 216
disequilibrium, 204
duration dependence, 299
dynamics, 189

EM algorithm, 284
encompassing, 180, 190, 212
Engle-Yoo 3-Step Estimator, 98
error correction, (ECM) 175, 206, 211, 216
  ECM test, 239
exchange rate dynamics, 299
exogeneity, 180

Fisher effect, 151
fixed format models, 86
forecasting, 53
fully-modified OLS, 97, 106, 144, 145

generalised method of moments, 271
German hyperinflation, 259, 271

Hamilton model, 297
Hannan-Rissanen recursion, 55
HEGY test, 166

idealised models, 226
inference, 189
inflation rate, 194
integrated, 184, 236, 243
  at the annual frequency, 160, 165
  at the long frequency, 154, 165
  at the semiannual frequency, 154, 165
interest rate, 194

Johansen Estimator, 100

Kalman filter, 55
Keynesian models, 246
Kullback-Liebler Information Criterion, 233

LBI test, 138
LBUI Test, 138

likelihood ratio, 57
linear filters, 181
long run, 175, 201, 211

Markov switching model, 284
MDL, 59
mean equality, 185
measures of fit, 234
model based procedures, 175
modified Box-Tiao estimator, 104
money demand, 193, 204
Monte Carlo experiment, 106
multiplier-accelerator model, 245
multivariate invariance principle, 140

Newey-West variant of White's tests, 271
non-cointegrating restrictions, 230
null hypothesis of cointegration, 133

OLS, 93
opportunity cost, 194
order estimator, 58

parameter encompassing, 180, 190
parsimony, 84
Parzen filter, 97, 123
Phillips-Perron correction, 277
PMDL, 59
predictive distributions, 54

quantity theory, 194

Radon Nikodym derivative, 57
rational bubble, 262
rational expectations, 260
RBC model, 247
real business cycles, 225
Recursive least squares, 207
Root mean squared errors, 84

seasonality, 198, 213
  causes for the seasonal variation, 161
  definition of seasonality, 154, 164
  seasonal adjustment, 180, 213
  seasonal cointegration, 154, 169
  seasonal component, 194
  seasonal dummies, 189
  seasonal economic time series, 154
  seasonal integration, 161
  seasonal unit roots, 164
    semiannual frequency, 175
    annual frequency, 175
sequential reduction, 205, 216
single-equation modelling, 205
Solow residual, 226
stochastic linear regression model, 55
structural breaks, 214
stylised facts, 227
super exogeneity, 211
super exogenous, 212

testable implication, 230
Tests for a stochastic seasonal pattern, 164
Tests for Cointegration, 138
total final expenditure, 194
transactions demand, 193
transformations, 216
Two-Step estimator, 98

Vector autoregression, VAR, 186, 227, 231, 247
variance dominance, 180, 193

Wald statistics, 274
weakly exogenous, 189, 205
White's autocorrelation correction, 271
White's heteroskedasticity correction, 271

X–11 seasonal adjustment, 172, 181, 190